NEUROSCIENCE RESEARCH PROGRESS

NEURAL NETWORK LEARNING IN HUMANS

NEUROSCIENCE RESEARCH PROGRESS

Additional books in this series can be found on Nova's website under the Series tab.

Additional e-books in this series can be found on Nova's website under the e-book tab.

NEUROSCIENCE RESEARCH PROGRESS

NEURAL NETWORK LEARNING IN HUMANS

GISELHER SCHALOW
DR. MED. HABIL., DR. RER. NAT., DIPL. ING.

New York

NOTICE TO THE READER

Additional color graphics may be available in the e-book version of this book.

Library of Congress Cataloging-in-Publication Data

ISBN: 978-1-63482-468-2

Library of Congress Control Number: 2015936105

Published by Nova Science Publishers, Inc. † New York

Contents

Preface

The study of learning the human neural network is considered medicine. The scientific research to improve neural network learning starts with an introduction to human neurophysiology, especially electrophysiology. The human nervous system is explored under physiologic and pathophysiologic conditions. This knowledge is used to development repair strategy for the human nervous system and other body functions. Treatment is achieved through learning, especially through movement-based learning.

If one wants to repair the neural networks of the human central nervous system (CNS) through learning, improve its functioning in the healthy case or to rebuild it artificially, basic knowledge of human anatomy, human neurophysiology and human neural network learning is needed. This is because the main difference between humans and animals is the capacity to learn. With learning, the injured, malfunctioning, or degenerating nervous system can be repaired. With learning, we can improve neural network functioning and improve intelligence. With the understanding of how the human neuronal network learns, it may become possible to partly rebuild the human brain artificially. For medicine, learning, and artificial network building, a basic knowledge of the human CNS is necessary. With behavioral human sciences alone, we will never understand how the human neuronal networks generate certain functions including learning.

One possibility to gain the necessary knowledge concerning human neuronal network learning is to measure, based on human anatomy research, how the human neuronal networks function under physiologic and pathophysiologic conditions on the single-neuron level. We might then try to change the pathologic functioning networks into physiologic functioning by learning. Movement-based learning for repairing the injured CNS is used to understand learning in the healthy cases. This strategy is achieved by measuring neural network functioning in the healthy human CNS and the injured one in brain-dead humans (somewhat healthy CNS) and patients with CNS injury. The learning method is "Coordination Dynamics Therapy", which repairs by movement-based learning and includes visual and auditory biofeedback. The methods for measuring single-neuron firing of the human neuronal networks under natural conditions is the single-nerve fiber action potential recording method, and the single-motor unit surface electromyography. Integrative CNS organization is measured with the so-called "Coordination Dynamics", which is based on the System Theory of Pattern Formation. These recording methods are shown in Figure 1. The basis for human neuronal network learning is human neurophysiology [1]. Neuronal network learning bears similarity to changing electronic circuits.

Learning to repair includes only movement-based learning, at first. The learning can be extended to include vision-induced, speech-induced, and auditory-induced learning. The learning is volitionally and un-volitionally.

This research project was started with human neurophysiology [1]. This knowledge was used for the repair of the human brain and the spinal cord [2]. Here it is shown that the repair is mainly achieved by learning.

Chapter I

Theory of Neural Network Learning

Abstract

Neural network learning starts with knowledge of basic functions of human neural networks and their communication with the outside world via specific information-coding impulse patterns, running into and out of the spinal cord. Neural network functions can only be explored thoroughly if it is partly known what impulse patterns run into and out of the networks. Even though the gained knowledge is rudimentary, it has immediate consequences for learning in order to repair the human central nervous system.

It starts with learning the anatomy of the human CNS and the nerve roots connected to it. Natural impulse patterns are recorded from the thin long nerve roots of the cauda equina with the single-nerve fiber action potential recording method. Specific natural impulse patterns are given, which code the changes in the outside world, similar to a touch or urinary bladder filling, and represent it in the CNS. To identify the nerve fibers in which the impulse patterns are conducted, a classification scheme for human peripheral nerve fibers is introduced. The firing patterns of the different premotor spinal oscillators are analyzed. They are important neural sub-networks of the spinal cord contributing to the generation of the motor programs for the body in order to move in the outside world. The firing of these oscillators changes from the occasional to the oscillatory firing mode. It is shown to be the mechanism that develops the motor programs for moving. A fundamental organization principle of the human nervous system, the phase and frequency coordination among neuron and neural sub-network firings, is measured. This phase and frequency coordination becomes impaired following injury. It can be improved by movement-based learning, when exercising on a special CDT device. For further repair of the CNS, automatisms and other functions have to be trained, like crawling, to stimulate other network parts to take over these functions. Since the single-nerve fiber action potential recording is an invasive method and difficult to apply, surface electromyography (sEMG) is used to study the output from the CNS. Single motor unit impulse patterns offer the knowledge of how motor programs are generated. It is shown that the patterns of single-motor unit neuron firings are the same as those recorded with the single-nerve fiber action potential recording method.

The system theory of pattern formation is introduced and related to human neurophysiology and movement-based learning. The stability of movements and other patterns can be understood. The quality of CNS functioning can be assessed non-invasively by pattern change when exercising the special CDT device. Quantifying CNS functioning is a requisite to judge learning, repair, and treatment in medicine. Physiologic and pathologic CNS functioning can now also partly be measured at the neuron level with the single-nerve fiber action potential recording method. The single-motor unit EMG correlates neural network functioning at the neuron level with the movement level. Changes of the geographical landscapes of movement patterns (attractor layouts) inform the learning process. The quality of CNS functioning can exactly be assessed by pattern changes when exercising on a special CDT device. One value, the coordination dynamics value, quantifies the quality of CNS functioning and is used for measuring the learning progress. The medical treatment, called CDT, is then developed out of the difference between physiologic and pathophysiologic CNS functioning. Phase and frequency coordination has to be improved and other neural networks or brain parts must function for repair by learning. By including the autonomic nervous system division into the learning process, cardio-vascular performance and urinary bladder function can be improved.

External loops of premotor spinal oscillators are shown to be a mechanism of communication with the outside world. But external loops of oscillators can also be used in the CNS neural networks for large-scale integration. By training many coordinated movements, getting them into the CNS via external loop of premotor spinal oscillators, large-scale integration of sub-network functioning in the CNS networks can be trained by movement-based learning. By this large-scale integration, which is coordinated up to milliseconds, CNS functioning improves by learning in the physiologic case, following injury or degeneration and in aging.

If the inhibitory system is reduced in function there is a devastating degeneration of the CNS. Imbalance between excitatory versus inhibitory levels occurs. As one consequence, the premotor spinal oscillators cannot coordinate their firing sufficiently anymore and a partial synchronization will give rise to tremors and other dysfunctions. Different kinds of pathologic oscillator organizations are demonstrated and analyzed in patients with Parkinson's disease. By movement-based learning, these pathologic network organizations can be 'caught' and trained to function more physiologically again. At the neuron level this entrainment means that we optimize the inhibitory system by increasing the weight of inhibitory synapses, when training at limits by neurogenesis of inhibitory neurons. By movement-based learning, these newly built neurons are integrated in the existing adult neural networks. By using animal data of hippocampus learning, neural network learning can partly be understood at the neuron level. The movement-based learning triggers excitation-neurogenesis coupling via Ca^{++} channels and NMDA receptors of neural stem/progenitor cells to change gene expression, especially at the progenitor level.

The learning for repair of the CNS was started with movement-based learning. If vision, hearing, and speech are included in the learning process for better management of the outside world, the neuronal networks learn in an integrative manner, and consequently better and faster, supported by models for neural network learning. New neurons enhance the accuracy of stored patterns especially when networks were more active and many different patterns were trained.

1. Introduction to Neural Network Learning

1.1. From Repair to Learning

Based on human neurophysiology [1], it has been shown that the human brain and spinal cord can partly be repaired [2] by movement-based learning. It seems that even to a very limited extent new nerve cells can be built anew in the human central nervous system (CNS) [3]. With the movement-based learning therapy "Coordination Dynamics Therapy" (CDT) it has been shown that CNS functioning could be improved or repaired after stroke [4], traumatic brain injury [5,10], spinal cord injury [6,7,13,16], cerebellar injury [12], cerebral palsy [9], hypoxic brain injury [11], in Parkinson's disease [8], spina bifida (myelomeningocele) [17] and scoliosis [18]. Speech had been induced and improved in a patient with severe cerebral palsy [1] and urinary bladder functions were repaired in spinal cord injured patients [1,16].

An important step in repairing the CNS was the learning transfer [15]. By training coordinated movements, vegetative [16] and cognitive functions [2] could be improved or repaired.

Therefore, it seems that by movement-based learning the neural networks of the whole CNS can be changed for repair. Since the human nervous system is involved in nearly all functions of the body, the overall health can be improved, which is important in CNS repair and in later life.

As in animal research, we learn from the movement-based learning in severe CNS injuries for the improvement of CNS functioning in mild injuries and in healthy CNS.

1.2. Tools and Strategy to Study Human Neural Network learning

To measure human neuronal network functioning and learning, we have to start from scratch, which means from the human anatomy, to see where in the human nervous system would it be possible to record from single neurons under natural conditions. With the simultaneous natural afferent and efferent impulse patterns generated by natural stimulation (human neurophysiology), neural network organization can be analyzed and the learning process of neuronal networks studied. Since the CNS can be seen as one large neuronal network, which is more or less activated in general, we can understand the measured learning transfer from movements to higher mental functions. If movement-based learning changes cognitive functions by learning transfer, it is possible that cognitive learning uses similar neural network organization changes than motor learning. One contribution to understand cognitive learning is therefore to understand motor learning.

The analysis of motor learning in the framework of the System Theory of Pattern Formation may help to understand the integrative aspects of learning.

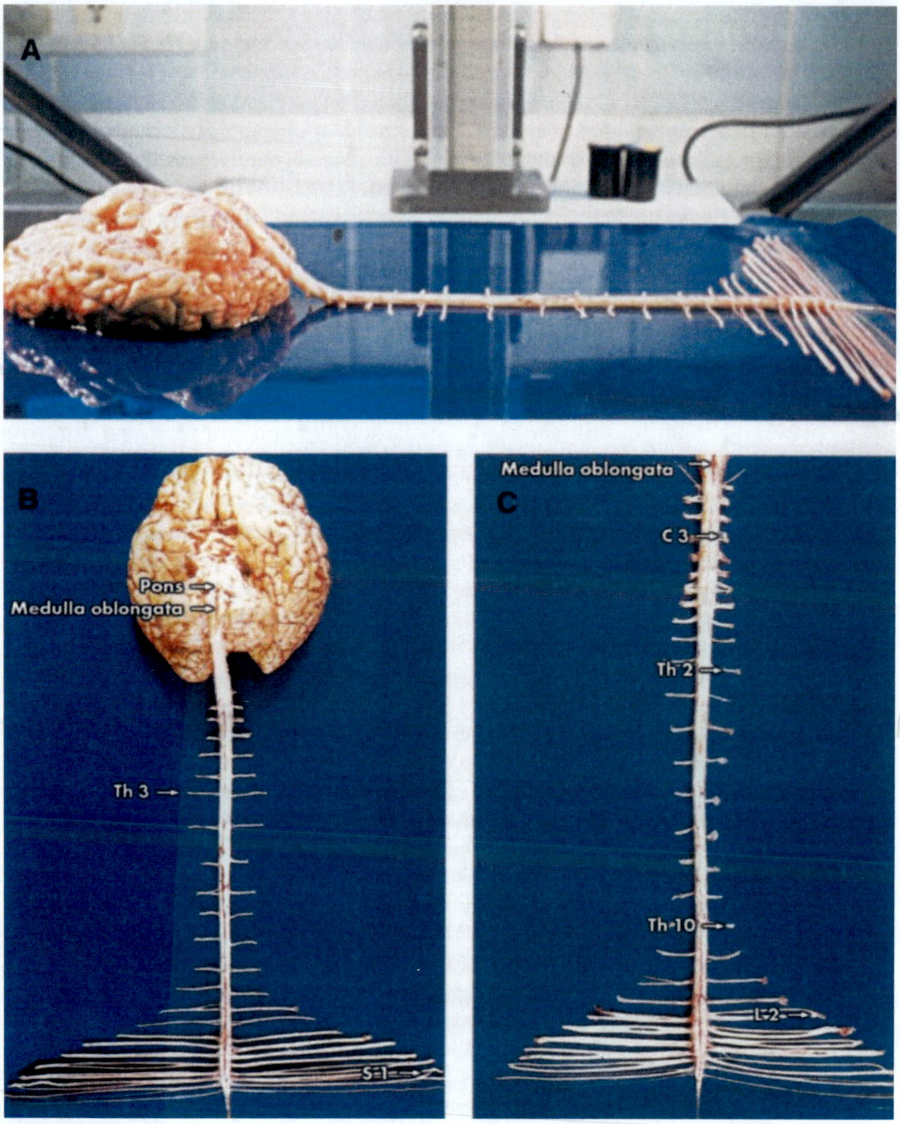

Figure 1. The human central nervous system. Some cervical (C3), thoracic (Th2, Th3, Th10), lumbar (L2) and sacral (S1) nerve roots are indicated. Dissection of the Author.

2. Anatomy of the Spinal Cord and Cauda Equina Nerve Roots

2.1. Anatomy of the Cauda Equina: A Site Where CNS Functioning Can Be Measured

Figure 1 shows the human CNS, which means the brain, brain stem, spinal cord, and nerve roots. To recognize the different lengths of the nerve roots, the nerve roots are separated. Figure 2A shows a cadaver dissection of the cauda equina, when the spinal canal and the dura mater are opened. This is the layout for recording natural impulse patterns running in and out of the CNS in brain-dead humans. Figure 2B shows the splitting in ventral and dorsal roots. The ventral (motor) roots are thinner than the dorsal roots. The last thick root in caudal direction is thc S1 root. The S4 and S5 nerve roots are thin, especially their ventral roots. In Figure 3A the principle recording layout is shown. An actual recording layout, when recording from a human patient with a spinal cord injury, is shown in Figure 3B. Because of the small operational field, it is obvious that detailed anatomical knowledge of the human cauda equina is necessary; to know what roots one is recording in the operation.

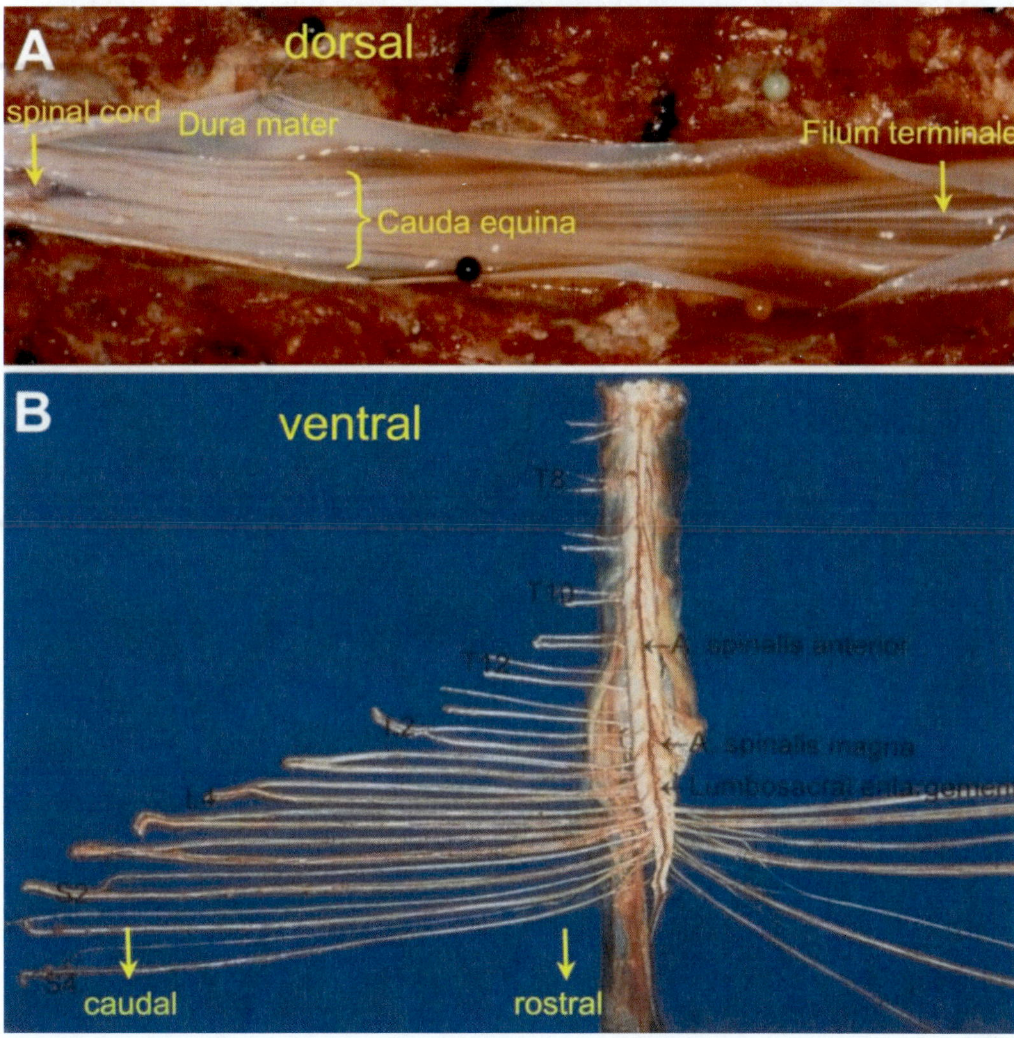

Figure 2. Schematic layout for the measurements of single nerve fiber action potentials. After opening the spinal canal (A) and rcmoving thc lower spinal cord with the cauda equina nerve roots and the dura mater (B), length and thickness of the roots can be seen. Some thoracic, lumbar, and sacral nerve roots are designated.

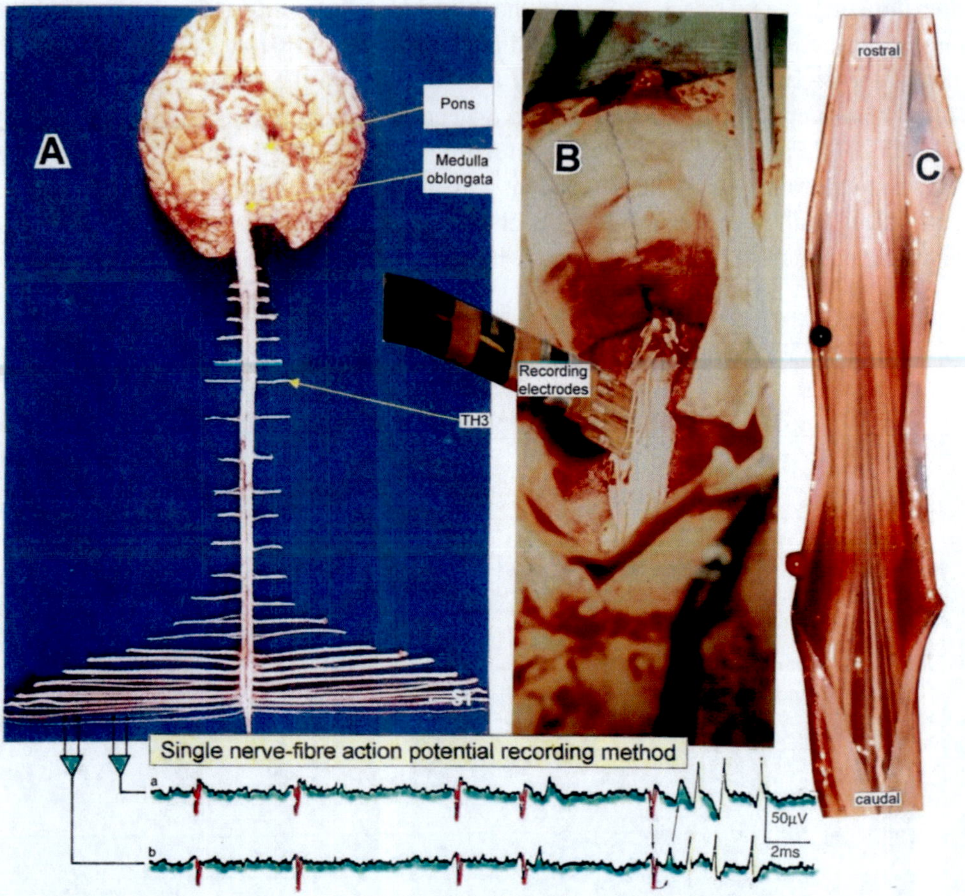

Figure 3. Anatomical layout for the recording of single-nerve fiber action potentials to analyze the self-organization of neuronal networks of the human CNS under physiologic and pathophysiologic conditions. A,B,C. By recording with two pairs of platinum wire electrodes (B) from sacral nerve roots (cauda equina, C) containing between 200 and 500 myelinated nerve fibers, records were obtained in which single nerve-fiber action potentials (APs) were identified from motoneurons (main AP phase downwards) and afferents (main AP phase upwards).

3. Recording of Single-Nerve Fiber Action Potentials (Electrophysiology)

3.1. Principle of Recording Single Afferent and Efferent Nerve Fiber Action Potentials

The development of the single-nerve fiber action potential recording method was possible because of the unique anatomical landscape of the human spinal canal. Because of the ascensus of the spinal cord, the lumbosacral nerve roots have become very long and form the cauda equina (Figure 2A). Since the caudal sacral nerve roots are very thin (Figure 2B) and nerve roots are only sheathed by a thin layer of cells (Figure 9), they are ideal for recording single-nerve fiber action potentials (APs) from undissected nerve roots. Since humans have no tail, continence (mainly S2 to S5) and sexual functions are mainly located in the conus medullaris only. Those functions are therefore represented in the lower sacral nerve roots and they are not intermingled with tail functions as they are, for example, in rats, cats and dogs.

The schematic recording layout is shown in Figure 4 to record single-nerve fiber action potentials (APs), extracellularly from undissected nerve roots (in this case from two fibers) with two pairs of platinum wire electrodes (electrode pair distance 10 mm; electrode distance in each pair 4 mm), pre-amplified (x1000), filtered (RC-filter, passing frequency 100 Hz - 10 kHz), and displayed on a digital storage oscilloscope, and also stored using a PCM-processor and a video recorder. APs from the afferent and the efferent fibers can clearly be distinguished since in the electrode arrangements, the main phase (second phase) from the afferent fiber is upwards and that of the efferent fiber is downwards (Figure 4A). E.g., the AP of a skin afferent fiber reaches a pair of electrodes first as negative and then as positive. According to the electrode setting used, the main phase is upwards. An AP of a motoneuron, coming from the opposite direction, would reach the electrodes in the order of positive-negative. The potential changes are therefore opposite and the main triphasic AP will point downwards. An AP in an afferent fiber first reaches the caudal electrode pair and then the rostral pair, whereas an AP of the efferent fiber first reaches the rostral electrode pair and then the caudal one. The conduction times are therefore also opposite. The reversing of the inputs to both preamplifiers does not change the ability to differentiate between afferent and efferent APs (Figure 4B).

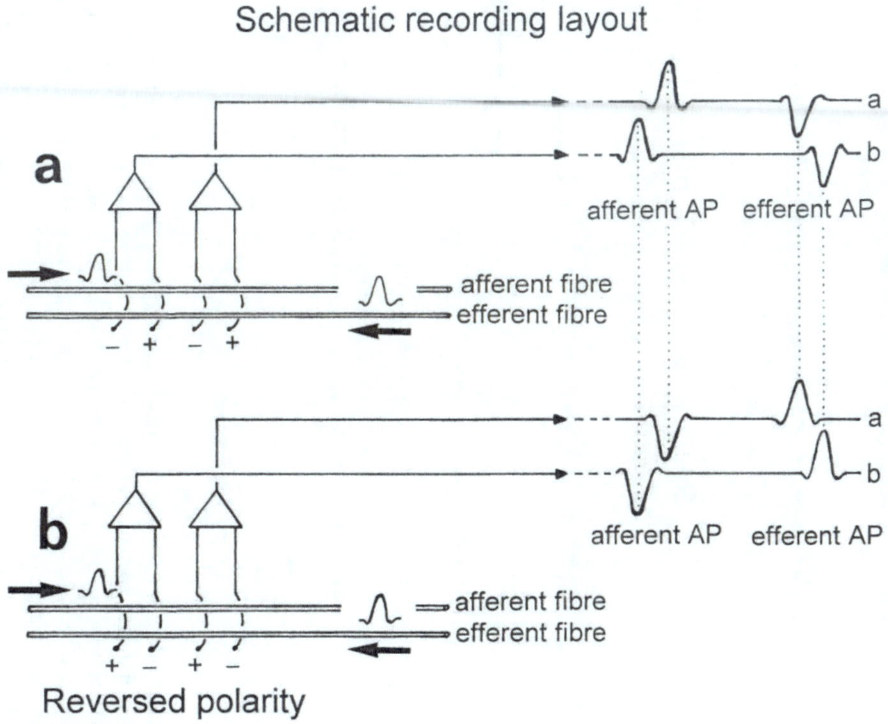

Figure 4. Schematic layout for recording single-nerve fiber action potentials (APs) from an afferent and an efferent nerve fiber. The reversing of the inputs to both preamplifiers does not change the ability to differentiate between afferent and efferent APs.

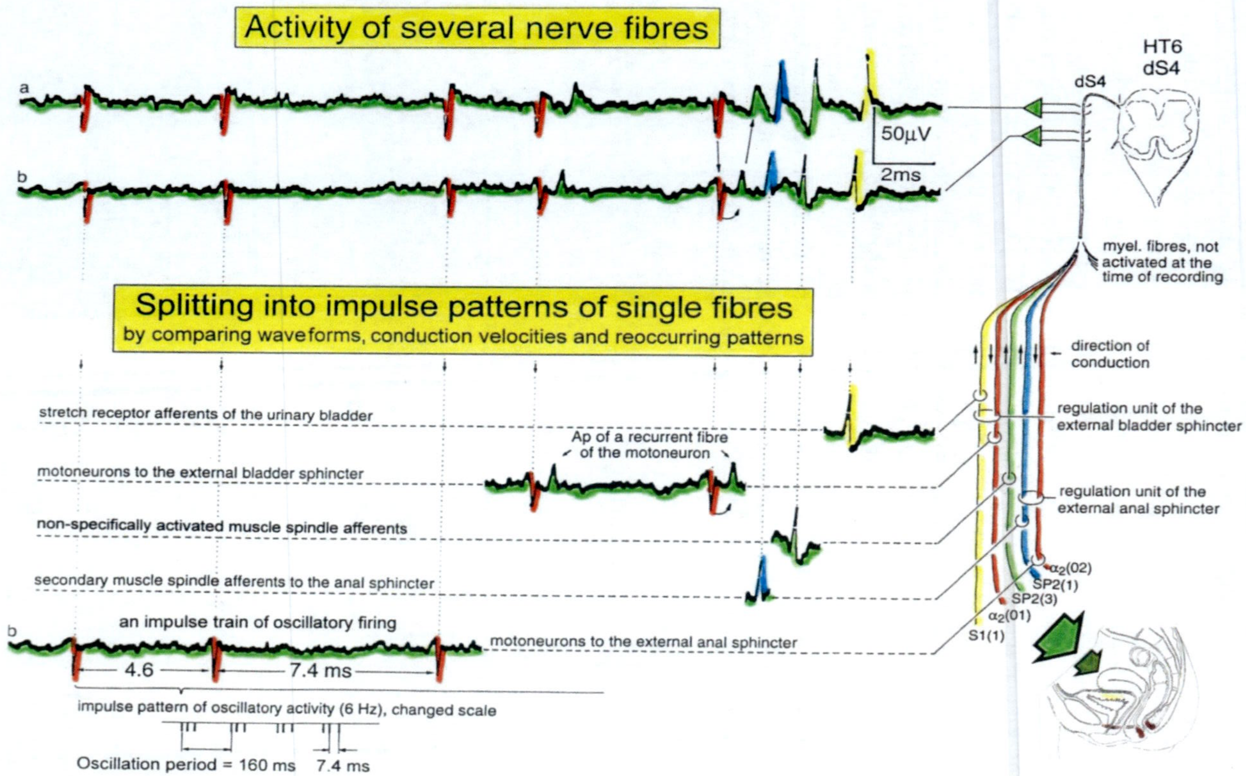

Figure 5. Schematic layout of the classification scheme for the human peripheral nervous system. By recording with two pairs of platinum wire electrodes from a nerve root containing approx. 500 myelinated nerve fibers, a recording is obtained in which 3 action potentials (APs) from 3 motoneurons (main AP phase downwards) can be seen. By measuring the conduction times and with the known electrode pair distance (10 mm) a conduction velocity distribution histogram is constructed in which the nerve fiber groups are characterized by ranges of conduction velocity values and peaks in asymmetrical distributions. After recording, the root was removed, fixated, embedded, and stained. Light microscope cross-sections were prepared and used to measure the mean diameter and the myelin sheath thickness (d). Distributions of nerve fiber diameters were constructed for four different ranges of myelin sheath thickness. Nerve fiber groups were characterized by the peak values of asymmetrical distributions. By correlating the peak values of the velocity distributions with those of the diameter distributions obtained from the same root, a classification scheme was constructed of the human peripheral nervous system. Brain-dead human HT6.

3.2. Recording of Single-nerve Fiber Action Potentials from Nerve Roots and Splitting of the Multiunit Recording into Natural Impulse Patterns of Several Single Afferent and Efferent Fibers

A real recording arrangement during an operation is shown in Figure 3A, B. To obtain natural impulse patterns simultaneously from several single afferent and efferent nerve fibers to analyze CNS functioning, the summed impulse traffic of several afferent and efferent fibers of a nerve root must be split into the impulse patterns of single fibers (Figure 5). The splitting is achieved by recognizing the APs from certain single fibers on the basis of wave form comparisons on the two recording traces and the conduction time, which an AP needs to travel from one electrode pair to the other one (10mm), and selecting these APs out. The summed impulse traffic of the recording in Figure 3A is split into the impulse patterns of 5 single afferent and efferent nerve fibers (Figure 5). In the thin lower sacral nerve roots, there are afferent and efferent fibers in the ventral and dorsal roots.

The splitting of the skin afferent activity, upon touch or pinprick into the natural impulse patterns of the different single touches and receptors, is shown in Figure 6. Such messages inform the CNS about changes in the periphery. Similar natural impulse patterns inform the CNS about changes in the urinary bladder or anal canal such as bladder filling or bladder or anal canal catheter pulling (Figure 7).

As we can measure the natural impulse patterns, generated by certain single receptors in the periphery, which run into the spinal cord (CNS) (afferents) and those patterns that leave the cord (efferents) in ensembles of single fibers simultaneously, it becomes possible to analyze the integrative properties of the largely unchanged CNS of brain-dead humans (HTs) at a cellular level. This also means that the change in function, caused by a CNS injury, can be identified.

4. Classification of Peripheral Human Nerve Fibers (Electrophysiology Combined with Morphometry)

4.1. Classification of Human Peripheral Nerve Fibers by the Group Conduction Velocity and the Group Nerve Fiber Diameter

For the analysis of CNS functioning and neural network learning, we must first identify the kind of nerve fibers from which the recordings are taken. A classification scheme of human peripheral nerve fibers is needed.

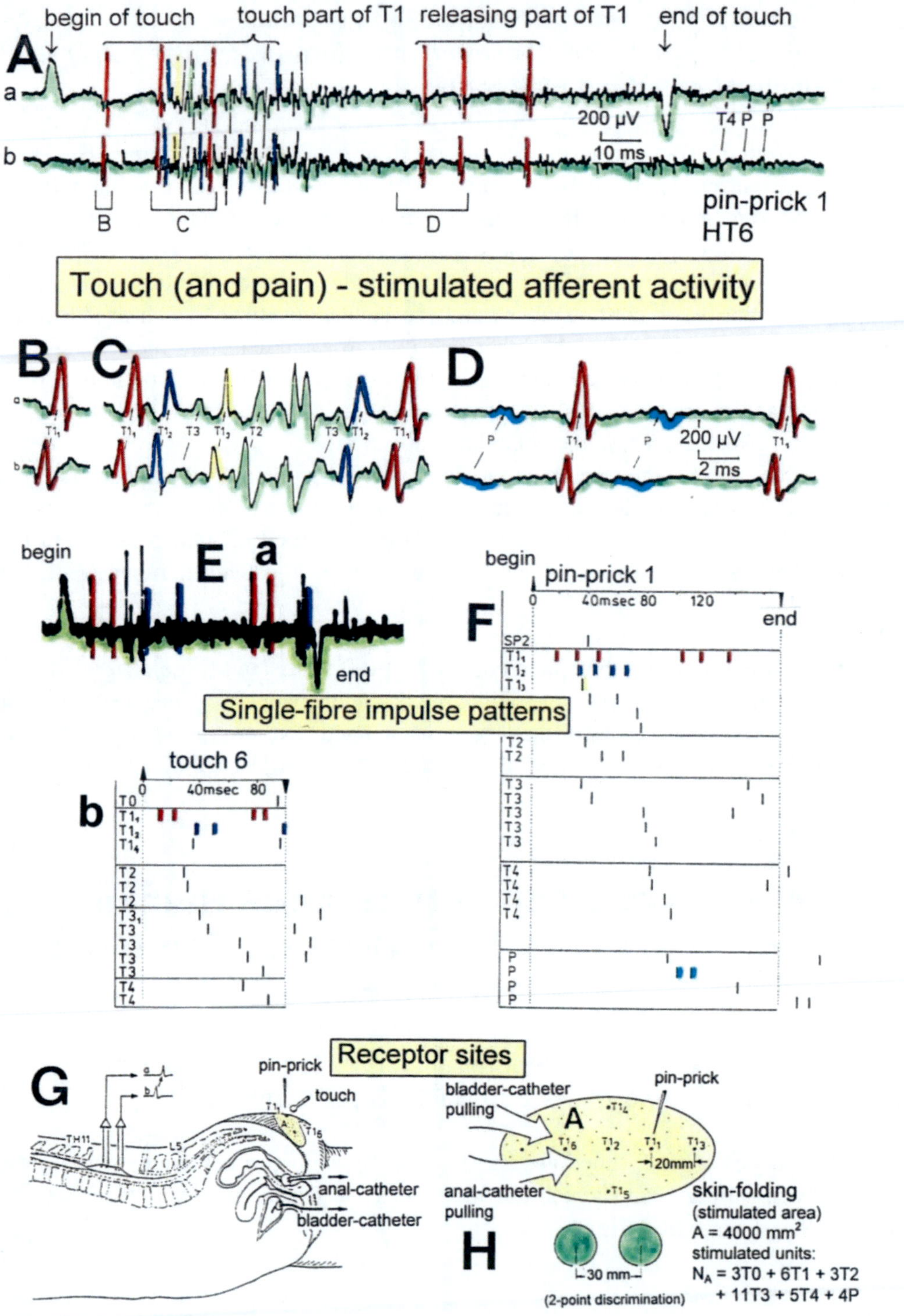

Figure 6. Touch (and pain)-stimulated afferent activity. Touch and pain activity stimulated by pin-pricking (A) and touching (Ea) at S5 or Co dermatomes and recorded extra-cellularly from a dorsal coccygeal root (brain-dead human HT6). T1, T2, T3, T4, P = mark action potentials (APs) from single touch and pain fibers. Subscripts 1, 2, 3 mark single fibers.

A. Whole sweep following pinprick 1 shown at a slow time base. The large upward artifact on trace ‘a’

electronically marks the beginning of the pinprick. The large downward artifact on trace 'a' marks the end of the pinprick. Note that 2 intervals of high activity of large APs occur, one after the beginning of the pin-prick with 1 AP in front, and a second before the end of the pinprick; potentials with small amplitude follow potentials of large amplitude. Time intervals B, C, and D are shown in a time-expanded form in Figures B, C, and D.
B, C, D. Time expanded sweep pieces of A. Identified APs are indicated. Note that the APs from the T11 touch unit can be safely identified by the waveforms in B, C, D.
Eb, F. AP occurrence patterns of single touch and pain fibers following short touch 6 and pin-prick 1. No pain afferents are stimulated upon touch 6. Upon pinprick 1, the single-fiber AP activity of the different touch and pain groups is identified by the AP waveforms on traces 'a' and 'b', and by the conduction times. The single touch afferents of the T1 group are marked with subscripts. One active secondary muscle spindle afferent fiber (SP2) could always be identified in F. Note that for pinprick 1, touch and pain afferents are stimulated whereas for touch 6 only touch afferents are stimulated.
G. Recording and stimulation arrangement for simultaneous recording of several single touch and pain units. A = area stimulated by skin folding, drawn in H in more detail. $T1_1$, $T1_6$ = suggested touch points of the $T1_1$ and $T1_6$-units.
H. Drawing of the very approximate skin area stimulated by skin folding. $T1_{1\text{-}6}$ = suggested focal T1 touch points. Two-point discrimination indicated for the sake of comparison. N_A = number of stimulated units in the dorsal coccygeal root. Skin tractions evoked by anal and bladder-catheter pulling are indicated by the large open arrows.

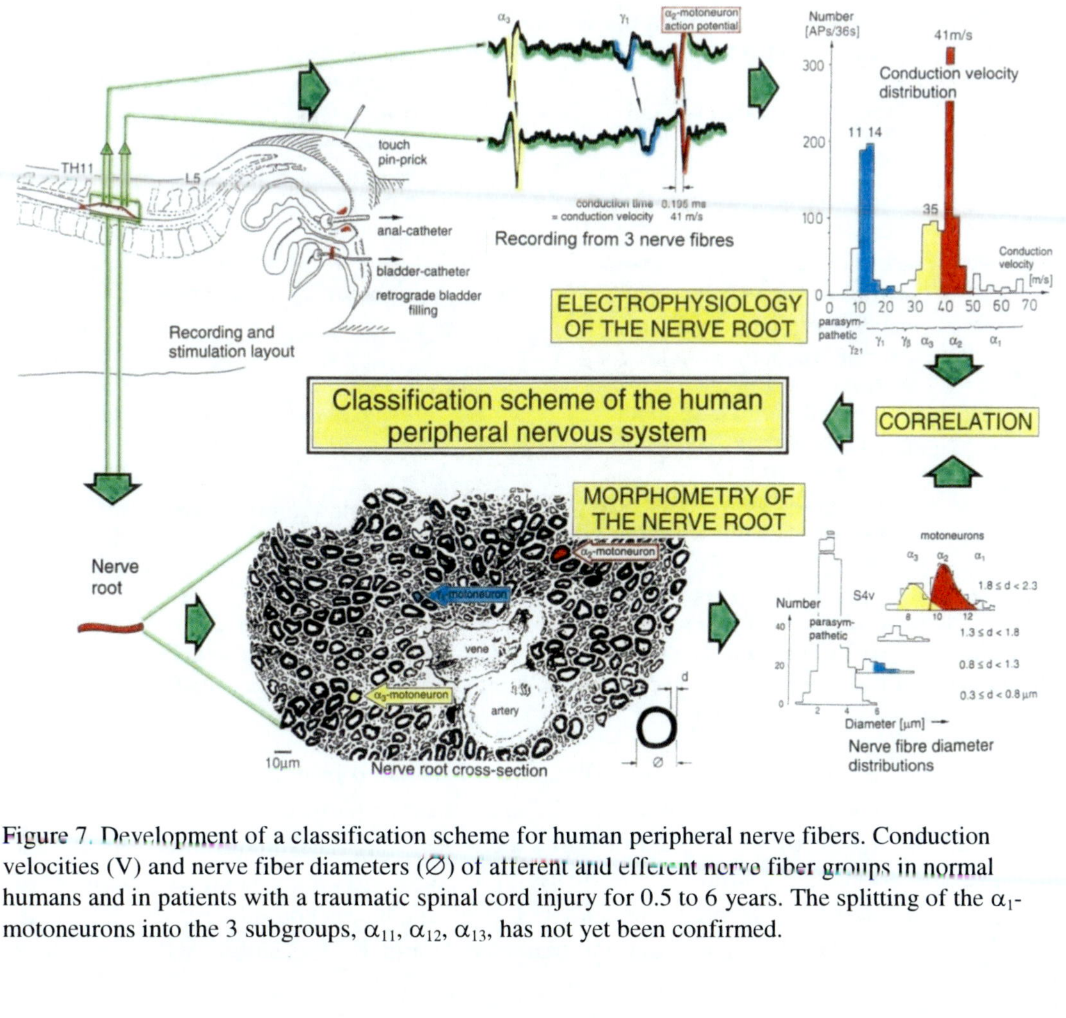

Figure 7. Development of a classification scheme for human peripheral nerve fibers. Conduction velocities (V) and nerve fiber diameters (∅) of afferent and efferent nerve fiber groups in normal humans and in patients with a traumatic spinal cord injury for 0.5 to 6 years. The splitting of the α_1-motoneurons into the 3 subgroups, α_{11}, α_{12}, α_{13}, has not yet been confirmed.

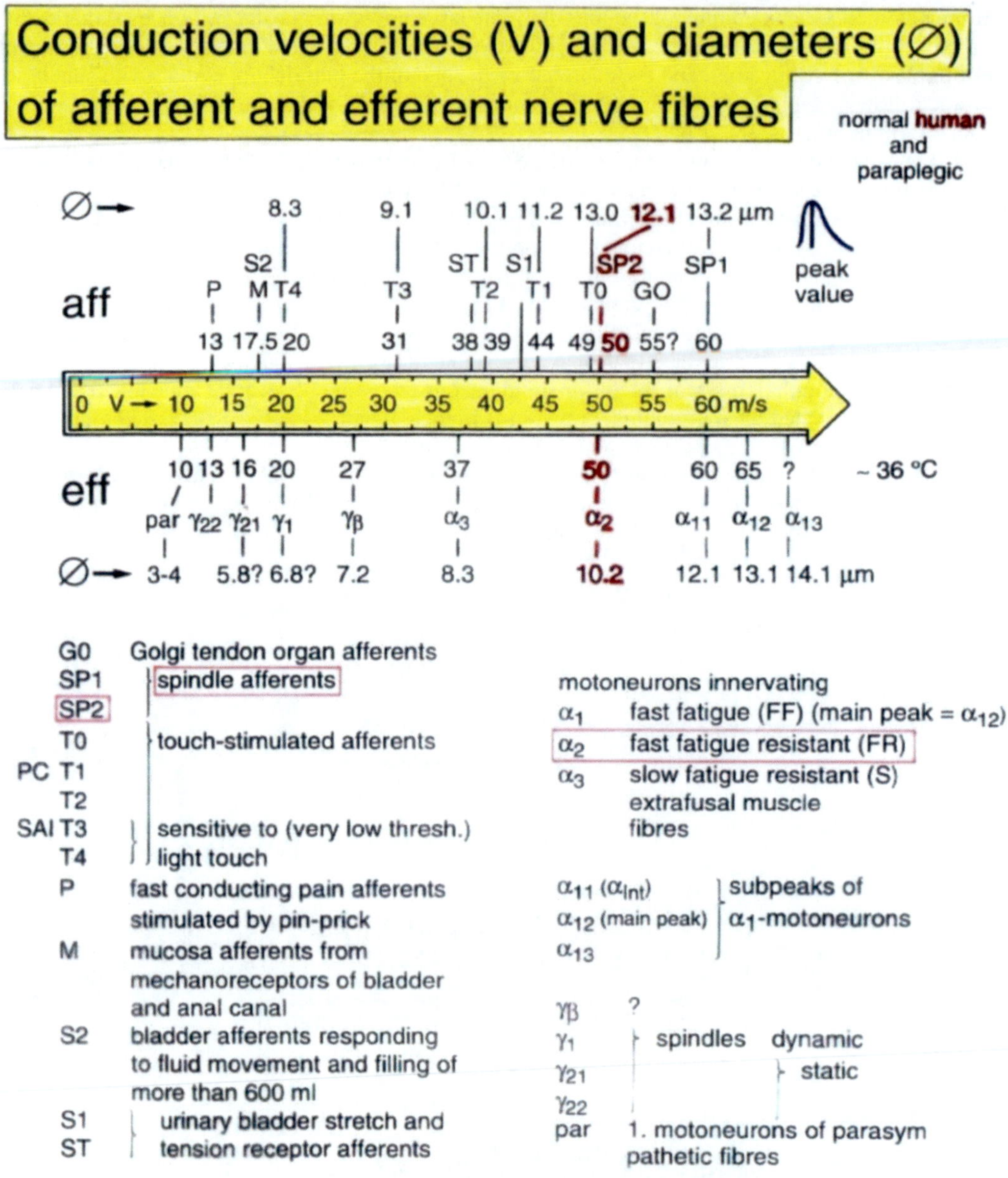

Figure 8. Classification scheme for human peripheral nerve fibers. Conduction velocities (V) and nerve fiber diameters (∅) of afferent and efferent nerve fiber groups in normal humans and in patients with a traumatic spinal cord lesion for 0.5 to 6 years. The splitting of the α_1-motoneurons into the 3 subgroups, α_{11}, α_{12}, α_{13}, has not yet been confirmed.

Conduction velocities of single nerve fibers were calculated from the conduction distance (electrode pair distance = 10 mm) and the conduction times (the time difference of a certain AP between the traces from two pairs of wire electrodes). Velocity distributions of afferent and efferent fibers were constructed, and distribution peaks were correlated to certain nerve fiber groups. The nerve roots from brain dead and surgical patients could be removed, fixated, embedded, and stained. Mean nerve fiber diameters could be measured, and nerve fiber diameter distributions constructed for different myelin sheath thicknesses (morphometry). By correlating the identified conduction velocity peaks with nerve fiber diameter peaks (Figure 7), a classification scheme for the human Peripheral Nervous System (PNS) was constructed, in which individual groups of nerve fibers are characterized by group conduction velocities and group nerve fiber diameters (Figure 8). This classification scheme is still incomplete and only holds for nerve fibers thicker than approx. 3.5μm. The classification schemes for animals do not apply to humans. For example, conduction velocities in rats, cats and dogs (max ≈ 120 m/s) are much higher than those in humans (max ≈ 70 m/s).

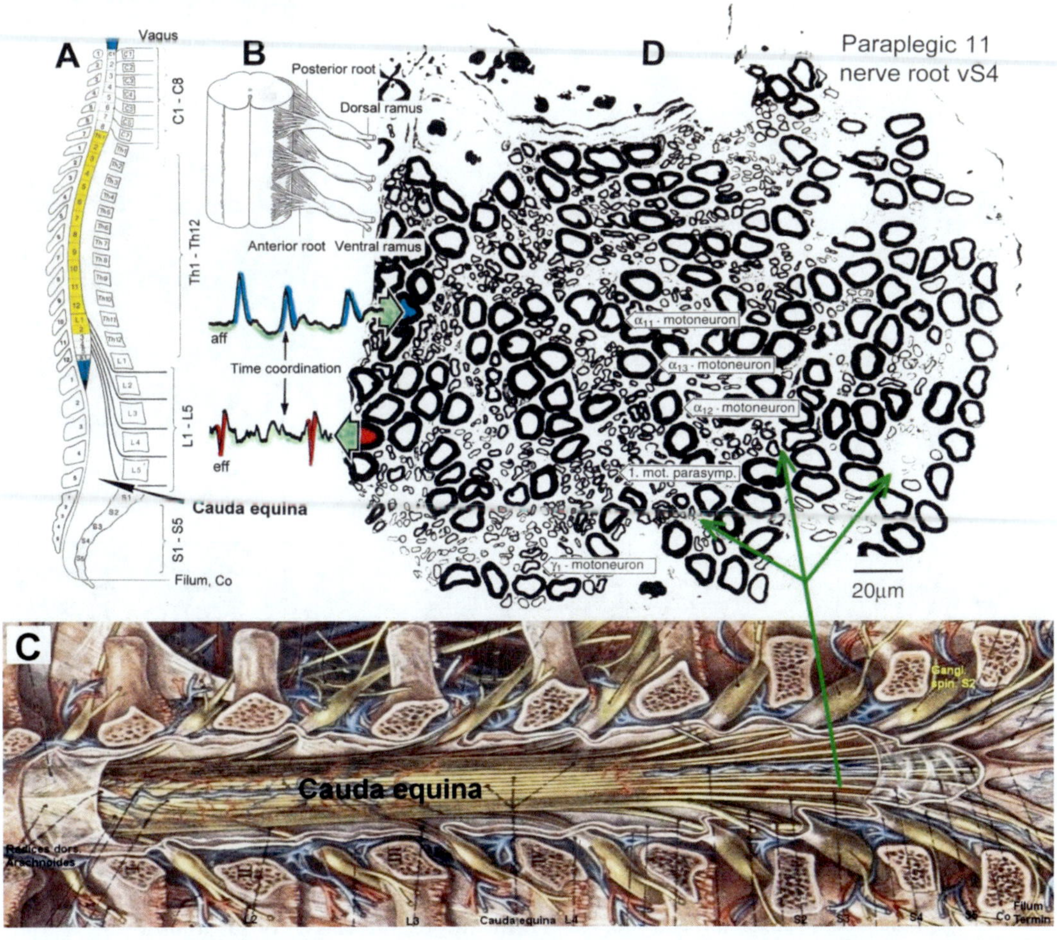

Figure 9. Anatomy to record single-nerve fiber action potentials. A. Ascensus of the human spinal cord gives rise to long nerve roots in the lumbar and sacral range. B. The nerve roots in the cervical range are short. C. Picture of the opened spinal canal with the cauda equina nerve roots, ganglions and spinal nerves. D. Real ventral S4 nerve root cross section with single-nerve fiber action potentials of afferent (aff) and efferent (eff) nerve fibers and their time coordination. Principle sizes of different nerve fiber cross sections are indicated.

It will thus become possible to record natural impulse patterns simultaneously from identified single afferent and efferent nerve fibers, and to analyze self-organizing mechanisms of the human CNS under physiologic and pathologic conditions.

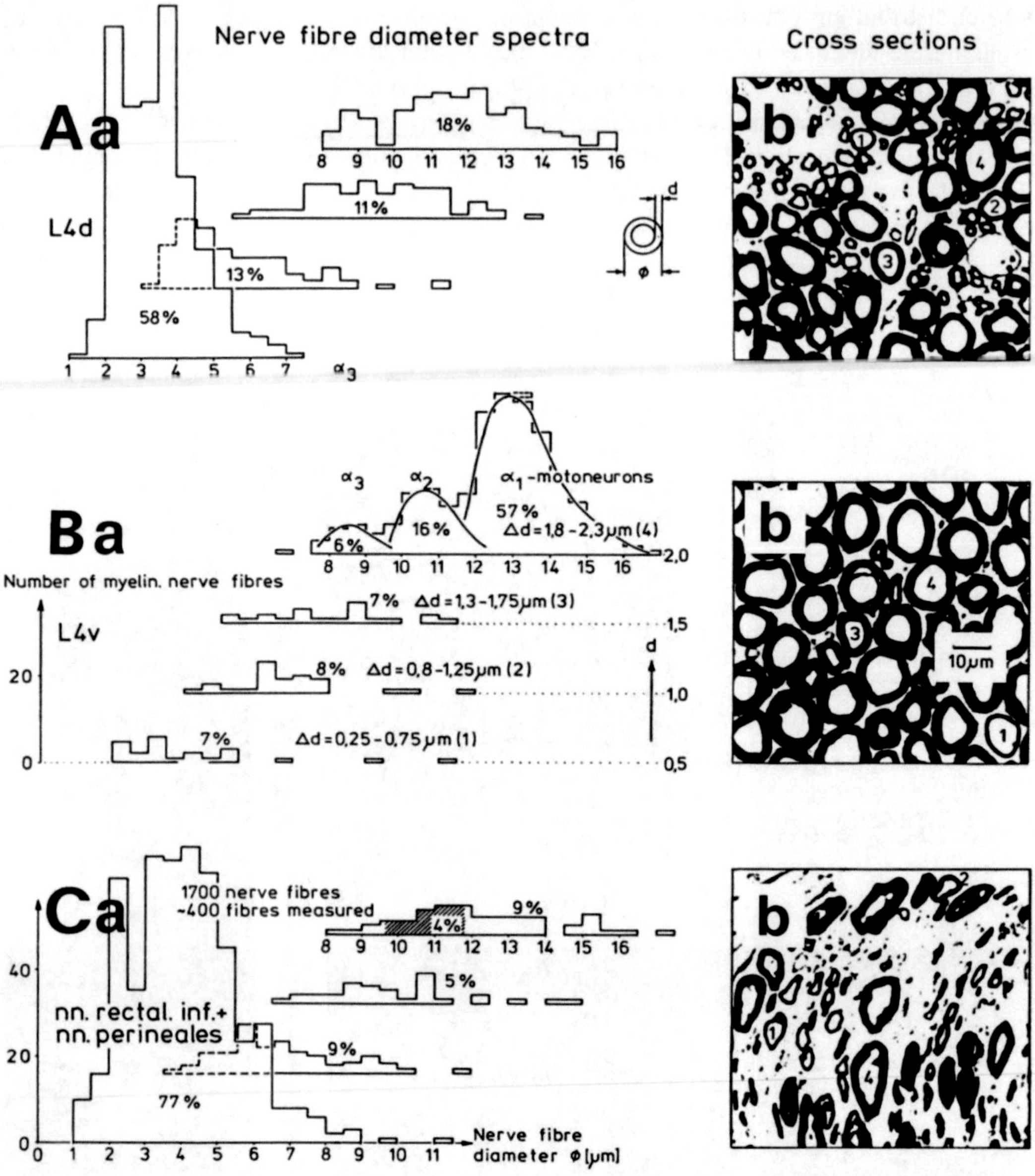

Figure 10. a. Nerve fibre diameter distribution histograms classified by 4 classes of myelin sheath thicknesses as indicated in Ba. % indicates the percentage of fibres in classes or subgroups. b. Corresponding characteristic cross sections. A few fibers are numbered by the myelin sheath thickness range they belong to. Dimension scale for A, B, and C are drawn in Bb. 8% shrinkage correction. For the definition of fibre diameter Ø and myelin sheath thickness d see insertion in Aa. – A. Nerve fibre diameter spectrum of a 4th dorsal lumbar root of a 47-year-old female human cadaver removed 2 to 5 hours after death, 660 fibres were measured. B. Spectra of a 4th ventral lumbar root (same case as in A). 320 fibres were measured. In the myelin sheath thickness range $1.8 \leq d < 2.3\mu m$ the distribution curves of the 3 α-motoneuron classes are drawn into the histogram. C. Spectra of the nervi rectales inferiors and perineales. Note, the majority of fibres (77%) have a very thin myelin sheath ($0.3 \leq d < 0.8\mu m$) with a relatively large amount of thick fibres. In the histogram of very thick myelin sheaths ($1.8 \leq d < 2.3\mu m$) the diameter range of α_2-motoneurons, to which sphincter motoneurons belong, is crosshatched (4% of the fibre).

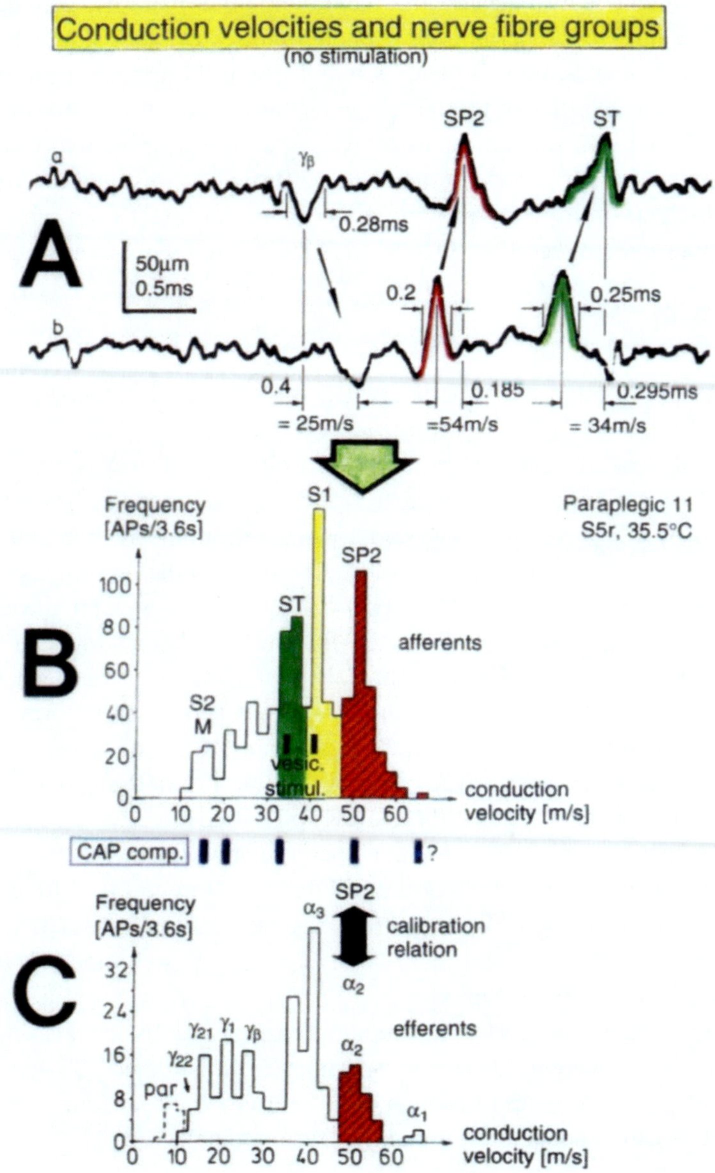

Figure 11. A. Sweep piece of recording; conduction times and corresponding conduction velocities are indicated. Root temperature at recording, 35.5°C.

B, C. Conduction velocity distributions of afferents (B) and efferents (C) obtained for a time interval of 3.6s with no additional stimulation. SP2 = secondary muscle spindle afferents, S1 = stretch receptor afferents of bladder, ST = tension receptor afferents, M = mucosal afferents, S2 = afferents responding to fluid movement; α_1 = α_1-motoneurons (FF), α_2 = α_2-motoneurons (FR), α_3 = α_3-motoneurons (S), γ_β = γ_β-motoneurons, γ_1 = γ_1-fusimotors (dynamic), γ_{21} = γ_{21}-fusimotors (static), γ_{22} = γ_{22}- fusimotors (static), par = preganglionic parasympathetic motoneurons. CAP comp. = group conduction velocities obtained from the components of compound action potentials (CAPs). Vesic. stimul. = group conduction velocities of bladder afferents obtained upon electrical intravesical stimulation (see Figures 8, 9). Calibration relation indicates the same peak group conduction velocity of secondary spindle afferents and α_2-motoneurons (cross-hatched). Velocity histogram classes ≤ and < (half closed (left) interval).

As we can simultaneously measure the natural impulse patterns, generated by certain identified single receptors in the periphery by touch, pinprick or other stimulations, which run into the spinal cord (CNS) and those patterns which leave the cord, it becomes possible to analyze the integrative properties of the largely unchanged CNS of brain-dead humans (HTs) at a cellular level. Figure 9 shows the anatomy (A,B,C) and measuring layout which opens the possibility to study neural network learning. The ascensus of the spinal cord in humans (A) gives rise to the long and thin nerve roots with no epineurium (D) from which single-nerve fiber action potentials can be recorded (D). The real recording layout in an operation is shown in Figure 3B.

4.2. The Neuron Microenvironment Influences Neural Network Functioning

As we will see below, the proliferation of stem cells depends on the microenvironment. Also, the function of nerve cells is related to the microenvironment. The classification of human nerve fibers by a group conduction velocity and a group nerve fiber diameter is only a first step towards a very exact classification. More exactness in the morphology of nerve fiber and nerve cell groups seem to exist with respect to classes. The myelin sheath thickness of nerve fibers roughly increases with the diameter. But the correlations indicate that there may be different populations of nerve fibers with their own correlation between myelin sheath thickness and diameter. From Figure 10 it can be seen that certain nerve fiber groups lie in a certain myelin sheath thickness range. The motoneurons, for example, lie in the myelin thickness range between 1.8 and 2.2μm (Figure 10Ba). From the point of view of the morphology, nerve fiber groups could still be further classified.

Also, the conduction velocities do not increase linearly with the diameter. The temperature dependence of the conduction velocity is different for different nerve fiber groups. But as can be seen from Figure 11, at least one calibration relation (between the secondary muscle spindle afferents and the α_2-motoneurons) exists to identify the conduction velocity groups in the velocity distributions.

In the learning of neuronal networks, the environmental cells are also included in the changes. It is not just the weights of synapses, which are modulated. The question remains why there is already such exactness in the size of neurons, their surrounding cells, and the conduction velocities with respect to grouping. For further details see [1].

5. Self-organization of Neuronal Networks of the Human Central Nervous System

5.1. Self-Organization of Premotor Spinal Network Oscillators

Typical firing patterns of motoneurons can be observed when motoneurons are activated with increasing strength of adequate afferent input. With low afferent input, the motoneurons fire occasionally. With increasing input they fire intermittently in an oscillatory manner and then continuously in an oscillatory manner (Figure 12). The demonstration that neurons of the CNS, in this case motoneurons, can fire both in an oscillatory manner and a non-oscillatory manner, which is very important for understanding the function of the human CNS. To describe the functioning of the CNS merely by reflex pathways and loops or coupling of rigid oscillators (of cellular or network origin) is in contradiction to empirical human data, namely that premotor spinal oscillators self-organize, as was concluded from measurements of simultaneous natural impulse patterns of afferent and efferent fibers (Figure 12).

As will be shown below, these self-organized premotor spinal network oscillators, of which the motoneuron is most likely a part, are sub-neural networks that coordinate their functioning. When this coordinated communication becomes impaired due to insufficient inhibition, they synchronize their firing with the consequence of pathologic tremors occurring in patients with Parkinson's disease [2,19].

In what follows, I shall concentrate mainly on the oscillatory firing of motoneurons, which takes place for high activation. In this high activation mode, these self-organized network oscillators can also be used as a reference basis when defining phase relations and thus phase and frequency coordination can be measured among neuron firing. For high and rather constant afferent input it was found that α-motoneurons fire repeatedly with impulse trains according to their type (Figure 13). The α_1-motoneurons (FF) fire rhythmically at around 10 Hz (range 8 to 20) with an impulse train consisting of 1 AP; α_2-motoneurons (FR) fire at approx. 6 to 9 Hz with 2 to 5 APs per impulse train, and α_3-motoneurons (S) fire with a frequency in the range of 1 Hz and with long impulse trains consisting of up to 40 APs (and more). The rhythmic firing patterns of α-motoneurons are probably generated by local neuronal networks of the spinal cord since oscillatory firing can be recorded from motoneurons of the disconnected spinal cord. The motoneuron is probably a part of the spinal network oscillator. The oscillation period (T) is roughly related to the number of action potentials (APs) per impulse train (n_{AP}), and this can be expressed by the formula: $T = 70ms + 30ms \bullet n_{AP}$. A typical premotor α_2-oscillator fires with 3 APs every 160ms ($T = 70ms + 30ms \bullet 3 = 160ms$) (Figures 5,12,14), and can change its firing pattern to 2 APs every 130ms for less activation or to 4 APs every 190ms for higher activation.

The α_1-oscillators respond very dynamically, but have few oscillator network properties. Their firing is absolutely correlated to the firing of primary spindle afferent fibers (Figures 65,66 of [2]). The α_2-oscillators respond less dynamically, have strong oscillatory properties and self-organize by the adequate afferent input patterns from several kinds of receptors, including secondary muscle spindle and urinary bladder afferents. The behavior of α_3-motoneurons is more static and their input is polymodal. The dynamics of responding to inputs (Figure 68 of [2]) increases from α_3 to α_2 to α_1-oscillator, in accordance with the dynamics of the 3 muscle fiber types the α-motoneurons innervate. The slow (S), medium fast (FR) (fast fatigue-resistant) and fast contracting muscle fibers (FF) (fast fatigable) have their own corresponding premotor networks in the spinal cord, namely that in which the α_1, α_2 and α_3-networks are integrated (Figure 13).

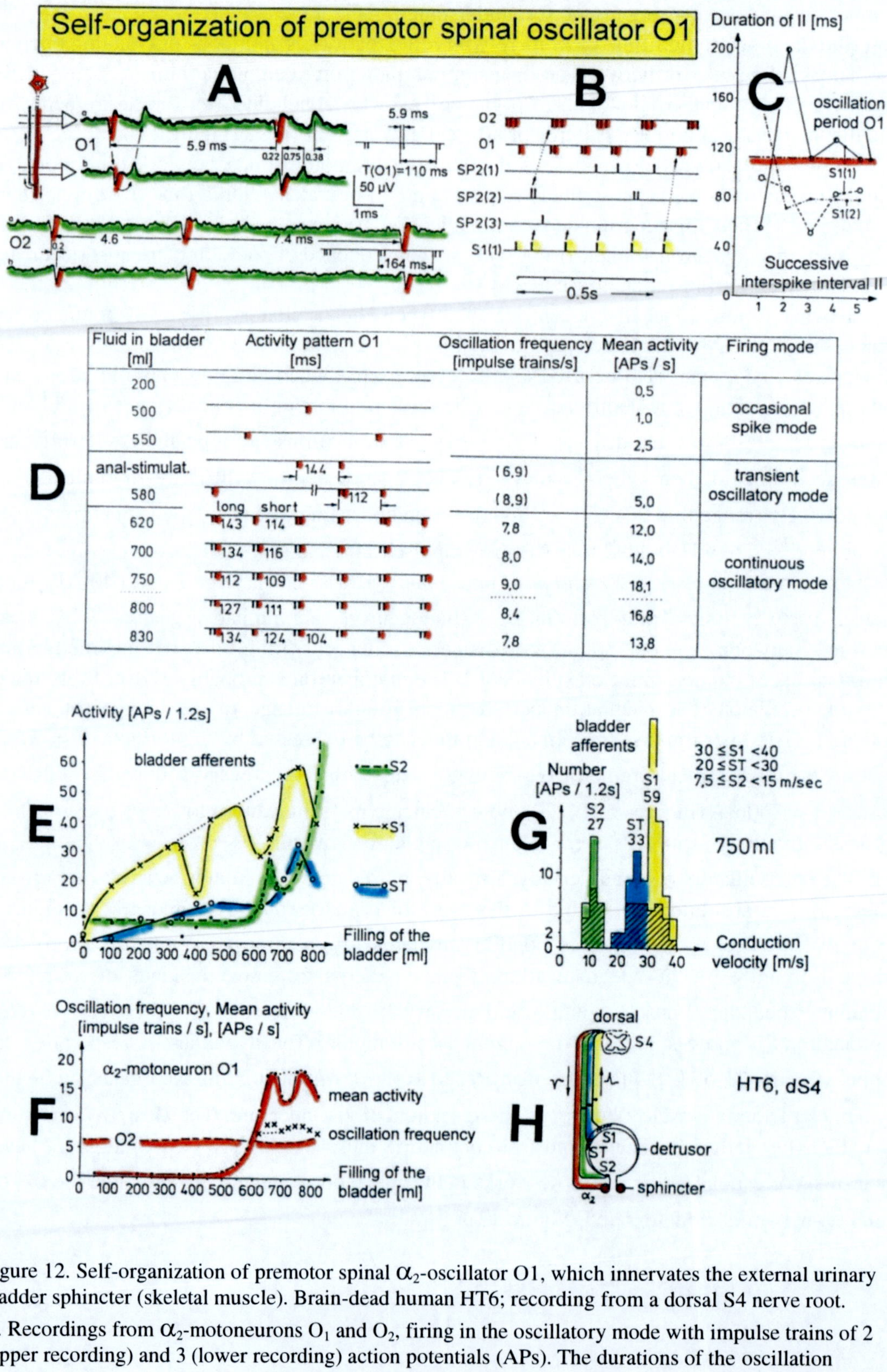

Figure 12. Self-organization of premotor spinal α_2-oscillator O1, which innervates the external urinary bladder sphincter (skeletal muscle). Brain-dead human HT6; recording from a dorsal S4 nerve root.

A. Recordings from α_2-motoneurons O_1 and O_2, firing in the oscillatory mode with impulse trains of 2 (upper recording) and 3 (lower recording) action potentials (APs). The durations of the oscillation periods were 110 (O_1) and 164ms (O_2). The interspike intervals of the impulse trains were 5.9ms (O_1) and 4.6 and 7.4ms (O_2). Motoneuron O_1 conducted at 36 m/s; its recurrent fiber conducted at 21 m/s. The measurement layout is shown schematically. The inserts show the oscillatory firing modes; they have not been drawn to scale.

B. Impulse patterns of oscillatory firing α_2-motoneuron O_2 innervating the external anal sphincter, in relation to the muscle spindle afferent activity SP2(1 to 3), activated by the stretch of the anal sphincter by the anal catheter, and impulse patterns of oscillatory firing α_2-motoneuron O1 innervating the external urethral sphincter, in relation to the stretch receptor afferent activity (S1(1)) of the urinary bladder, activated by 750 ml bladder filling. Phase relations between APs of SP2(2) and O_2 and between APs of S1(1) and O_1 are indicated by the small arrows.

C. Three series of successive interspike intervals of the 2 stretch receptor afferent fibers S1(1) and S1(2) activated by retrograde bladder filling. The oscillation period of oscillatory firing motoneuron O1, activated only by bladder filling, is shown.

D. The firing in the occasional spike mode, the transient, and the constant oscillatory firing mode of α_2-motoneuron O_1 in response to filling of the bladder. In the 'activity pattern' column changing durations of oscillation periods are given. The oscillation frequencies in the brackets give the frequencies at the moment of oscillation for the transient oscillatory mode. Downward deflections are schematized APs.

Interspike intervals of the close APs $\approx$ 6.0ms (A).

E. Activity levels of stretch (S1) and tension (ST) and flow receptor afferents (S2) (E) and of sphincter α_2-motoneuron O_1 (F) in response to retrograde filling of the bladder. The activity values of the S1, ST and S2 afferents are taken from histograms like the one in G. Filling of the bladder was stopped once between 600 and 650 ml.

F. The small dotted lines represent mean activity (APs/s) and oscillation frequency (impulse trains/s) of α_2-motoneuron O_1 if bladder filling were not stopped in between. Note that the mean activity increases continuously with the filling of the bladder from 550 to 650 ml, even though motoneuron O_1 started to fire in the oscillatory mode from 620 ml on (D). Note further that the oscillatory firing motoneuron O_2 (frequency of firing with impulse trains is shown) is almost not affected by the filling of the bladder and by the start of the oscillatory firing of motoneuron O_1.

G. Conduction velocity frequency distribution histogram of stretch, tension and flow receptor afferent activity at 750 ml. The activities of afferents S1, ST and S2 are quantified by counting the afferent conduction velocities under the peaks (open plus hatched part), with the conduction velocity limits given in the insert. The counts (27, 33, 59) are given below the peak labeled S1, ST and S2 and plotted into E for the afferent activity at 750 ml.

H. Schematic drawing of the anatomical arrangement of the afferents and the motoneuron O_1.

5.2. Phase and Frequency Coordination among Neuron Firing for Human CNS Self-Organization

Now, measuring the organization principles of the human CNS is attempted. It will be shown that neurons and sub-neuronal networks coordinate their firing up to a few milliseconds. When this phase and frequency coordination becomes impaired, organization patterns of the CNS become impaired, instable or are lost. Every functional or structural modulation of the neuronal networks changes this phase and the frequency coordination among neuron firing. Learning is related to the exactness and complexity of the many coordinations among single neuron firings or sub-neuronal networks. One strategy for repair is to improve the injury impaired phase and frequency coordination among neuron firing by movement-based learning. The coordinated movements must activate the CNS integratively, so that as many phase and frequencies coordinations as possible are trained simultaneously to improve the exactness and complexity of CNS self-organization. By exercising very coordinated movements on the special coordination dynamics therapy (CDT) device (Figures 14,15), the CNS learns from the device via the movement induced afferent input to improve its coordinated firing of the neurons and sub-neuronal networks.

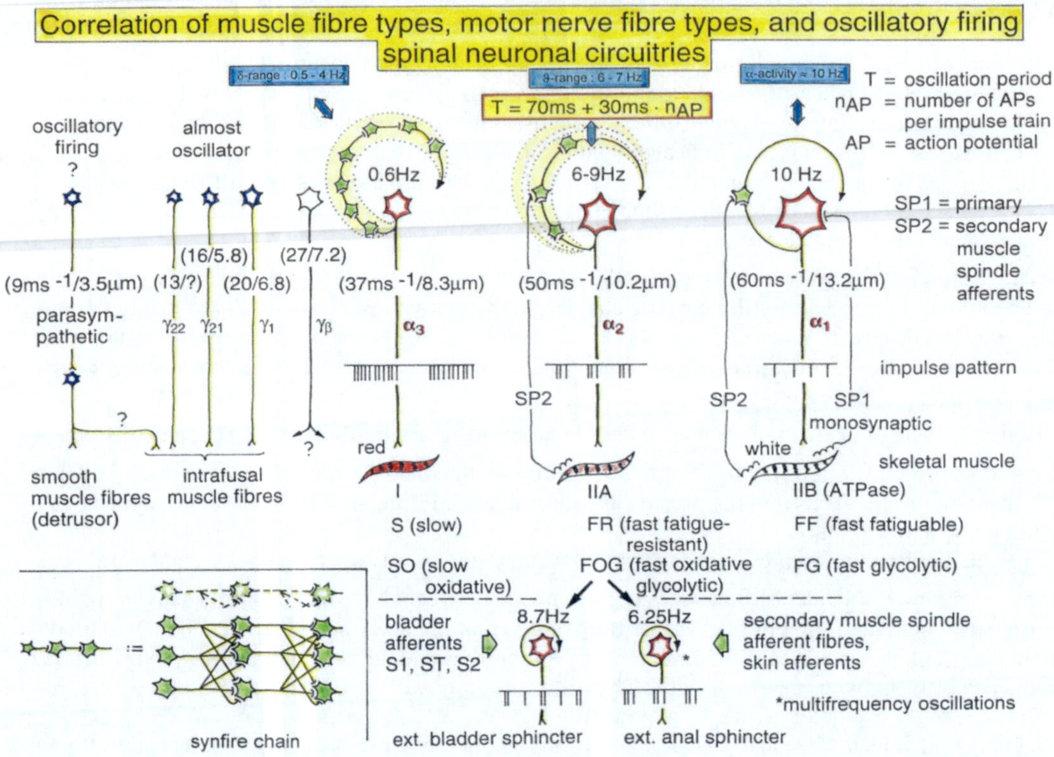

Figure 13. Correlation of muscle fiber types, motor nerve fiber types, and oscillatory firing spinal neuronal networks, based on histochemical, morphological, and physiological properties. This figure provides a simplified correlation between muscle fiber, motoneuron and sacral oscillator types. No additional subtypes have been included. The existence of α_1-motoneuron (FF) oscillators firing at 10 Hz has been predicted and they have been identified in paraplegics (unpublished observation). α = motoneuron, γ_1, γ_2 = dynamic and static fusimotors, parasympathetic = parasympathetic preganglionic motoneuron. S1, ST, S2 = stretch, tension and flow receptor afferents.

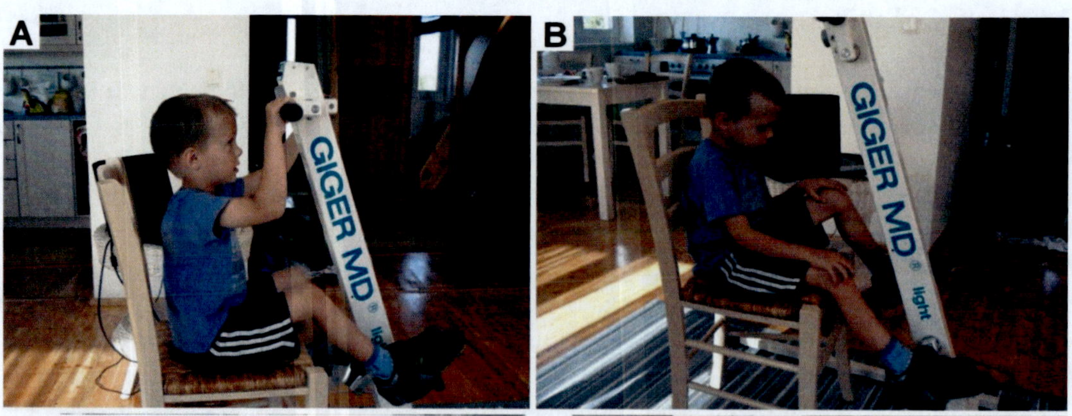

Figure 14. (Continued).

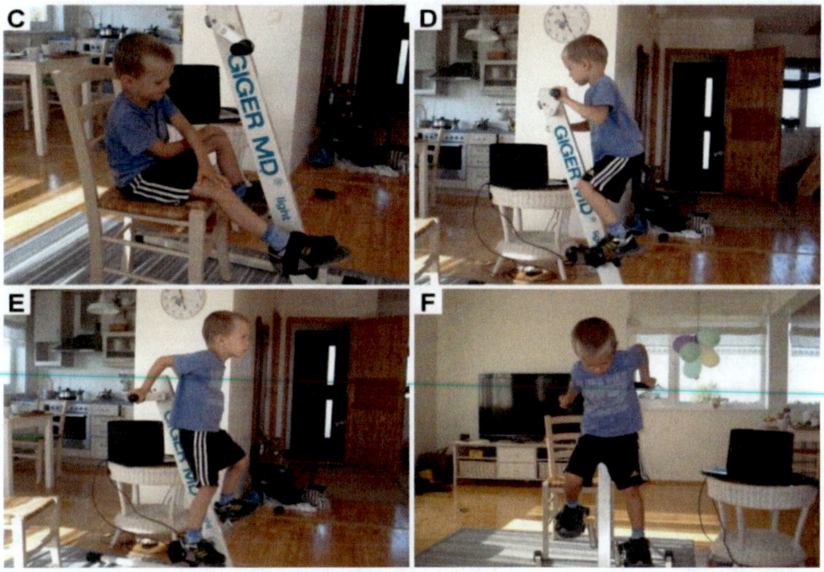

Figure 14. Different training patterns realized when exercising on the special CDT device. The healthy 4.5-year-old boy Jonas exercises in the normal sitting position with pattern change (A), the pace gait (B) and trot gait pattern (C) with no pattern change and two standing positions with pattern change from pace to trot gait (D,E,F).

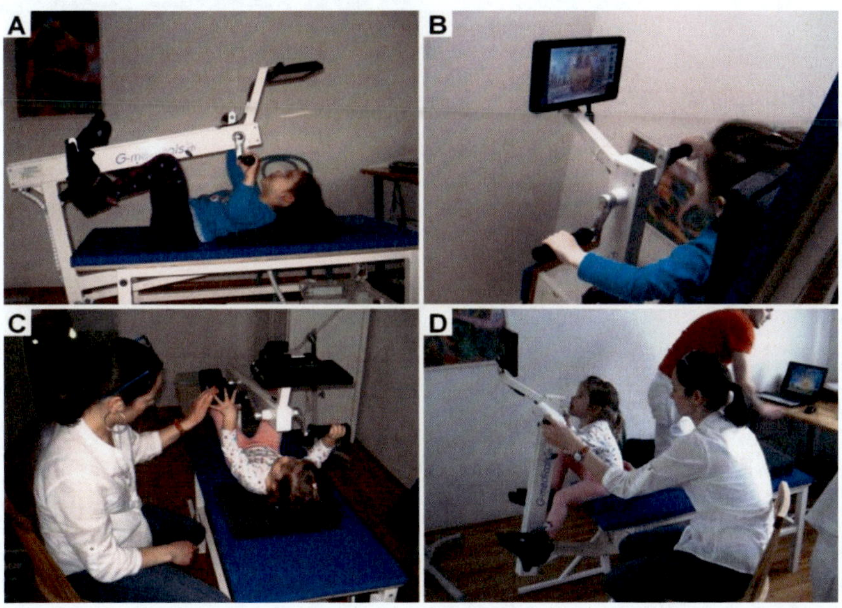

Figure 15. Administration of movement-based learning therapy with coordinated arm, leg, and trunk movements to children with cerebral palsy. The coordinated movements on the special CDT device are administered to patients in the age range between 3 and 6 years. Some patients turn volitionally; in other ones the movement has to be supported. To increase the motivation of exercising, the children can watch video films. If they stop turning, the video film stops. The position of exercising can be changed from the lying to the sitting position. If the hand is spastic (C) and cannot hold the leaver, the hand is fixed (D). This special CDT device for measuring and therapy (int. pat.) is produced by the firm: Giger Engineering, Martin Giger dipl.Ing.ETH/SIA, Herrenweg 1, 4500 Solothurn, Switzerland, www.g-medicals.ch.

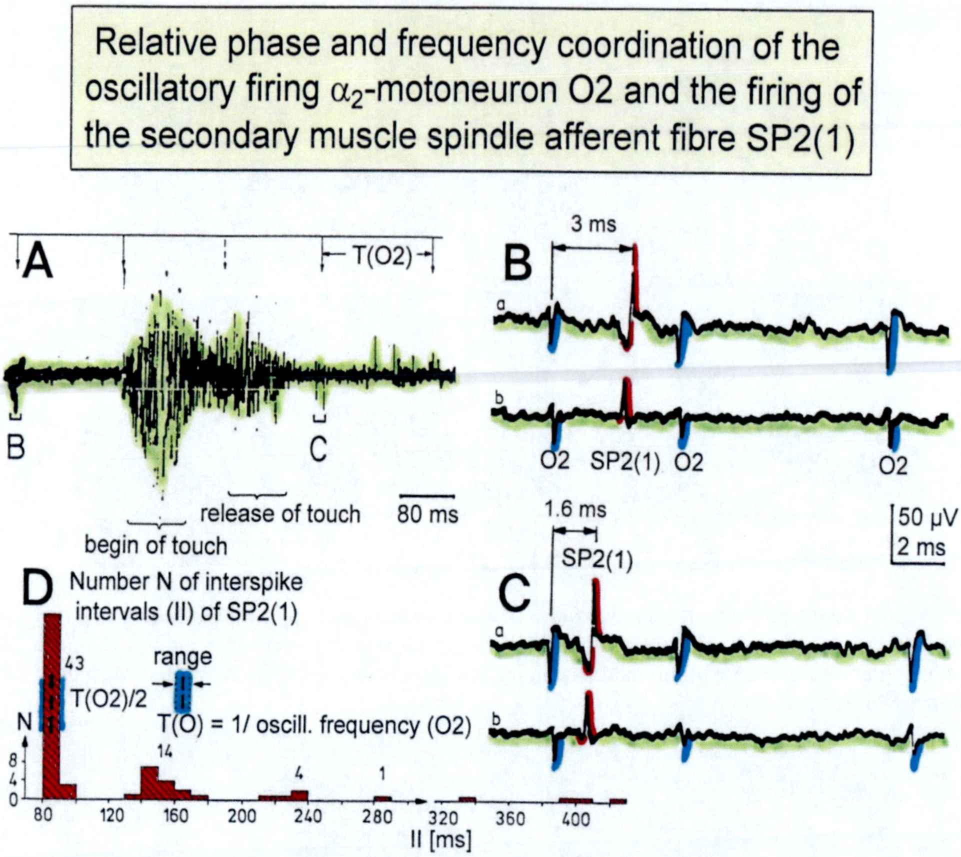

Figure 16. Time relation between the occurrence of the action potentials (APs) of oscillatory firing α_2-motoneuron O2 and the firing of the secondary muscle spindle afferent fiber SP2(1). HT6. S4 dorsal root recording.

A. Overall view of the used sweep piece; only trace "a" shown. Four oscillation cycle periods of motoneuron O2 are indicated (T(O2)). The APs of the impulse trains can be only partly recognized, because of the slow time base and poor digitalization. One impulse train (dashed arrow) is lost in the touch-stimulated activity, which consists of a touch (large overall activity) and a release part (lower overall amplitude).

B,C. Sweep pieces from A, time stretched. In B, motoneuron impulse train APs is marked O2, spindle afferent APs are marked SP2(1). Note that the APs of the spindle afferent fiber are not time-locked to the first AP of the impulse train of the rhythmically firing motoneuron (relative phase coordination). Digitalization 4 times better than in A, but still rather poor, as can be seen from the low amplitudes of the motoneuron APs on trace "b" in C.

D. Occurrence of interspike intervals of the secondary muscle spindle afferent fiber SP2(1). The numbers give the amount of IIs in each distribution peak. The oscillation period of motoneuron O2 (and the range of variation) and the half period are indicated by short dashed lines. Note that the IIs of fiber SP2(1) are very similar to the oscillation period (or the half of it) of α_2-motoneuron O_2 (relative frequency coordination).

In Figure 14 the healthy boy, Jonas, improves his motor and higher mental functions by an instrumented supervised phase and frequency improvement. The exactly performed coordinated movements, given by the special CDT device, train the neurons to fire in a more coordinated manner.

In Figure 15, girls with cerebral palsy exercise on a special CDT device to re-learn exact phase and frequency coordination, which was strongly impaired by CNS damage or malformation. With the improvement of coordinated firing of the neurons and sub-neuronal networks, certain organization patterns re-appear or improve its performance.

Relative phase and frequency coordination between the APs of the oscillatory firing α_2-motoneuron O2 and the secondary muscle spindle afferent fiber SP2(1) can be seen directly in the original recordings in Figure 16B,C. The firing of the oscillator and the sweep pieces, which are shown time-expanded are indicated at the summary trace "D". Figure 9B,C shows the AP-impulse train of oscillator O2 in connection with one of its driving spindle afferent AP. Because of the duration of the phase relation of around zero milliseconds between the firing of the driving SP2(1)-fiber (firing mostly every 80ms) and the impulse train of the oscillatory firing motoneuron O2 (T(O2) ≈ 160ms), the SP2(1)-fiber AP (every second AP) appeared at a similar time as the impulse train. Because the AP of the spindle afferent fiber had a characteristic waveform, it was easy to extract its impulse pattern from the summed impulse traffic of this S4 dorsal root. During touch-induced skin afferent activity, as in Figure 6, the activities of the motoneuron and the spindle afferent fiber were covered by the skin afferent activity. After the cessation of the skin afferent activity, the afferent and efferent APs were found again at their expected time positions of the regular firings. The phase coordination between the firings of the oscillatory firing motoneuron O2 and the secondary muscle spindle afferent fiber SP2(1) at the time when records B,C were taken, was 1.6ms (3ms - 1.4ms, Figure 16B,C).

In Figure 16D, the relative frequency coordination between the firings of the SP2(1)-fiber and the impulse train of the oscillator is indicated. For the time period evaluated, the correlation between the firing of the motoneuron and the spindle afferent fiber was in the range of between 3 and 5ms (Figure 16D).

In Figure 17, considerations concerning the relative frequency coordination are extended to the activity of further afferent fibers and γ-motoneurons of the same root. "G" of Figure 17 shows sweep pieces of the original recordings; A through F shows the interspike interval distributions of spindle afferents and γ-motoneurons. It can be seen from the overlapping of the oscillator frequency T(O2) and T(O2)/2 distribution ranges and the interspike interval distributions of the afferents that, from the viewpoint of frequency coordination, fiber SP2(1) contributed strongly to the drive of oscillator O2, whereas there was a weaker contribution from other afferents, i.e. less overlapping between the distributions of the afferents and the range of the basic frequency or the first harmonic of the oscillator. Also, γ-motoneurons showed only little frequency correlation at that time period.

The fact that neurons fire in a relatively coordinated way of up to a few milliseconds is used for re-organizing the injured CNS by re-learning phase and frequency coordination between neuron firings when exercising movements coordinated with an exactness of up to a few milliseconds, using a special coordination dynamic therapy device, i.e. by instrumented supervised phase and frequency re-learning (Figure 15).

As will be shown below, phase and frequency coordination among α_1-motoneuron firings can directly be seen in recordings of single-motor unit surface electromyography.

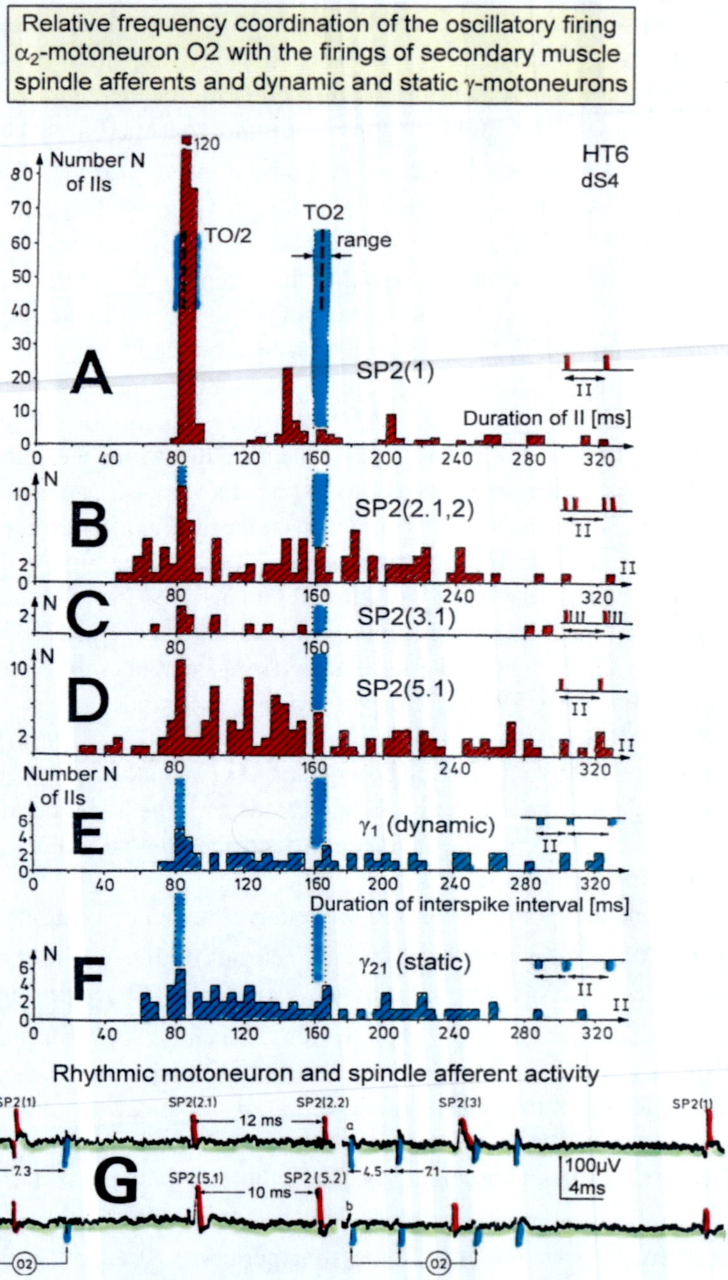

Figure 17. Interspike interval distributions of single endings of four secondary muscle spindle afferents (SP2) and two γ-motoneurons, recorded simultaneously. In A, the oscillation period TO2 (impulse train length = 3 APs) with its range of simultaneously recorded oscillatory firing α_2-motoneuron O2 (see G) is drawn for comparison; also, the halves of the oscillation period TO2/2 are indicated. Note that the interspike interval distributions of spindle afferents and γ-motoneurons have the shortest interspike interval, nearly identical to the half of the oscillation period (relative frequency coordination). The schematic impulse pattern in A to F shows the procedure for measuring the interspike intervals. Original records of the firing patterns of α_2-motoneuron O2 and the secondary muscle spindle afferents SP2(1), SP2(2), SP2(3) and SP2(5) are shown in G. Brain-dead human HT6, dS4 root.

5.3. Relative Phase and Frequency Coordination between the Firings of α and γ-Motoneurons and Secondary Muscle Spindle Afferents Recorded with the Single-nerve Fiber Action Potential Recording Method

With the single-nerve fiber action potential recording, it was shown that the neurons of the human CNS organize themselves by phase and frequency coordination (Figures 16, 17). Following injury, this organization principle is disrupted (see below). Figure 17a schematically shows a recording from a dorsal S4 root of a brain-dead human. Of the summed afferent and efferent impulse traffic the natural impulse patterns of one α_2-motoneuron, a dynamic (γ_1) and a static γ-motoneuron (γ_{21}) and two to three secondary muscle spindle afferent fibers could be extracted (for classification see Figure 8). The natural stimulations performed were pin-pricking (pain) of sacral dermatomes inside the continence automatism zone and urinary bladder catheter pulling [1,2]. It can be seen from Figure 17a that the α_2-motoneuron (α_2(O2)) fired in an oscillatory manner with 2 to 3 impulses per impulse train, and sometimes there was a break in the oscillatory firing. The impulse train for α_1-motor units consisted of only one action potential. Phase coordination's between the α_2-motoneuron, the γ-motoneurons and the secondary muscle spindle afferent fibers are indicated by different arrows and the dotted and dashed lines. It can be seen from Figure 17a that there were many coordination's between the different neurons. The relative phase and frequency coordination seems to hold for all neurons and is an integrative mechanism for the self-organization of the neuronal networks of the human CNS.

6. Surface Electromyography to Record Motor Programs, Oscillatory Firing, and Phase and Frequency Coordination among Motor Units (Electrophysiology)

6.1. Recording of Single-Motor Units

Another human electro-physiologic tool to measure natural impulse patterns of neurons is the surface electromyography (sEMG). With the same recording system used to record single-nerve fiber APs, just replacing the wire electrodes with EMG surface electrodes, single-motor unit firing and motor programs can be recorded non-invasively. The sEMG recording arrangement is shown in Figure 18 for recording motor programs from an infant.

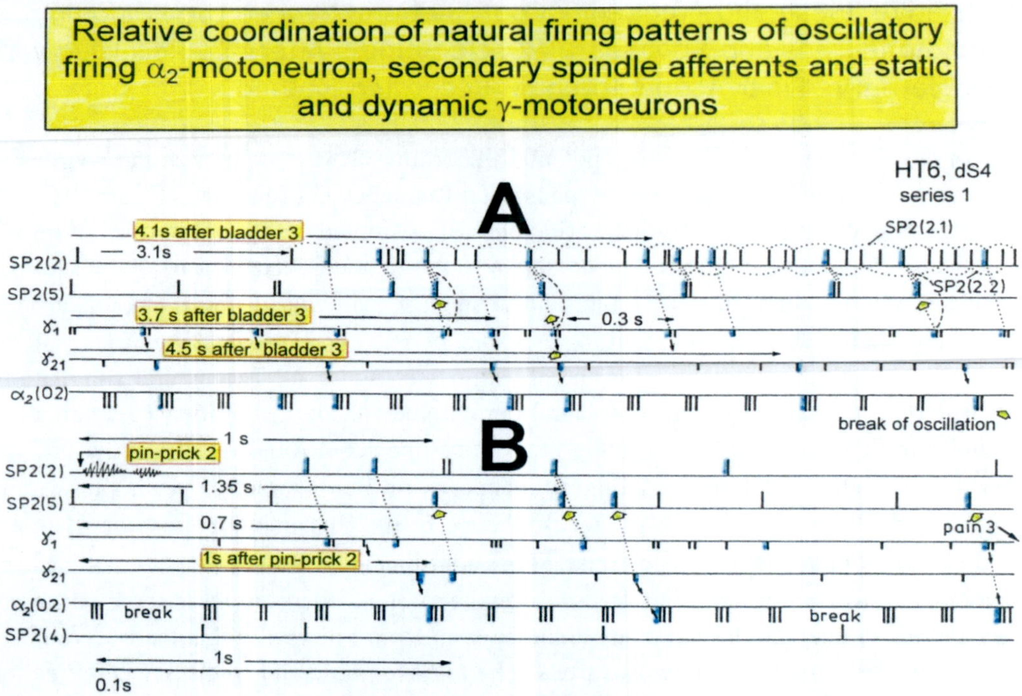

Figure 17a. Phase and frequency coordination between the extracellular recorded action potentials of simultaneously recorded γ-motoneurons (γ_1 and γ_{21}), secondary spindle afferent fibers (SP2(2), SP2(4), SP2(5)) and oscillatory firing α_2-motoneuron O2 following bladder catheter pulling (bladder 3) (A) and pin-prick 2 (B). B was recorded before A. In A the impulse patterns of the 2 encoding sites SP2(2.1) and SP2(2.2) of the single parent fiber SP2(2) are indicated by the dotted curves. Times to the activity increases of γ-motoneurons and secondary spindle afferents following stimulation are indicated. Similar time intervals of the occurrence of γ-motoneuron APs and SP2(5) fiber APs (phase coordination) are indicated by the open arrows, and the similar time intervals of γ-motoneuron APs and α-motoneuron APs are indicated by small arrows. Similar time intervals of the APs of fibers SP2(2) and SP2(5) are indicated by the double dotted lines, those of γ_1-APs and the SP2(2) fiber APs by a dotted line, and those of γ_1-APs and the SP2(2)-SP2(5) correlation by a curved dashed line. HT6; dS4-root.

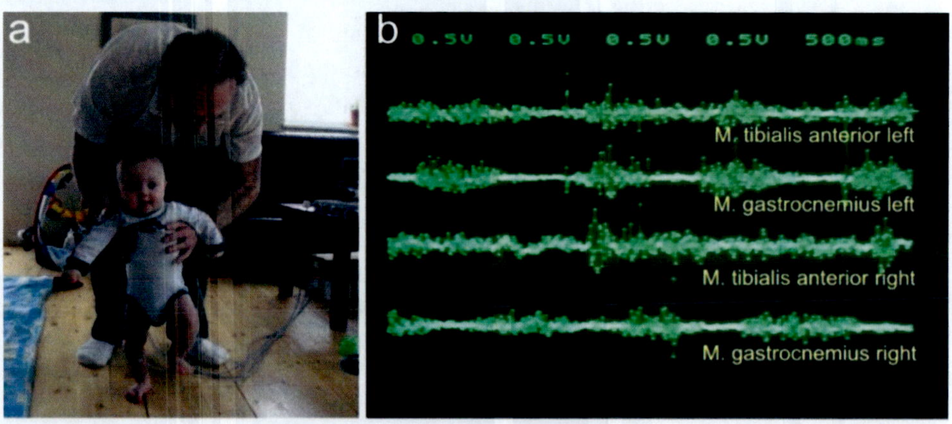

Figure 18. (Continued).

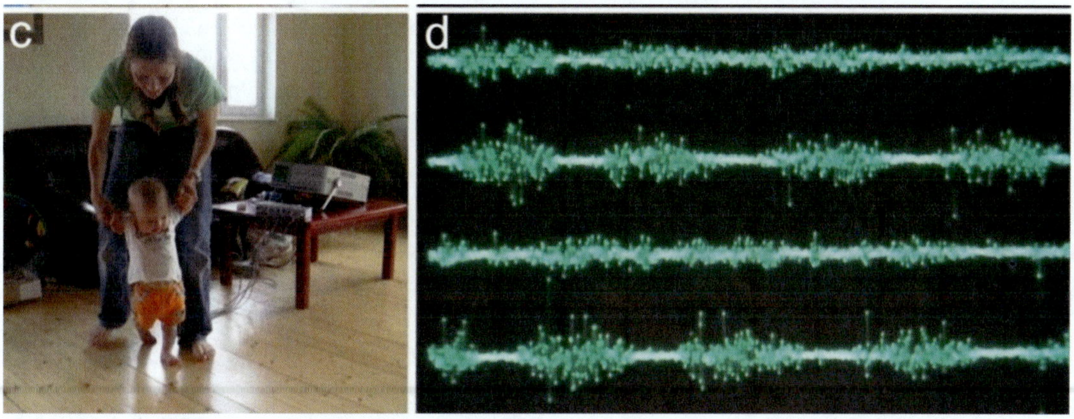

Figure 18. Surface EMG obtained from the healthy 5-months-old (a,b) and 8-months-old old "Jürgen" (c,d) during supported walking. a. Walking resembles automatic stepping, because of the strong lifting of the left knee. The toes of the right foot are plantar flexed, which is not physiologic. b. Surface EMG motor programs of left and right tibialis anterior and gastrocnemius muscles. Note that there is no antagonistic action between the tibialis anterior and gastrocnemius muscles. The right tibialis anterior muscle shows no motor program. c. The walking is more walking-like and not so much automatic-stepping-like. d. Better motor programs then 3 months earlier (b). Still there exists no antagonistic action between the tibialis anterior and gastrocnemius muscles. The activation of the right tibialis anterior muscle is a bit better than 3 months ago (b).

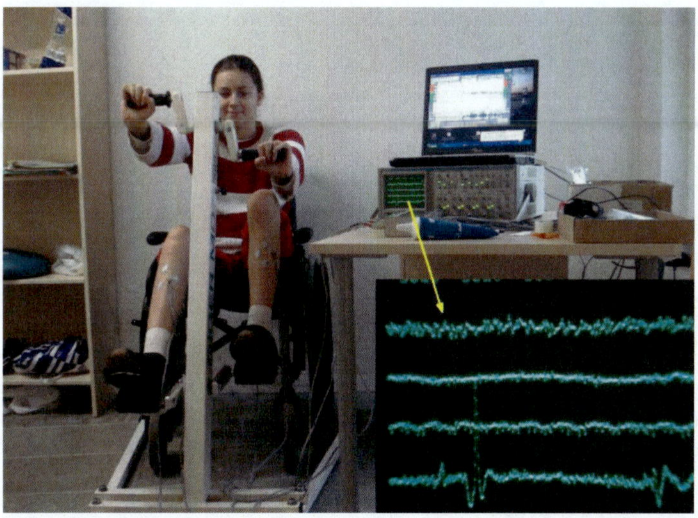

Figure 19. Layout for measuring coordination dynamics (arrhythmicity of exercising) between arm and leg movements, displayed on the laptop; for the intermediate coordination's between pace and trot gait, the fluctuation of the network states is larger. The recording of sEMG activity (displayed on the oscilloscope) from the tibialis anterior and other muscles is also shown. The inset shows single motor unit action potentials on the lowest trace. The recordings are taken from a patient with a motoric complete cervical spinal cord injury C5/6.

When surface EMG is performed from a healthy person or child, coordinated motor programs can be recorded from the different muscles (Figure 18). The patterns of recruitment of motor units cannot be seen in such a motor program, because the number of activated motor units is so high that single motor units cannot be followed. However, when only a few motor units can be activated in a certain muscle, then the pattern of activation of the motor units and the coordination between them can be seen. If the CNS of a patient is functioning rather physiologically because of a long lasting intensive coordination dynamics therapy, then an analysis of the generation of the motor program becomes possible based on single motor unit firing. Figure 19 shows the sEMG recording layout of such a patient from whom single motor unit firing can be recorded.

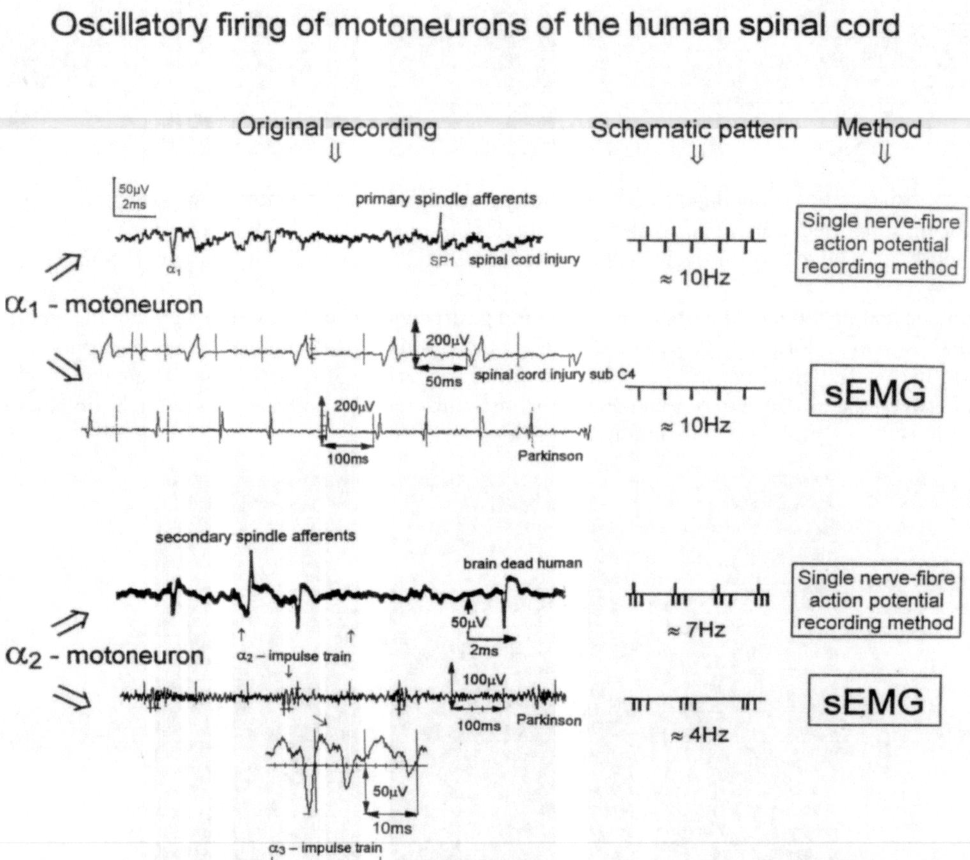

Figure 20. Oscillatory firing patterns of α_1, α_2, and α_3-motoneurons recorded from motoneuron axons with the single-nerve fiber action potential recording method and by surface electromyography (sEMG) from FF, FR, and S-type motor units. The left panel shows original recordings, the middle panel the schematic patterns; the recording methods are indicated on the right side. The recordings were taken from patients with spinal cord injury and Parkinson's disease and from brain-dead humans.

6.2. Oscillatory Firing of Motoneurons and Motor Units

In Figure 17, the different frequency patterns of oscillatory firing of motoneurons are shown. Original records were taken with the single-nerve fiber action potential recording method from motoneuron axons and surface electromyography (sEMG) from single motor units. α_1-Motoneurons innervate FF-type muscle fibers and fire rhythmically with impulse trains consisting of 1 action potential in the order of 10Hz (Figures 13,20). α_2-Motoneurons innervate FR-type muscle fibers and fire rhythmically with impulse trains consisting of 2 to 5 action potentials in the range of 4 to 7 Hz. The amplitude of the extracellular action potential of the α_2-motoneurons (axon group diameter = 10.2μm, axon group conduction velocity = 50m/s) is on average smaller than that of the α_1-motoneurons (axon group diameter = 13.1μm, axon group conduction velocity = 65m/s) (Figure 8), depending on the position of the axon in the nerve root with respect to the recording electrodes. FR-type motor unit potentials have much smaller amplitudes than the motor unit potentials of FF-type muscle fibers. The α_3-motoneurons (axon group diameter = 8.3μm, axon group conduction velocity = 37m/s) innervate S-type muscle fibers and fire oscillatory at a frequency of around 1 Hz with long impulse trains (up to 50 action potentials per impulse train). The motor unit firing of single S-type muscle fiber motor units could not be safely identified by sEMG because their amplitudes are still smaller than those of FR-type motor units and are thus difficult to identify. The impulse patterns of oscillatory firing motoneurons obtained with sEMG are similar or the same as those obtained with the single-nerve fiber action potential recording method (Figure 20). This confirms the accuracy of the single-nerve fiber action potential recording method. Since sEMG is a non-invasive recording method, oscillatory firing can be recorded easily when using appropriate patients.

6.3. Motor Program Generation, Oscillatory Firing and Coordination Among α_1-Motor Units (FF-type)

When a healthy person exercises on the special coordination dynamics therapy device (Figure 14), coordinated motor programs can be recorded from the different muscles (Figure 15). The patterns of recruitment of motor units cannot be seen in such a motor program. The number of activated motor units is so high that single motor units cannot be followed. But when only a few motor units can be activated in a certain muscle, then the pattern of activation of the motor units and the coordination between them can be seen. If the CNS of a patient is functioning rather physiologically as a result of a long lasting intensive coordination dynamics therapy, then an analysis of the generation of the motor program becomes possible based on single motor unit firing (Figure 19).

It can be seen from Figures 21 and 22 how the CNS generates a motor program in a muscle, if only a few α_1-motor units can be activated in that muscle. By increasing the load on the special CDT device (in Newton's), more muscle power is required to perform the movements and the motor unit firing rates increase, in that the motor units start to fire more rhythmically and with increasing frequency. Motor units coordinate their firing in order to avoid tremor.

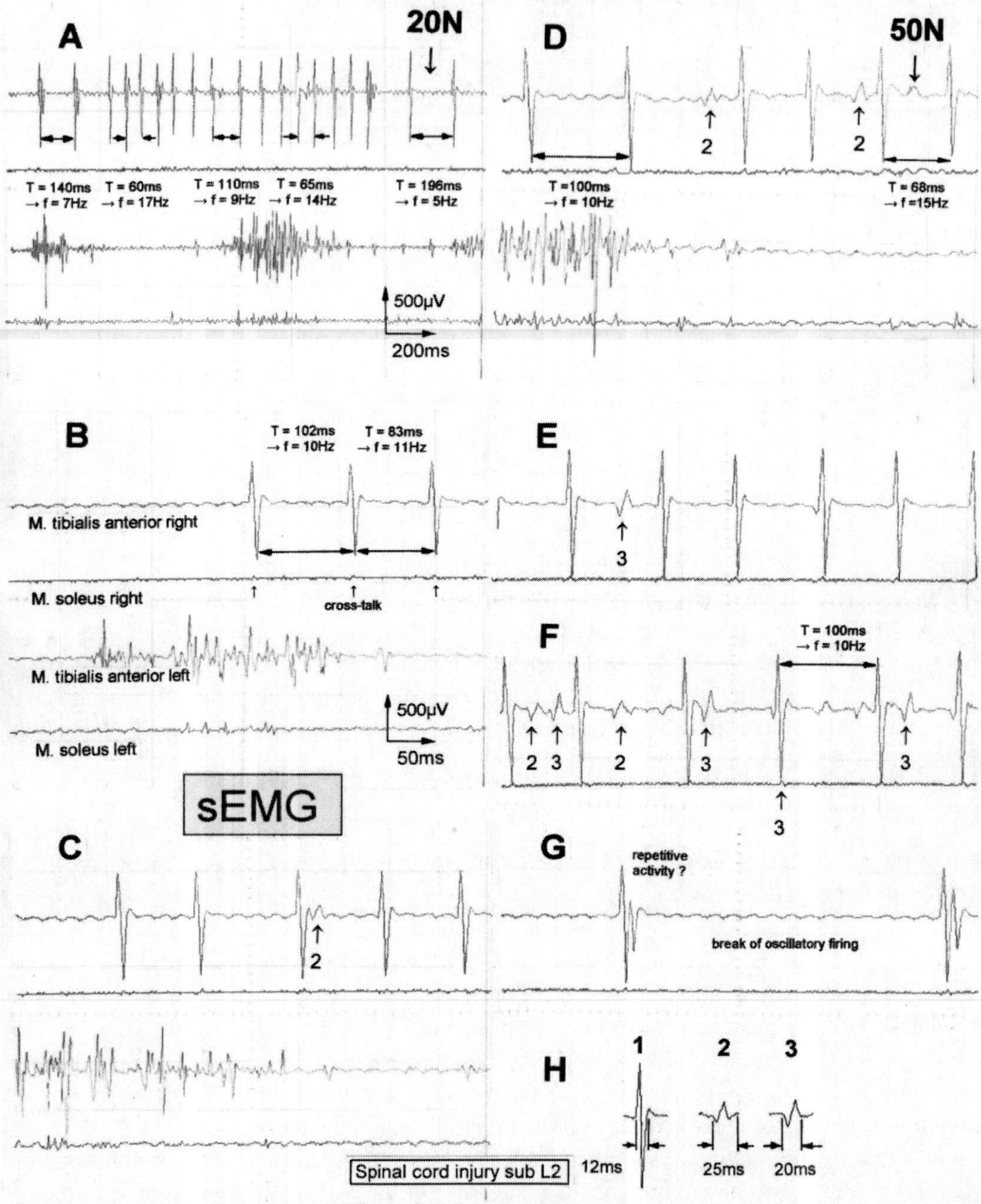

Figure 21. Non-synchronous partly coordinated firing of FF-type motor units of the right tibialis anterior muscle from a patient with spinal cord injury sub L2 suffered in an accident, measured by surface EMG; the firing was activated by exercising on the special coordination dynamics therapy device at the loads of 20 and 50N (Figure 1). Oscillation periods (T [ms]) and oscillation frequencies (f [Hz]) of oscillatory firing motor unit 1 are partly indicated. The waveforms of 3 motor units are identified in the right tibialis anterior muscle (top trace) and are shown in 'H'. Motor units 2 and 3 are marked in 'A' through 'G'; the identification of large amplitude motor unit 1 is obvious. The left tibialis anterior muscle (third trace) already shows a motor program; little activity in the soleus muscles for this low-load activation.

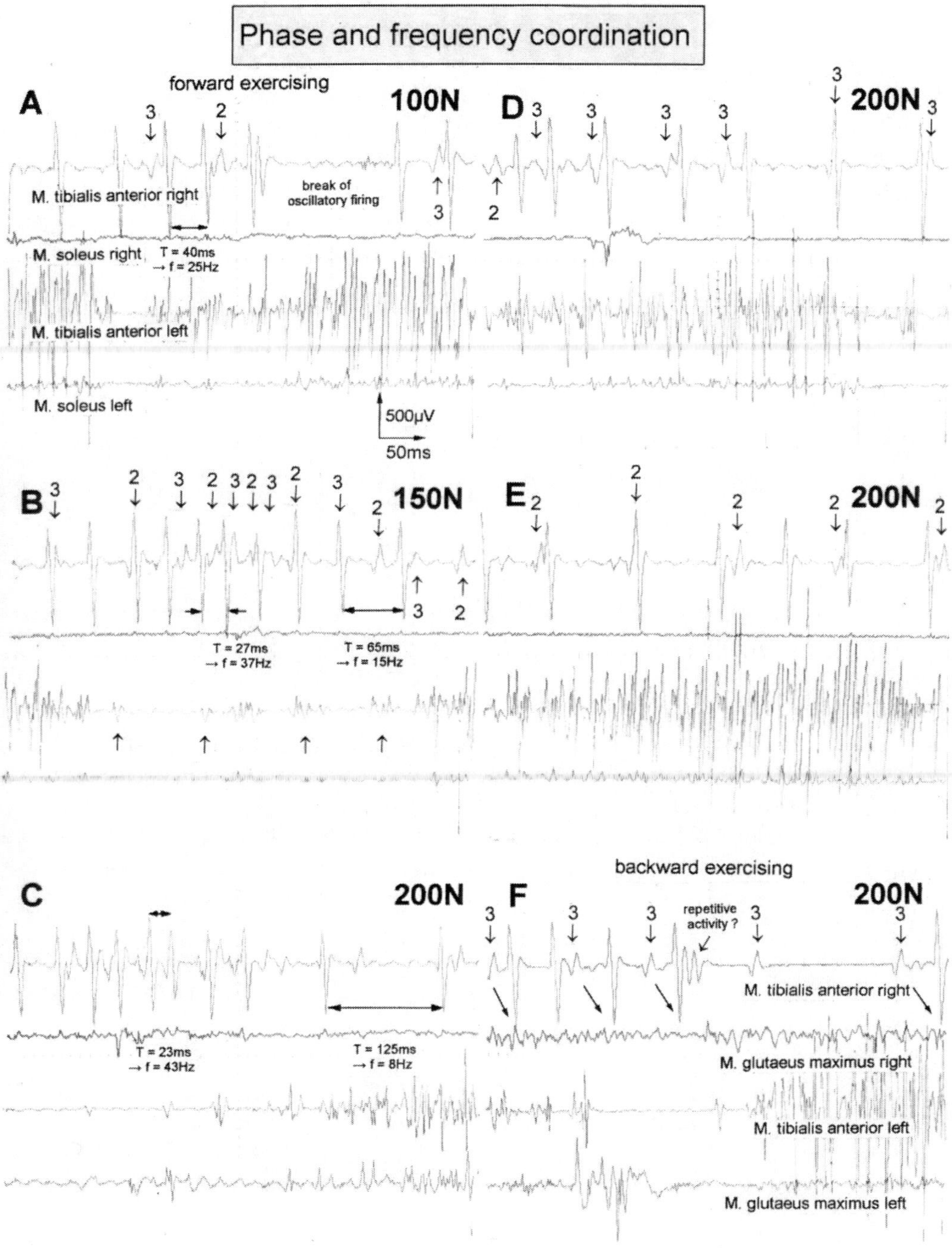

Figure 22. Phase and frequency coordination between oscillatory firing motor units (FF-type) during the generation of a motor program during exercise on the special coordination dynamics therapy device at loads increasing from 100 to 200N. Oscillation periods (T) and oscillation frequencies (f [Hz]) of oscillatory firing motor unit 1 are partly indicated. 'A,B,D,E' same recording situation as in Figure 3; 'C,F' soleus electrodes shifted to gluteus muscles to check early re-innervation upon therapy. The waveforms of the 3 identified FF-type motor unit potentials '1', '2', and '3' are the same as in Figure 3; motor units '2' and '3' are partly marked. In 'F', some coordination's between motor unit '3' and '1' are marked.

In a patient with a spinal cord injury (injury level sub L2 following an accident), the left tibialis anterior muscle (third trace in Figure 21) was already strongly re-innervated after two years of treatment. The overlapping of the motor unit potentials was so strong that single motor unit firing could only be seen occasionally. In the right tibialis anterior muscle (top trace in Figures 21,22) however, oscillatory and coordinated firing of mainly 3 single α_1-motor units could be identified. Figure 21H shows the waveforms of the 3 different motor unit potentials (1, 2, and 3). Motor unit 1 had the lowest threshold and the largest amplitude. In the majority of cases, the motor units can be identified safely, though not always. It is especially difficult when the motor units fire simultaneously or nearly simultaneously. It will now be shown how the firing frequencies and coordination between α_1-motor units firing (FF-type) changes for different movements and loads.

6.4. Firing Frequency Increases with Increasing Load

In Figure 21A, motor unit 1 fired oscillatory for a certain period to generate a motor program, when the patient exercised at the load of 20N on the special CDT device. The frequency fluctuated between 5 and 17Hz. Quite often, the frequency of this motor unit 1 was in the order of 10Hz for low load exercising at 20 or 50N (Figure 21). Motor units 2 and 3 fired mostly only occasionally. In Figure 21F, motor unit 3 fired transiently oscillatory at the frequency of approximately 6Hz.

For exercising at higher loads, the firing frequencies of the motor units increased. At 100N, motor unit 1 reached 25Hz (Figure 22A); the corresponding values at 150N and 200N being 37 Hz (Figure 22B) and 43 Hz (Figure 22C). Also, motor units 2 and 3 increased their firing rates. They started to fire also rhythmically with increasing frequencies for increasing load. At 150N (Figure 22B), motor unit 2 and 3 reached the frequency of 22Hz and 25Hz respectively. Exercising at 200N is a rather high load, even for a healthy person. The lowest frequencies measured at 200N were 9Hz and 10.5Hz for motor unit 2 (Figure 22E) and 3 (Figure 22F), respectively. For some time motor unit 1 stopped firing oscillatory (Figure 22F).

6.5. Motor Program Development

In the left tibialis anterior muscle (Figures 21,22, third trace), motor unit activity was recorded when muscle power was needed, and none or nearly none was recorded when no muscle power was needed, in some similarity to the physiologic motor programs. In the right tibialis anterior muscle (top trace) the activation pattern was different. When the muscle had to be activated, the motor units were more activated (Figures 21A, 22B) and when no muscle power was needed, less activity was recorded (Figures 21B, G, 22E, F). The activity of the motor units increased and decreased. For high load exercising, the motor units fired rhythmically with increasing and decreasing frequencies. During low load exercising, the motor units fired rhythmically with low frequencies or stopped firing in an oscillatory manner and fired only occasionally. The activity changes from motor unit firing to no motor unit firing during the motor program were not as pronounced in the right tibialis anterior muscle, where only a few motor units could be activated, as in the left tibialis anterior muscle. It was as if inhibition were missing. Inhibition is necessary to give motor unit firing patters their structure. As measured earlier, the high activity mode of α-motoneuron firing is the oscillatory firing (Figure 12).

6.6. Phase and Frequency Coordination between the Firing of the Motor Units

But how did the motor units coordinate their firing? As can be seen from Figures 21 and 22, the motor units did not synchronize their firing. In Figure 21F, motor unit 1 is firing in an oscillatory manner at around 10Hz, and motor unit 3 shows a phase lag of 22ms with respect to motor unit 1 (the 2 right potentials). But when motor unit 2 fired additionally (on the left part of the trace), motor unit 3 changed its phase with respect to motor unit 1 so that all 3 motor units fired unequally to give rise to a smooth non-rhythmic muscle contraction. In Figure 22D (200N), motor unit 3 fired approximately at the same time as motor unit 1. The motor unit potentials partly added up, because of partial synchronized firing of the corresponding motoneurons. The distributed firing of motor units was poor at that time interval. In Figure 22E, motor unit 2 fired approximately at the same time as motor unit 1. In Figure 22F, motor unit 3 fired in phase correlation to motor unit 1, but it was not synchronized. When motor unit 1 transiently stopped to fire in an oscillatory manner, motor unit 3 continued firing in an oscillatory manner. When motor unit 1 started to fire in an oscillatory fashion again, the same phase coordination built up to the firing of motor unit 3. For the high activation parts in Figure 22B,C, motor units 1, 2 and 3, and some other unidentified units, fired very close to each other (but not in a synchronized manner) and the potentials overlapped, because the frequencies were very high to generate strong muscle power.

6.7. Phase and Frequency Coordination's between α_1-Motor Unit Firing of Different Muscles and Different Arms

Figures 21 and 22 show relative phase and frequency coordination for motor unit firings in the same muscle, namely the right tibialis anterior muscle, of a patient with a spinal cord injury sub L2 suffered during an accident. Figure 23 shows phase and frequency coordination among motor units between different muscles and arms in a spinal cord injury patient (sub C4/5). A single FF-type motor unit of the right flexor carpi radialis muscle fired at approximately 8Hz when activated on volition (Figure 23A). In Figure 23B, the same motor unit is seen firing at 12.5Hz. This motor unit fired in relative coordination (marked with the arrows) with another single FF-type motor unit in the left flexor carpi radialis muscle. The phase relations have different values in 'A' and 'B'. However, also note that the motor unit in the flexor changed its oscillatory firing from 6.7Hz to 10Hz, probably to develop more motor unit force. Not every motor unit potential is correlated. There may be correlations to motor units of other muscles, which were not recorded from. In Figure 23C, phase relations are indicated with the arrows from the right extensor carpi radialis to the left extensor carpi radialis. In Figure 23D, phase relations are indicated by the arrows, from the left extensor carpi radialis to the left flexor carpi radialis. Thus, there is coordinated firing between all motor units activated on volition, when the patient is not exercising on the special coordination dynamics therapy device, but is in position to do so.

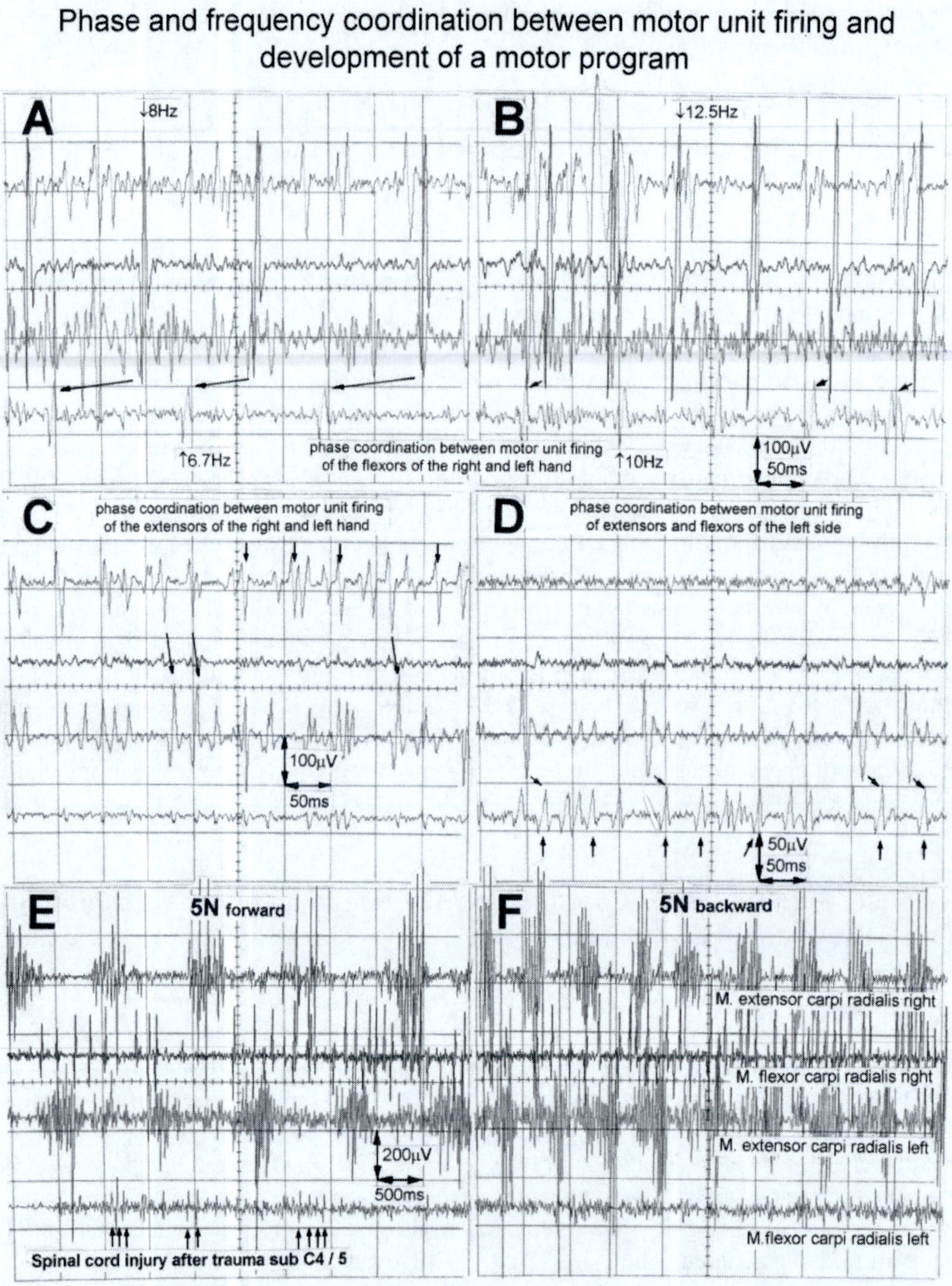

Figure 23. Phase and frequency coordination of single-α_1-motor motor unit firing between different muscles and between different arms. Phase coordination's are indicated by the arrows between motor unit potentials. A-D. The patient with a spinal cord injury was in position at the special coordination dynamics therapy device but activated muscles on volition upon looking onto the oscilloscope screen. E, F. Patient exercising on the special device. Note the beginning of the appearance of a motor program in the right flexor carpi radialis muscle for 5N forward turning. Recordings for a patient with a complete spinal cord injury sub C4/5.

During exercise on the special coordination dynamics therapy device at 5N, a motor program was partly generated in the right and left extensor carpi radialis muscles (Figure 23E,F). In the right flexor carpi radialis muscle, the CNS tried to build up a motor program with 2 or 3 innervated motor units. In the left flexor carpi radialis muscle, no motor program was generated, even though small motor unit potentials can be seen (marked by the arrows in Figure 23E).

It was measured in Figure 23 that phase and frequency coordination also takes place among motoneurons, innervating different muscles. Above it was shown that there is coordinated firing between afferents and efferents (Figures 16,17). It is likely that phase and frequency coordination is a general organization principle of the human CNS.

7. Similar Efferent Impulse Patterns Obtained with the Two Electrophysiological Methods Single-nerve Fiber Action Potential Recording Method and Single-Motor Unit sEMG

With the single-nerve fiber action potential recording method, natural impulse patterns were recorded simultaneously from several single afferent and efferent nerve fibers and functions of the human CNS were analyzed. The disadvantage is that this is an invasive method. With surface EMG, natural impulse patterns of several motor units and motor programs were recorded. The disadvantage is that one can obtain only natural impulse patterns from the motoneurons and not from afferent fibers. The advantage, however, is that it is a non-invasive recording method.

By correlating the natural impulse patterns obtained with the single-nerve action potential recording method with those obtained with the surface EMG, one can verify whether the natural impulse patterns are the same or similar and can also verify the accuracy of the measured patterns. Organization can be measured from the neuron level and followed right up to the muscle level.

As Figure 24 shows, the quality of the sEMG recordings can be much better than that of the single-nerve fiber recordings. The amplitude of the single-motor unit APs is in the range of 20 times greater than those of single-nerve fiber APs and the duration of the motor unit potentials is also in the range of 20 times longer (Figure 24). The precise amplitude and duration depends on the kind of fiber. It is very important that the natural firing patterns obtained with both electrophysiological recording methods are the same for the different kinds of motoneurons (Figure 20). Both human electrophysiological methods can therefore be used to analyze human CNS functioning. For study, learning only the non-invasive sEMG can be used.

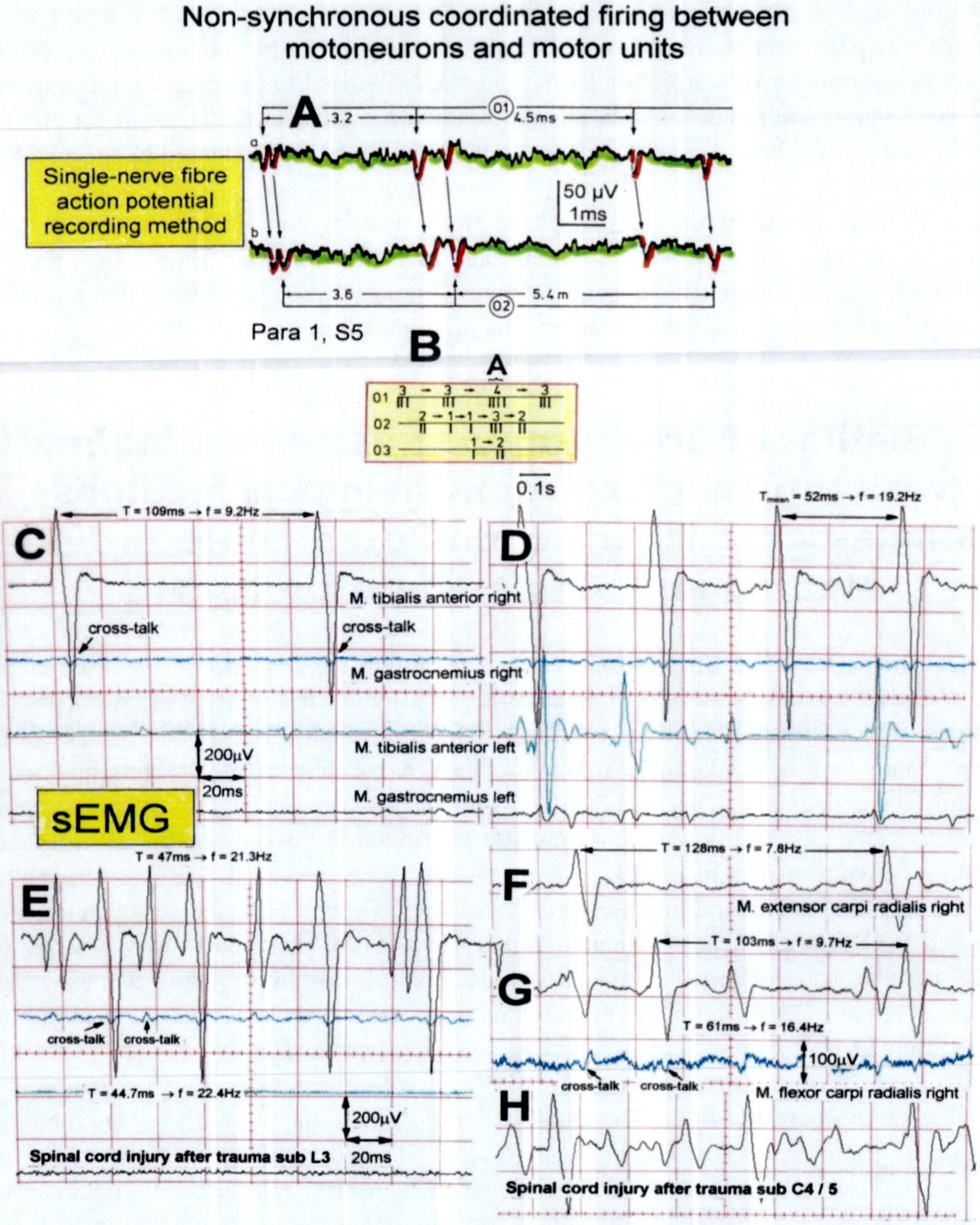

Figure 24. Non-synchronous relative coordinated firing of motoneurons from patients with spinal cord injury, recorded with the single nerve-fiber action potential recording method (A,B) and with sEMG from FF-type motor units (C-H) activated on volition. Oscillation periods (T) and oscillation frequencies (f [Hz]) are indicated. Note that the waveform identification of the motor unit action potentials in C, D, and E is safe. Note further that the α_1 – motoneuron, giving rise to the FF-type motor unit potential of 'C', fired in 'C' oscillatory at 9.2Hz, in 'D' at 19.2Hz, while in 'E' it fired in a non-synchronous relative coordination with another α_1 – motoneuron giving rise to another FF-type motor unit potential. Even the cross-talks from M. tibialis anterior to M. gastrocnemius were of such a high quality that they could be used for waveform identification and thus for firing pattern identification (C, D, E). The oscillatory firing patterns of motoneurons, recorded from motoneuron axons (A, B), showed normally no synchronization, as reported earlier.

Figure 18 shows that the motor program improves during development between the age of 5 to 8 months, and from Figure 27 it can be seen that the motor program improves, i.e. repairs following six months of CDT. However, a safe quantification of improvement is not entirely possible. As we need to measure CNS functioning and learning in patients reliably and non-invasively and want to characterize CNS functioning and learning by a single, authoritative parameter a further human CNS measuring tool is needed. By using the System Theory of Pattern Formation and making the step from the muscle and motor program level to the movement pattern level, it will become possible to measure the quality of CNS organization by one value, namely the coordination dynamics value, to better understand human CNS organization and learning.

8. Integrative Physiology: System Theory of Pattern Formation

8.1. The System Theory of Pattern Formation for Understanding Neuronal Network Organization and Learning

To understand the on-going changes of movement and other patterns in healthy humans and in patients with CNS injury, malformation, and degeneration (aging), the System Theory of Pattern Formation is used. In a complex system like the human CNS, patterns are generated by a nervous system, which seeks cooperative stability. Stability is what defines collective states. The system has the tendency to slip into the collective states to which it is attracted. When an infant crawls (Figure 25), its arms and legs are strongly attracted to the 'pace' and 'trot' gait patterns. The attraction is so strong that intermediate crawling patterns seemingly do not exist, as if the patterns are hard-wired. But with the help of the special CDT device (in the background of Figure 25) the CNS can generate intermediate coordination patterns. A patient with a CNS injury often crawls with intermediate arm and leg coordination patterns and has to re-learn the pace and trot gait coordination's for CNS repair and shifts in this way the attractors for crawling to the pace and trot gait coordination's. Attractive states and attractors of the CNS organization can be pictured as a ball in a potential well, or more generally in an attractor layout (Figure 27). Changes in CNS functioning are characterized as continuous stabilization and destabilization, over time, of preferred attractor states.

Figure 25 shows a girl with cerebral palsy who tries to relearn the attractor state patterns pace and trot gait crawling. A therapist is crawling in interpersonal coordination to speed up the learning process. The visual input from the exact crawling of the therapist into the CNS of the girl improves the performance of the trot gait pattern, in this case. For this supervised learning, the cerebral palsy girl does not need to concentrate on it. It is working automatically. This interpersonal coordination is something like when soldiers march together. Once they get the rhythm among each other, the marching coordination works automatically. It was even reported that soldiers could march together in interpersonal coordination when half sleeping.

Figure 25. Trot gate crawling of a cerebral palsy girl in interpersonal coordination with the therapist. The crawling performance of the therapist is not optimal. The right arm is leading with respect to the left knee.

Figure 26. Cerebral palsy girl and therapist are crawling in optimal interpersonal coordination; only the crawling performance of the girl is not optimal. In the therapist, the left knee and the right hand are lifting synchronizedly.

This supervised teaching of the therapist, so that the patient learns faster, needs a lot of concentration. The therapist must copy the pattern of the patient and then motivate her into a better performance. In doing so, the therapist sometimes is also losing her own movement pattern. With the concentration on the patient and adapting partly to the pattern of the patient, the stability of her own movement pattern is reducing strongly and is easily lost. With adaptation to the patient, her potential well of her movement pattern became more shallow and deformed and the ball jumps then easily out of the well (Figure 28).

In Figure 26, the therapist is having a good performance of the trot gait pattern. The interpersonal coordination between the therapist and the patient is that moment is optimal. Both are in a good trot gait pattern performance.

To reduce for understanding the complexity of human neural networks of the many billions of neurons, *order parameters* or *collective variables* are introduced for the generation of certain movements. An equation of motion describes the coordination patterns dynamics. However, coordination patterns are not only determined by the task or biological function. Patterns adjust continuously to requirements from the environment (transmitted by impulse patterns from stimulated receptors in the periphery), memory, intention, and support given by a therapist. The specific requirements are captured by the concept of behavioral information and are made part of a vector field that attracts toward the required patterns. The coordination pattern dynamics, characterized by equations of motion of collective variables (the vector X), takes the general following form [20]:

$$dX/dt = F_{intr}(X) + \sum c_{inf} F_{inf}(X,t) \quad (2)$$

where F_{intr} designates the *intrinsic dynamics* of the nervous system. These intrinsic dynamics capture the anatomical (neuronal network structure), physiological, and pathological states of the CNS, and its muscular-skeletal elements.

$\sum c_{inf} F_{inf}(X,t)$ represents the sum of external influences ($F_{inf}(X,t)$) with their relative strength (c_{inf}) pertaining to each influence. The so-called *behavioral information* $F_{inf}(X,t)$ includes cognitive states, emotional states, intentions, motivations, instructions, inter-personal coordination, movement support, etc. During motor learning, or while applying therapy to a patient, these extrinsic influences become extremely important, because the intrinsic (pattern) dynamics can be changed with these extrinsic influences by altering the equation of motion. By modulating the behavioral information, the intrinsic dynamics of the neuronal networks can be influenced further, that is if CDT is no longer efficient in repairing the injured CNS, requiring the therapy to be updated. With respect to a healthy athlete, the movement performance can be improved by modulating the behavioral information by, for example, including in the training program the exercising on a special CDT device to improve CNS functioning.

If the behavioral information includes the exercising of extremely coordinated, integrative movements, like exercising on the special CDT device, then the quality of CNS self-organization can be enhanced by improving the exactness of self-organization, namely the precision of phase and frequency coordination between neuron and neural assembly firings. By improving the precision of organization of the intrinsic dynamics, that is the specific variability of the injured networks, certain patterns do eventually re-appear in the case of repairing the injured CNS by movement-based learning.

8.2. Learning Implications for Treatment Derived from the Equations of Motion of the Collective Variables (Formula 2)

From the repair by learning in the severely injured CNS, we study learning in the healthy CNS, because the impact on the learning mechanisms is higher, like in animal experimentation.

1. Behavioral requirements $\mathbf{F}_{inf}$ (like intention, support, and instruction) affect the whole coordination dynamics, including stability, rather than only certain coordination patterns. The **change of the whole coordination pattern dynamics** of the CNS by the behavioral information **is one scientific basis for learning transfer** between different patterns and **stability changes of patterns** (as for example the reduction of spasticity). The other scientific basis for learning transfer is followed from human neurophysiology, namely that nerve cells or neural sub-networks are involved in different neural network organizations [1].
2. Intrinsic coordination tendencies captured by the intrinsic dynamics influence the performed pattern systematically because the degree to which intrinsic tendencies conflict or agree with the required patterns determines the variability of the performed coordination pattern.
3. A reduction in stability of movements and other patterns when intrinsic and informational requirements conflict, may lead to loss of stability and abrupt change while behavioral information is changing smoothly.
4. The intrinsic dynamics $\mathbf{F}_{intr}$ include vegetative and higher mental functions (these are also patterns of the coordination dynamics), which indicate that via exercising coordinated movements with support and/or instructions ($\mathbf{F}_{inf}$), urinary bladder function, intelligence and speech may be partly repaired or improved following CNS injury or malformation.
5. When in an injured CNS with a certain set of behavioral information ($\sum \mathbf{c}_{inf}\mathbf{F}_{inf}$) the intrinsic coordination dynamics ($\mathbf{F}_{intr}$) can no longer be influenced during coordination dynamics therapy, then this set of behavioral information must be changed (using different $\mathbf{F}_{inf}$), or balanced differently (using different $\mathbf{c}_{inf}$), to further improve CNS organization dynamics.
6. However, the equations of motion of the coordination pattern dynamics (formula 2) provide no information about the specific behavioral information (F_{inf}) and training intensity ($\mathbf{c}_{inf}$) with which the CNS can be efficiently repaired by learning in the patient. We need to have detailed knowledge of the human CNS on the single neuron and neural assembly level [1], as well as the knowledge at the integrative level, to find the specific behavioral information for the repair by learning the human CNS.

A *first novel step* in coordination dynamics therapy is the inference derived from formula 2 of the equation of motion. It suggests that movement learning not only improves the performance of that particular movement, but also improves the other non-trainable functions by *transfer of learning*. These functions include vegetative functions like bladder control, speech and higher mental functions.

Furthermore, we have a means by which the stability of physiological network states can be increased (e.g. movements, continence, continuous concentration in performing certain tasks, speech, etc.) and simultaneously the stability of pathological network states, like spasticity, are decreased. The coordination (pattern) dynamics therapy, partly based on the System Theory of Pattern Formation in combination with human neurophysiology, thus offers us an important theoretical basis and a practical tool to diagnose, quantify, and repair/improve the functioning human nervous system at the macroscopic level.

8.3. Geographical Landscape of Attractors

The drawback of the equation of motion of the order parameters (formula 2) is that it is normally not possible to find a mathematical solution to it. But by defining a potential function and by picturing the attractive states and attractors by a ball in a potential well or rather by a ball moving in a geographical landscape of attractors (Figure 26), we form a theoretical basis to understand and measure stability of certain coordinated movement patterns (i.e. the deepness of the potential well of an attractor) in patients with CNS injury who receive on-going therapy.

By studying the pattern change of certain highly coordinated arm and leg movements, while a subject is exercising on a special coordination dynamics therapy device (Figure 19), pattern stability can be made visible and the mean stability per one minute can be measured by the arrhythmicity of exercising (see below). Such value, so-called coordination dynamics value, quantifies CNS functioning objectively, integratively, and non-invasively. The *assessment of quality of CNS organization by pattern change is a second novel step* in CDT.

To make the strategy of pattern formation, pattern stability, pattern assessment, and pattern picturing understandable, the procedure is demonstrated for the simple movement 'jumping on springboard', which is used during CDT, especially for the repair of the urinary bladder, and training in the up-right weight-bearing posture (very important in patients with SCI).

8.4. Equation of Motion, Potential Function and Attractor Layout for the Movement 'Jumping on Springboard'

For the special movement 'jumping on springboard' with no behavioral information ($\sum c_{inf} F_{inf}(X,t) = 0$) the equations of motion (formula 2) take the form:

$$d\varphi/dt = f_{intr}(\varphi)$$

Where φ is the relative phase between the two moving legs and is the only collective variable of this special movement.

The mathematical solution of $d\varphi/dt = f_{intr}(\varphi)$ in the Haken-Kelso-Bunz model [20,21] (for the approximations being made, see Ref. 22) gives the equation of motion for jumping on a springboard for the symmetric case [23]:

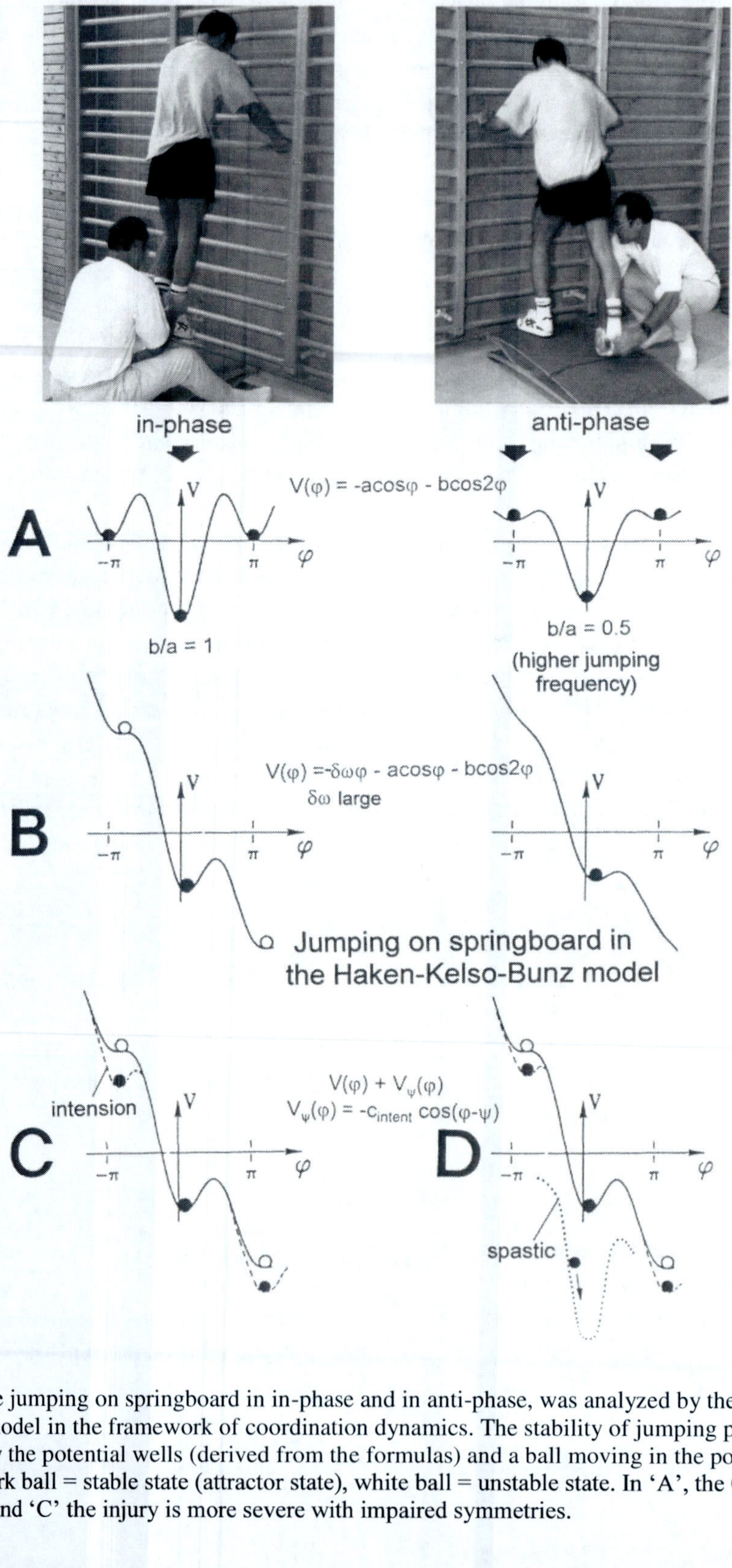

Figure 27. The jumping on springboard in in-phase and in anti-phase, was analyzed by the Haken-Kelso-Bunz model in the framework of coordination dynamics. The stability of jumping patterns is represented by the potential wells (derived from the formulas) and a ball moving in the potential landscape. Dark ball = stable state (attractor state), white ball = unstable state. In ‘A’, the CNS injury is small; in ‘B’ and ‘C’ the injury is more severe with impaired symmetries.

$d\varphi/dt = -a(t)\sin\varphi - 2b(t)\sin 2\varphi$

The so-called potential function is defined by

$d\varphi/dt = -\partial V(\varphi,t)/\partial\varphi$ (19).

By integration, we obtain the potential function for jumping on a springboard:

$V(\varphi,t) = -a(t)\cos\varphi - b(t)\cos 2\varphi$

The potential function $V(\varphi,t) = -a(t)\cos\varphi - b(t)\cos 2\varphi$ can be plotted for different φ and certain ratios of the parameters a and b, and is shown in Figure 27.

The potential function shows two attractor states, namely the jumping in in-phase ($\varphi = 0$) and the jumping in anti-phase ($\varphi = \pm\pi$). Especially for the higher frequencies (smaller b/a), the jumping in-phase has a higher stability (the potential well is deeper) than the jumping in anti-phase. Asymmetry (not tackled mathematically here) strongly changes the stabilities of the attractor states (depths of potential wells) (Figure 27).

The human CNS, seeking for cooperative stability, slips into the collective states to which it is attracted. For jumping on springboard, these attractive states are the jumping in in-phase and in anti-phase. For crawling (not creeping) the attractive states are the pace (in-phase) and in trot gait coordinations (anti-phase).

Since such a potential function can no longer be derived from more general movements, especially when the CNS is injured, malformed, or degenerating, the temporal stability of different movement patterns for a characterization of CNS functioning must be measured. This is partly possible by measuring the so-called coordination (pattern) dynamics (see below).

8.5. Including the Variability of Phase and Frequency Coordination among Neuron Firing into the Equation of Motion of the Collective Variables

Depending on the relationship between the initial coordination dynamics (so-called intrinsic dynamics, $F_{intr}(X)$, depending strongly on the severance of the injury) and the patterns to be re-learned (termed behavioral information, $\sum c_{inf} F_{inf}(X,t)$, which act as attractors of the coordination pattern dynamics toward the required patterns), qualitative changes in the attractor layout occur with learning, accompanied by qualitative evidence for loss (or change) of stability. The nature of change due to learning (e.g., abrupt versus gradual) arises from the cooperative and competitive interplay between the behavioral information (supported jumping or walking of the patient) and the intrinsic dynamics.

A completely different, additional nature of necessary learning is needed in the repair of CNS injury. The impaired phase and frequency coordination among neuron firing (Figures 13-15), must be repaired by re-learning for proper CNS self-organization. This perturbation of CNS self-organization produces deviations from the attractor states and changes the attractor layout because of altered hard wiring due to injury. In a first approximation, this tremendously increased variability of phase and frequency coordination can be included into the equations of motion of the collective variables and gives further understanding of pattern change in patients with CNS injury, as for example the switch from a movement pattern to a spastic pattern (Figure 16B).

In the Haken-Kelso-Bunz model, the jumping on springboard (Figure 27) can be described in terms of relative phase between the rhythmically moving legs. Without specific behavioral information the dynamical description is defined by a vector field (a differential equation) expressing the rate of change in relative phase, dφ/dt, as a function of the derivative of its potential, V(φ):

$$d\varphi/dt = -\,dV(\varphi)/d\varphi + (Q\xi_t)^{1/2} \tag{3}$$

where $V(\varphi) = -\,a\cos(\varphi) - b\cos(2\varphi)$ and $(Q\xi_t)^{1/2}$ is the phase and frequency variability of strength Q (where ξ_t is Gaussian white noise of unit variance). Zanone and Kelso [24] introduced noise in Equation 3 (from a logic point of view), because all real systems described by low-dimensional dynamics are coupled to many subsystems at a more microscopic level. One may view noise as a continuously applied perturbation that produces deviations from the attractor state. Such fluctuations are conceptionally important in dynamical modeling of phase transition or bifurcation phenomena and are essential in effecting transitions.

I included noise in Equation 3 (from the experimental point of view) because of the measured increased variability of phase and frequency coordination among the coordinated firing of neurons and neural assemblies in the human CNS. This at the neuron level, measured fluctuation of phase and frequency coordination is giving rise to phase transitions or bifurcation phenomena, which are essential in causing transitions among attractor states under physiologic (small fluctuation; Figure 15 (HT5 or normal eigenfrequency distributions); Figure 12B) and pathologic conditions (large range of the eigenfrequency (Figure 15) (Para 2 distribution) and consequently large variabilities of phase and frequency coordination (Figure 13)). The relative stability of an attractor state is, therefore, reflected by the depth of each potential well (I) and the strength Q of the variability of phase and frequency coordination (II), and the attraction of attractor states is reflected by the slope at each point of the potential curve.

The behavioral changes when jumping on springboard (Figure 28) are represented by the over-damped movement of a rolling ball in the potential landscape for the physiologic (Figure 28A, Q small = little fluctuation of phase and frequency coordination) and the pathologic case (Figure 28B,C, Q large = large variability). The increased fluctuation in the rather stable state, due to increased variability of phase and frequency coordination, will have greater probability of "kicking" the system out of attractor the basin (Figure 28B,C), especially in the asymmetric case.

In the healthy CNS, the phase and frequency variability is small (short arrows) and the jumping in-phase and anti-phase is stable (Figure 28A). Following injury, the potential landscape is deformed and the fluctuation of the network states, generating jumping, is high (Figure 28B). The in-phase jumping is still stable in spite of the increased fluctuation, because the basin of attraction is deep. The jumping anti-phase became unstable because the basin of attraction is shallow and the increased fluctuation in the state has a greater probability of "kicking" the system out of the basin. A switch into a spastic state is also possible. In severe CNS injury or malformation, the patient cannot jump any more in anti-phase because of the missing attractors for anti-phase jumping (Figure 28C). Support is needed for anti-phase jumping (Figure 27, upper right). The jumping in-phase is still possible but unstable (Figure 27, upper left).

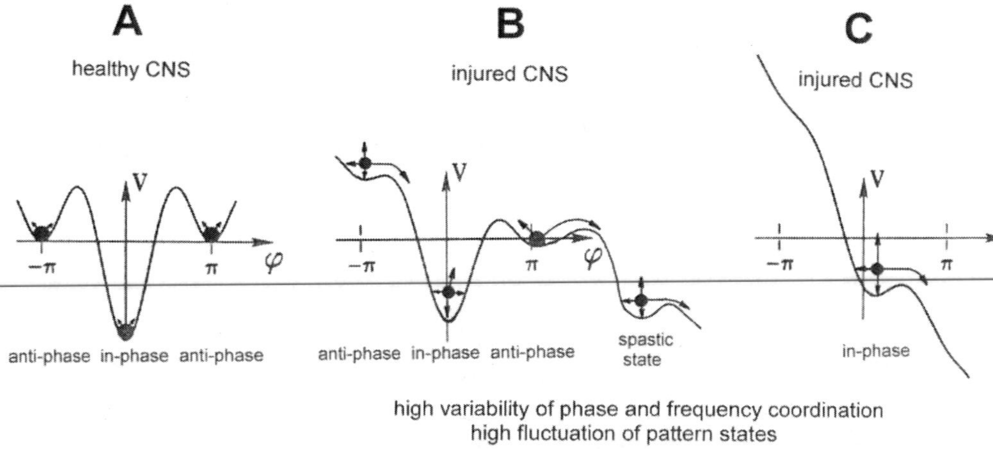

Figure 28. The potential, V(φ), of the coordination dynamics for jumping on springboard of a healthy (A) and injured CNS (B,C). The region around each local minimum acts like a well that weakly traps the system into a coordinated state. Behavioral changes are represented by the over-damped movement of a rolling ball in the potential "landscape". High fluctuations (indicated by long arrows attached to the ball (network state)) in the stable state, due to high variability of phase and frequency coordination (in the injured case), will have a greater probability of "kicking" the system out of the basins of attraction (B,C) than for low fluctuations (short arrows) (A), due to small variability of phase and frequency coordination (in A). In B, only the in-phase jumping is stable, even though the fluctuation is high. In C, there is only an attractor basin for the in-phase jumping, but the fluctuation is so high that there is a high probability that the system is kicked out of the basin of attraction. The patient can no longer jump in anti-phase and has difficulty with jumping in-phase. The stability of jumping depends on the motor program (deepness of basin of attraction) and the fluctuation of the pattern state (moving of the ball) caused by the increased variability of phase and frequency coordination due to the injury.

Upon performing very exact coordinated movements, imposed by devices (as for example shown in Figures 14 and 15), the nervous system of the patient learns to reduce the variability of phase and frequency coordination and achieves in this way a small fluctuation of the network states again, as shown in Figure 28A. The progress in treatment (learning) is that the in-phase jumping in Figure 28C and the anti-phase jumping in Figure 28B become stable (Figure 28A) again. Also, the potential landscape will change due to the reduction of the phase and frequency variability. The important consequence for treatment is that when exercising on special CDT devices and reducing, in this way, the variability of phase and frequency coordination, the patient can induce the formation of patterns again, without having trained them (learning transfer). Upon improving the coordinated firing of neurons, a cerebral palsy child may become able to speak or may develop social behaviors.

In conclusion, the impairment of phase and frequency coordination, measured at the neuron level in humans (see below), can be included in the coordination dynamics at the collective variable level. The decrease of the variability of phase and frequency coordination (one kind of coordination repair) is an essential part of CNS development and repair by movement-based learning.

8.6. Geographical Landscape of Attractors

The drawback of the equation of motion of the order parameters (Formula 2) is that it is normally not possible to find a mathematical solution to it. But by defining a potential function and by picturing the attractive states and attractors by a ball in a potential well or rather by a ball moving in a geographical landscape of attractors (Figures 27,28), we form a theoretical basis to understand and measure the stability of certain coordinated movement patterns (i.e. the depth of the potential well of an attractor) in patients with CNS injury who receive on-going therapy.

8.7. CNS Repair upon Stability Changes of Physiologic and Pathophysiologic Patterns: Improvement of Geographical Landscape of Attractors

Before showing, with human neurophysiology, that the phase and frequency coordination becomes impaired following injury, and that even in the healthy CNS the coordinated firing of neurons is sub-optimal, the integrative aspects of CNS learning should be finished.

When jumping on a springboard (Figures 29,30), the pattern changes can be represented by the over damped movement of a rolling ball in the potential landscape for the physiologic (Figure 28A, little fluctuation of phase and frequency coordination) and the pathologic case (Figure 28B,C, large variability of phase and frequency coordination). In the healthy CNS, the phase and frequency variability is small (short arrows of the moving ball) and the jumping in in-phase and in anti-phase is stable (Figure 28A). Following injury, the potential landscape is deformed and the fluctuation of the network states, generating jumping, is high (Figure 28B). The in-phase jumping is still stable in spite of the increased fluctuation (larger fluctuation arrows), because the basin of attraction is deep. The jumping in anti-phase became unstable because the basin of attraction is shallow and the increased fluctuation in the state has a greater probability of "kicking" the system out of the basin. In CNS injury, a switch into a spastic state is also possible. In severe CNS injury or malformation, the patient can no longer jump in anti-phase because of the missing attractors for anti-phase jumping (Figure 28C). The attractor layout is asymmetrical and deformed. Support is needed for anti-phase jumping. Jumping in-phase is still possible but unstable.

Upon performing very exact coordinated movements, imposed by devices, the nervous system of the patient learns to reduce the variability of phase and frequency coordination and achieves in this way a small fluctuation of the network states again as shown in Figure 28A. The progress in treatment (learning) is that the in-phase jumping in Figure 28C and the anti-phase jumping in Figure 28B become stable (Figure 28A) again. Also, the potential landscape will change due to the reduction of the phase and frequency variability. The important consequence for learning/treatment is that when exercising on special devices and reducing in this way the variability of phase and frequency coordination, the patient can induce the formation of patterns again, without having trained them (learning transfer). But exercising only on the special device is not enough, in itself, to improve the attractor layout sufficiently. Regrettably, there is no miracle device. Many different movements must be trained to repair the CNS.

8.8. Pattern Stability When Jumping on Springboard

Jumping on springboard (Figures 29,30) and exercising on the special CDT device (Figure 15) are the most important movements to repair the urinary bladder by learning. Figure 29 shows the supported jumping on springboard. The therapist (Author) is supporting the in-phase jumping to realize the movement in the patient with a motoric complete cervical spinal cord injury (SCI) (C5/6) [2]. The hands are fixed because there is little handgrip power. The legs are supported so that they cannot slip to the side, and because there is very little leg function to manage the gravity, further trunk support is needed. The Author is actually supporting the jumping movement. Such jumping movement became possible following three years of coordination dynamics therapy (CDT) when some regeneration of the spinal cord took place.

Figure 29. In-phase jumping on springboard of a patient with a formally motoric complete spinal cord injury C5/6 supported by fixations and the Author.

Re-learning of up-right movements is very important for patient with SCI. With an Eigenfrequency of 1Hz of the springboard, the α_3-motoneuron (S) oscillators activating leg muscles of slow type (S) (Figure 13) are especially improved in its firing. The jumping on springboard is oscillator formation learning. Attempt is made for the patient to re-learn jumping on springboard, which changes the geographical landscape of attractors for jumping from the pathologic case (Figure 28C) to the physiologic case (Figure 28A). First, the in-phase jumping can be achieved because the stability of the in-phase jumping is higher (Figure 28A) or plainly speaking, the in-phase jumping is the easier movement.

Figure 30. Supported jumping on springboard of a girl following tumor removal and damage of the lumbar spinal cord to repair the lumbar cord by learning. The Author supports the jumping in anti-phase.

In Figure 30, a girl is trying to re-learn jumping in anti-phase after cancer treatment. Following extirpation of the tumor, located beside the lumbar spinal cord at the ganglion and foramen intervertebrale, the lumbar spinal cord got damaged and probably also the blood supply to the cord (Figure 2). The blood supply of the spinal cord [25] is critical in spinal cord injury and tumor operations. Leg functions and continence, located in the caudal spinal cord, became impaired in the girl. The relearning mainly of jumping, walking and exercising on the special CDT device should repair the functions of the lumbar spinal cord (including continence) and its blood supply.

8.9. Reduction of Spasticity

When performing movements like walking, running, crawling, or exercising on a special CDT device (Figure 32), which imposes highly coordinated movements on the patient, the coordination dynamics can be changed in the way that the stability of spastic states decrease, and the stability of the movement states increase. Such changes of coordination pattern dynamics can be pictured again by means of an attractor layout. An attractor is pictured as a potential well (attractor valley) into which a rolling ball is attracted. The position of the ball represents the momentary state of the system. Figure 31 shows schematically such an attractor layout with the two attractor's *spasticity* and *coordinated movement*. When exercise is commenced (A), the spastic state is very stable (the attractor valley is deep) and the state of the system is attracted towards the attractor state *spasticity*. With exercise, the attractor layout is changing the short-term memory in the way that the attractor *spasticity* is getting shallower and the attractor *physiologic movement* is getting deeper (B). Because of fluctuation due to variability of phase and frequency coordination, the position of the ball, which represents the momentary state of the system, is switching between the attractor states *spasticity* and *movement*. Spasticity and movements are present simultaneously in the patient. With further exercise, the attractor *movement* becomes deeper (more stable) than the attractor *spasticity*. The patient can now perform the movements with little or no spasticity. The transient reduction of spasticity in the short-term memory, achieved by many hundreds of coordinated movements, can last from a few minutes up to several hours; this is indicated in Figure 31 by the two long arrows. The shorter backward arrow (from right to left) indicates that spasticity has slightly reduced the long-term memory. The coordination dynamics have changed. Repeated exercising will further reduce the stability of *spasticity* and increase the stability of the coordinated *movement* and will further change intrinsic coordination tendencies in the long-term memory.

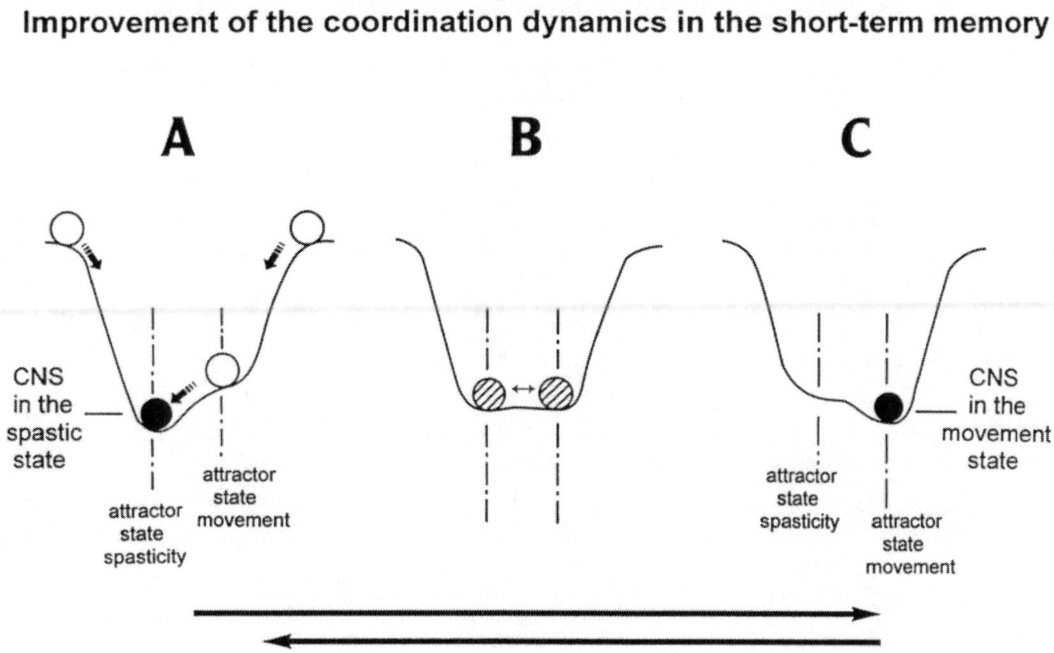

Figure 31. Therapy-induced spasticity reduction in the short-term memory. The position of the ball represents the state of the system and the potential well, the attractor. The ball is attracted to the stable position in the deepness of the whole, called attractor state. The attractor layout, consisting of two attractive states of different stability, is changing upon exercising with very coordinated rhythmic movements. Black ball = stable state, open ball = very unstable state, hatched ball = spasticity and movement co-exist. Variability of phase and frequency coordination not indicated.

Figure 32. A girl, following cancer treatment, is exercising on the special CDT device. Same girl as in figure 30. She is exercising in a rather flexed position of the legs to avoid the pathologic pattern extensor spasticity of the legs.

In Figure 32, the girl is exercising on the special CDT device to reduce the side effects of the cancer treatment, which were in this case the damage of the lower spinal cord. By training on the special CDT device, she is improving the impaired phase and frequency coordination to enhance pattern stability via improving the variability of phase and frequency coordination. Because of the damage of the lumbar spinal cord, she got the pathologic pattern extensor spasticity, similar then in traumatic spinal cord injury. When she trains the coordinated movements, she has to avoid the pathologic pattern not to deepen the potential well (the stability) of the extensor spasticity. The girl is therefore exercising in a rather flexed position to be far away from the extension. If she would train more in a stretched position, she would slip partly into the extensor spasticity. In the picture of system theory of pattern formation, the ball representing the state of the system is partly in the attractor extensor spasticity.

8.10. Quantifying CNS Function by Measuring Pattern Stability upon Pattern Change When Exercising on the Special CDT Device

Experimentally, the underlying dynamics of coordinated movements can be found in the temporal stability of coordination patterns and can be assessed through pattern change. A change of the coordinated movement patterns is generated, when a subject is exercising on the special CDT and recording device (Figures 19,32), where the coordination between arms and legs, imposed by the device, changes continuously between pace (P) and trot gait (K) and backwards. The stability of the intrinsic coordination tendencies is measured by the deviations and differential stability during the performance of these rhythmic movements (Figure 33F). When the differential stability of the movement pattern is high, the arrhythmicity of exercising is small, and when the stability is low, the arrhythmicity of exercising in that pattern is high. From the coordination dynamics trace in Figure 33F it can be seen that in the healthy case the arrhythmicity is high for the pace and trot gait coordination, and is low for the intermediate coordination between pace and trot gate. The pace and trot gait coordination, between arms and leg movements, have a high stability and the intermediate coordination patterns have a low stability.

The plotting of the differential stability, over time of the frequency of exercising, generates an attractor layout for this special movement (Figure 33G), and the mean stability per minute can be measured by the arrhythmia of exercising (df/dt:f, f = frequency; or dv/dt, v = angular velocity). Such differential stability value per minute, the so-called coordination dynamics value, quantifies CNS functioning objectively, integratively, and non-invasively. The assessment of quality of CNS organization by pattern change is a *third novel step in CDT.*

8.11. Forward-backward Symmetry Impairment

As the coordination dynamics traces in Figure 20 show, the lack of symmetry between forward and backward exercising can be made visible and quantified in a patient with cerebellum and cerebrum injuries by measuring the coordination dynamics (temporal stability) through pattern change. When exercising at a load of between 100 and 200N, the values for the exercising frequency and the coordination dynamics increased and decreased rhythmically both for forward (Figure 20A) and backward turning (Figure 20B). But when turning forward, the arrhythmicity was small to the right side of the pace (P) and trot gait (K) coordinations (Figure 20A) and when turning backwards (Figure 20B), the arrhythmia was small to the left side of the pace (P) and trot gait (K) coordinations. Therefore, there seems to be a mirror picture shift of the arrhythmia between exercising forward and backward with respect to the pace and trot gait coordination's.

In the case of a healthy patient, the lowest arrhythmia of exercising (the highest pattern stability) was positioned at the pace and trot gait coordination's (Figure 33F). To make this mirror image change with respect to forward and backward exercising better visible in the patient, the coordination dynamics traces for forward and backward exercising are arranged together in Figure 33C (lower part). The recordings were taken one week later to show that patterns of the highest stability were at the same place with respect to the pace and trot gait coordinations, to show that there was no phase drifting. It can clearly be seen in Figure 33C that the periods of lowest arrhythmia shifted rather symmetrically away from the pace (P) and trot gait (K) coordination positions with respect to forward and backward exercising. When picturing this rather opposite symmetric shift of the arrhythmia of exercising in an attractor layout of pattern formation in the framework of the system theory of pattern formation, the attractor states (patterns with highest stability) shifted away from the pace and trot gait coordination patterns (Figure 33E). In the attractor layout of a healthy, physically active girl (Figure 33G), drawn from the coordination dynamics recording of Figure 33F, the stable movement patterns (when exercising on the special device) are lying at the pace and trot gait coordinations. For the patient with the brain injury, the stable movement patterns (small arrhythmia of exercising) shifted with respect to the cycle of pattern change (lower part of Figure 33G) between $\varphi = 45°$ and 65° forwards for forward exercising, and between $\varphi = 50°$ and 65° backwards for backward exercising (Figure 33E). This rather opposite shift of the attractor states shows similarity to the rather mirror picture change of the antagonicity impairment of the antagonistic muscle activation of the tibialis anterior and gastrocnemius muscles, measured by surface EMG in this patient (Figure 34). We can therefore measure symmetry impairment of CNS organization with respect to forward and backward exercising with the special CDT device. Symmetry diagnostics, with the special CDT device, can also be performed with respect to right and left, and rostral and caudal.

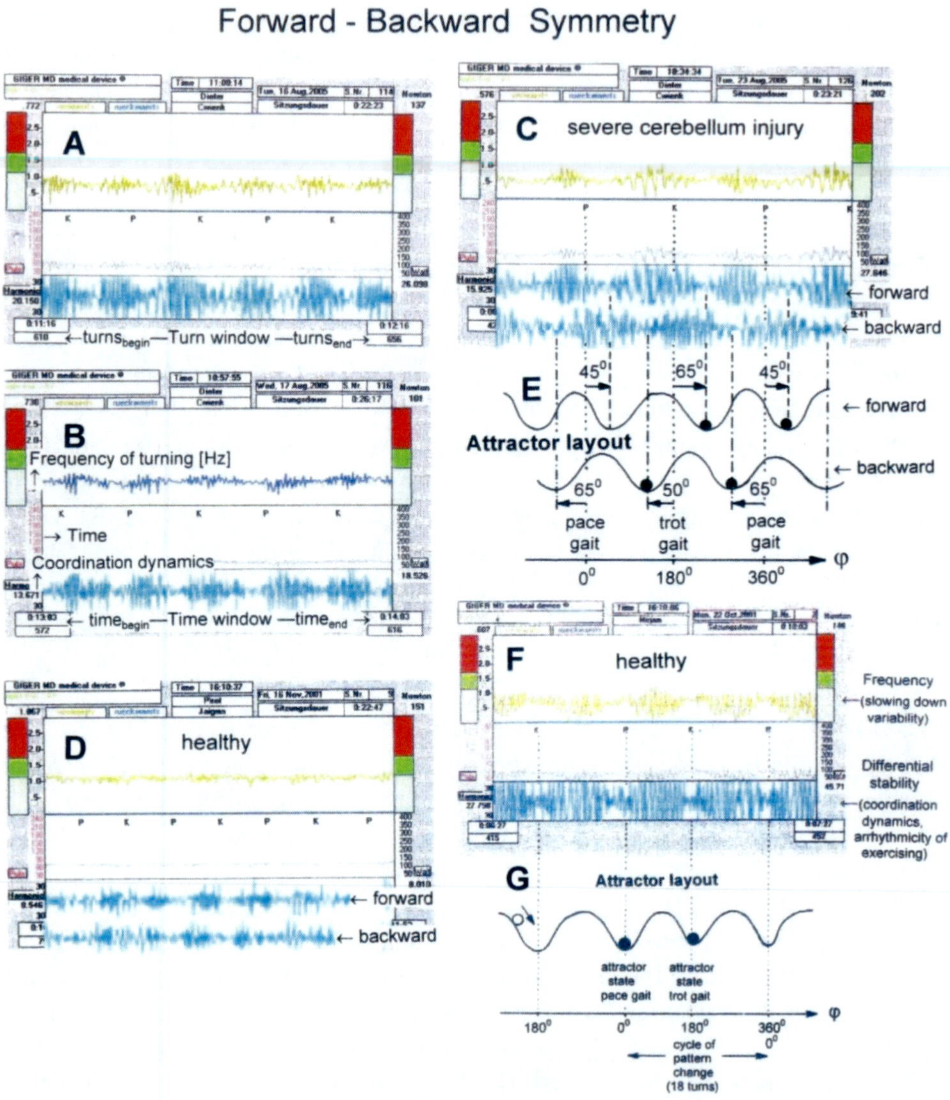

Figure 33. Forward-backward movement symmetry impairment in a patient with severe cerebellum injury for exercising on the special coordination dynamics therapy device (A-C). Coordination dynamics figures of a healthy person are shown for comparison in D and F. In G the attractor layout is constructed for the healthy case. The attractors are the pace and trot gait movements. The state of the system (the white ball) is attracted to the ground of the potential well. Being in the attractor state, the ball is pictured black. In E, the attractors of a brain-injured patient are not at the pace and trot gait coordinations, and are at different coordinations for forward and backward exercising.

8.12. Motor Pattern Diagnostic by sEMG for Updating Coordination Dynamics Therapy (CDT)

In Figures 21 to 23 sEMG demonstrated coordinated firing of single motor units. But sEMG can also be used to improve or update coordination dynamics therapy (CDT). In the above coordination pattern dynamics, characterized by equations of motion of collective variables, it was emphasized that the intrinsic dynamics $F_{intr}(X)$ (CNS functioning) can be modulated by the treatment (behavioral information $\sum c_{inf} F_{inf}(X,t)$ upon changing the equation of motion, if CDT is no longer efficient in repairing the injured CNS. The system theory of pattern formation, however, is giving no information on how to change the therapy for further improvement of CNS organization. But such information can come from human neurophysiology. It will be demonstrated now how motor patterns, recorded with sEMG, from different movements can show us which movements have the best motor patterns and should be trained most because they offer the highest rate of learning. We shall first analyze recorded motor patterns when the patient with the brain injury is exercising forward and backward on the special CDT device. The coordination pattern dynamics give us information about CNS functioning in general. The sEMG motor patterns show what is wrong in the different motor patterns and gives indications as to what to change in the administered therapy. Natural impulse patterns at the neuron level give us information about CNS functioning at the neural level.

8.13. Symmetry Improvement of Motor Programs of Antagonistic Muscles by Increasing the Integrativity of Movement Learning

In a patient with cerebrum and cerebellum injury, surface electrodes were placed on the right and left tibialis anterior and the right and left lateral gastrocnemius muscles. When the patient only exercised with the legs, like on a fitness bicycle, the left gastrocnemius muscle was activated too late with respect to the left tibialis anterior muscle (Figure 34A, at $\varphi = 200^0$ instead of 180^0 (antagonistic muscles) for exercising in the forward direction, and it was activated too early for exercising in the backward direction (Figure 34B, at 150^0 instead of 180^0). The antagonicity of the motor programs thus showed pathologic symmetry between exercising in the forward and backward direction, namely $\Delta\varphi_{forward} = 20^0$ against $\Delta\varphi_{backward} = -30^0$. When exercising with arms and legs (Figure 34C,D) this asymmetric coordination did not change in the stronger left leg.

However, an improvement of the symmetry of antagonistic coordination between the gastrocnemius and the tibialis anterior muscles could be observed in the right weaker leg when exercising in the forward or backward direction using arms and legs as compared to only using the legs. Upon exercising in the forward direction using only legs, the right gastrocnemius muscle was activated too late with respect to the right tibialis anterior muscle (Figure 34A, at 230^0 instead of 180^0; $\Delta\varphi_{forward} = 50^0$), and too early for exercising in the backward direction (Figure 34B, at 65^0 instead of 180^0; $\Delta\varphi_{backward} = -115^0$). When exercising in the forward direction with arms and legs, the right gastrocnemius muscle was only activated a little late (Figure 34C, only at 210^0 instead of 180^0; $\Delta\varphi_{forward} = 30^0$). When exercising in the backward direction, the right gastrocnemius muscle was not activated as early with respect to the right tibialis anterior muscle (only at 105^0 instead of 180^0; $\Delta\varphi_{backward} = -75^0$). Thus, it may be stated that the antagonistic coordination between the right tibialis anterior and the right gastrocnemius muscles improved by exercising in the forward and backward direction by 20^0 and 40^0, respectively, upon exercising with arms and legs as compared to only using the legs, like on a fitness bicycle. Generally speaking, this antagonistic symmetry between the tibialis anterior and the right gastrocnemius muscles improved by 60^0 for exercising in the forward and backward direction following exercise with both arms and legs. During this more integrative exercising on the special CDT device using both arms and legs, the antagonistic activation of the tibialis anterior and the gastrocnemius muscles improved in the short-term memory. Therefore, in this patient, the antagonicity impairment could be treated by exercising very coordinated movements, with arms and legs on the special device. It is likely that, other, very coordinated four-limb movements could also improve the antagonicity of muscle activation.

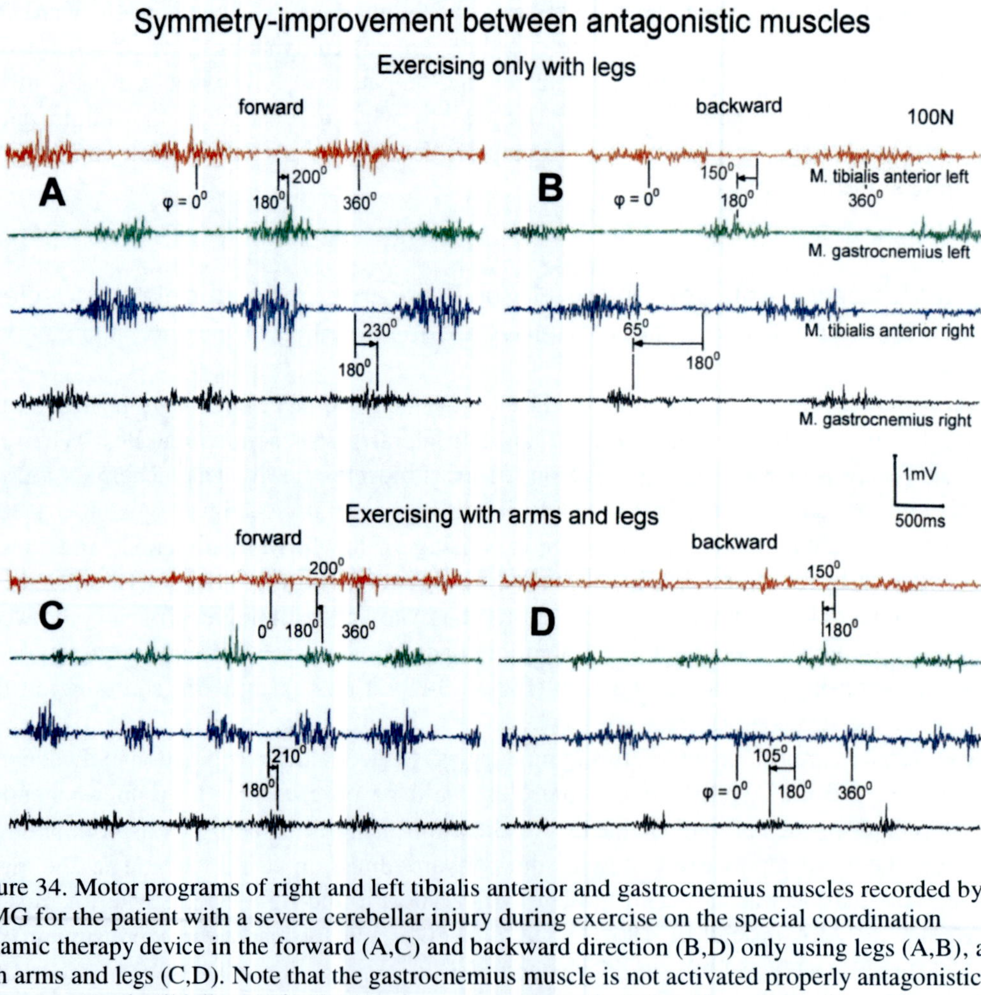

Figure 34. Motor programs of right and left tibialis anterior and gastrocnemius muscles recorded by sEMG for the patient with a severe cerebellar injury during exercise on the special coordination dynamic therapy device in the forward (A,C) and backward direction (B,D) only using legs (A,B), and with arms and legs (C,D). Note that the gastrocnemius muscle is not activated properly antagonistically with respect to the tibialis anterior (at 180 degrees): it gets activated earlier or later, depending upon whether movements are performed in the forward or backward direction and whether only legs, or arms and legs, are used.

8.14. Learning in the Short-term Memory from the Better Opposite Side

The better, in terms of antagonistic activation, left side did not improve in this patient, whereas the poor right side improved by 36% ($\Delta\varphi_{forward}$ - $\Delta\varphi_{backward}$ reduced from 165^0 to 105^0). With the integrativity increasing from coordinated leg movements to coordinated arm and leg movements, the better left side remained unchanged, whereas the poor right side improved by 36%.

This one-sided symmetry improvement can be interpreted in the way that the poor right side learned in the short-term memory from the good left side. This network, responsible for this learning effect, may be located in the spinocerebellum and/or in the spinal cord and is called co-movement.

8.15. Neural Network Learning when Exercising on the Special CDT Device

When exercising pace and trot gait coordinations during walking, running, and crawling, mainly the neural networks of the spinal cord are trained, because these stereotyped automatic movements are mainly located in the intumescentia lumbosacralis and cervicalis and the connections between them. But when exercising on the special CDT device, the intermediate coordination's between pace and trot gait are also trained. To perform these intermediate coordinations, the spinal cord neuronal networks must be varied by the neuronal networks of the brain.

Therefore, by exercising on the special CDT device the brain networks are trained and must learn to improve their functioning in the healthy case and following CNS injury. The CNS learns these complicated coordinations between pace and trot gait, and learns in this way movement patterns in the deep complexity of CNS organization. But for getting deeply into the complexity of the CNS organization, the device must impose the movements on the CNS exactly, so that the CNS neuronal networks can improve the phase and frequency coordination of self-organization.

Human neurophysiology is needed to clarify and quantify what exactness means, and what the consequences are for CNS organization. Therefore, to understand learning in the healthy CNS, and repair it by learning following the injury, we need to understand the reason and consequences of the coordinated firing of neurons. We need to understand why phase and frequency coordination among impulse patterns is important and why neurons work as coordination detectors.

Coordination of action potentials, among action potential impulse patterns, seem to be important for CNS self-organization because CNS functioning is achieved by digitalized and analogized communication. Therefore, we turn back to the coordinated firing of neurons.

9. The Improvement of Coordinated Firing of Neurons by and for Learning

9.1. Impaired Phase and Frequency Following CNS Injury and Its Repair by Learning

It has been shown above that neuronal networks self-organize by phase and frequency coordination of neuron firing (Figures 16,17,22-24) by a few milliseconds.

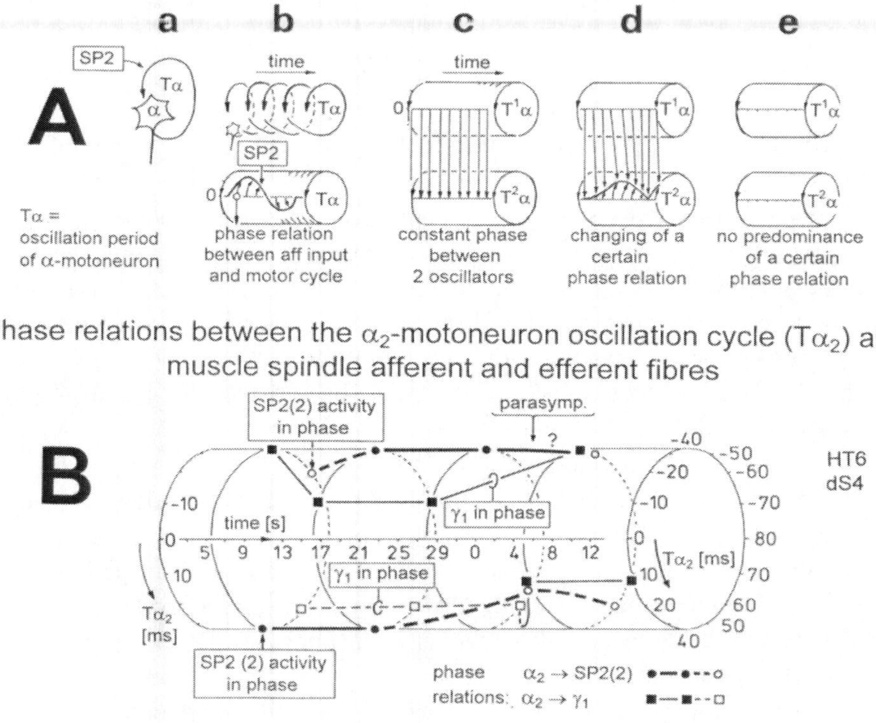

Figure 35. (A) Derivation of the simultaneous description of interspike intervals and phase relations. (a,b) The oscillation period of an oscillatory firing α-motoneuron is schematically characterized by the length of the loop (perimeter). Successive oscillation periods with ongoing time yield a cylinder. Flashing a stroboscope on such a cylinder with the same frequency as that of the rotation of the cylinder would make a black spot on the turning cylinder and not move up or down. If the frequency of the cylinder or the stroboscope changed slowly, the black spot would move up or down. If the black spot moves from left to right with ongoing time, a curve is obtained. By replacing the flash of the stroboscope by the occurrence of the APs of the spindle afferent fiber (or another oscillatory firing motoneuron) with respect to the APs of the oscillatory firing motoneuron, phase relation changes are made visible in the lower part of 'b' for a constant oscillation period (cylinder with no diameter changes). (c) A constant phase between two oscillatory firing motoneurons results in a constant line on the cylinder with ongoing time. (d) A changing phase creates a curve on the cylinder circumference. (e) If there is a loss of predominance of a certain phase between two motoneurons (the black spot gets diffused with ongoing time and is then lost) there is no line or curve.
(B) Interspike interval and phase data from the brain-dead human HT6 (root dS4) are plotted in the representation of A. Filled dots and squares represent average phases (phase relations); thick and thin lines connect the dots to show trends. Note that the phase relations change little; the frequency of the sphincter α_2-motoneuron ($1/T\alpha_2$) changes little - the cylinder does not change its diameter.

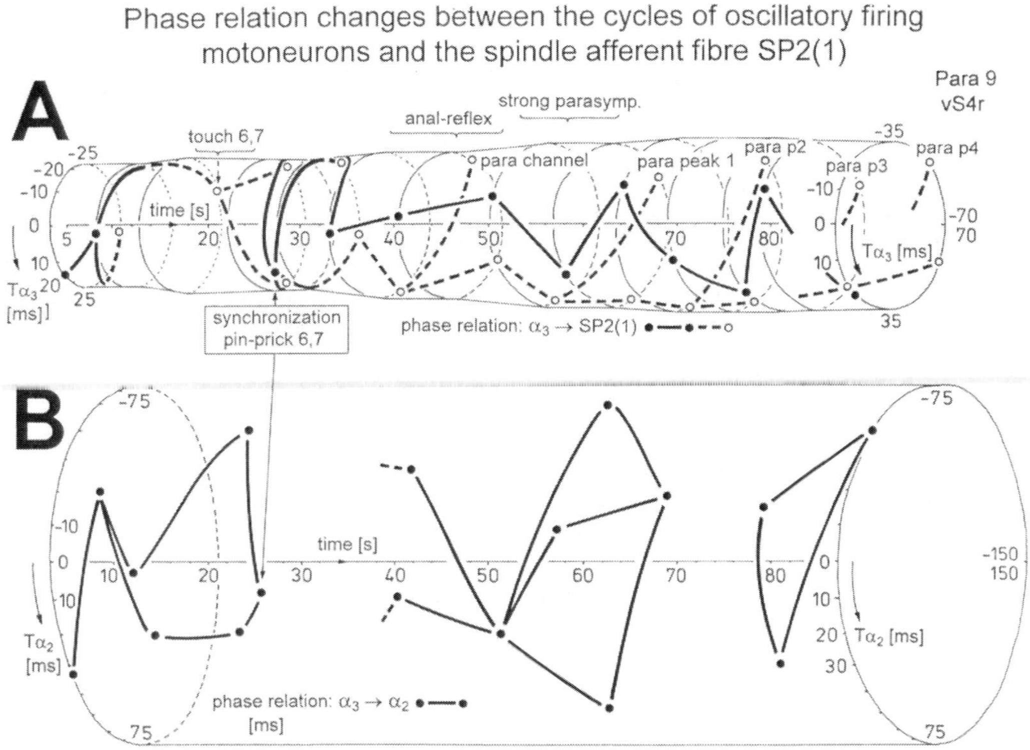

Figure 36. (A) Phase relations between the secondary muscle spindle afferent fiber SP2(1) and the oscillatory firing α_3-motoneuron, taken from Figures 163-165 (and additional data), are plotted on the oscillation period cylinder $T\alpha_3$ (mean oscillation periods are taken from Figures 164,165) according to Figure 48a. The cylinder is changing its diameter (perimeter) because the oscillation period changes. Phase changes in ms are scaled on the cylinder circumference. The ongoing time (to the right) is scaled on the axis of the cylinder (time intervals are taken from Figure 10A of [12]). Existing phase relations are represented by dots (filled and open (back-side)); lines (filled and dashed (back-side)) only connect the phase relations to show trends. para peak 1, para p2, para p3, para p4 = activity peaks of the SP2(1) fiber due to parasympathetic activation (see Figure 10A right). (B) Phase relations between the α_3 and α_2-motoneurons plotted onto the oscillation period cylinder of the α_2-motoneuron. Dots represent phase relations, taken from Figures 11B,12B of [12]. Note that the phase relations of the paraplegic 9 are much more variable than those of the brain-dead human HT6 (Figure 48aB); also, the number of phase relations changes.

Following CNS injury, this coordination becomes impaired, or more precisely, the variability of phase and frequency coordination sharply increases. In the somewhat healthy case (with respect to the spinal cord) of a brain dead patient, the variability of phase coordination is small (Figure 35). Following spinal cord injury for example, the variation of phase coordination increases (Figure 36). For further details, see Refs [1,2]. Following injury the variability of phase coordination therefore increases strongly.

A tremendous increase of the Eigenfrequency band of α_2-oscillators following injury is shown in Figure 37. In the case of a brain-dead human the Eigenfrequency band is small. But following spinal cord injury, the Eigen-frequencies of the α_2-premotor spinal oscillators vary strongly (Figure 37). In the healthy case, the variation of the Eigenfrequency is only little as is suggested in Figure 37. With such an Eigenfrequency variation, frequency coordination becomes strongly impaired, and continuous synchronization easily occurs and the specificity of network organization becomes strongly impaired. In Figure 28, the stability of the movement pattern jumping on springboard was given by the deepness of the potential well, and the variability of phase and frequency coordination, indicated by an arrow attached to the ball representing the network state. In the case of a high variability of phase and frequency coordination in the injury case, the fluctuation of the network state is high, indicated by a long arrow attached to the ball, and there is a high probability that the system is "kicked" out of the basins of attraction especially when the potential well is shallow. The patient may no longer be able to jump in a certain jumping mode. If we want to improve the stability of a movement pattern we have to increase by learning the deepness of the potential well and to decrease the variability of phase and frequency coordination. As will be shown below, when exercising on the special CDT device the variability of phase and frequency coordination can be reduced. This improved coordination among neuron firings is also valid for other patterns when generated by the same networks. Therefore, network states deep in the complexity of CNS organization will also become stable and are relearned in this way.

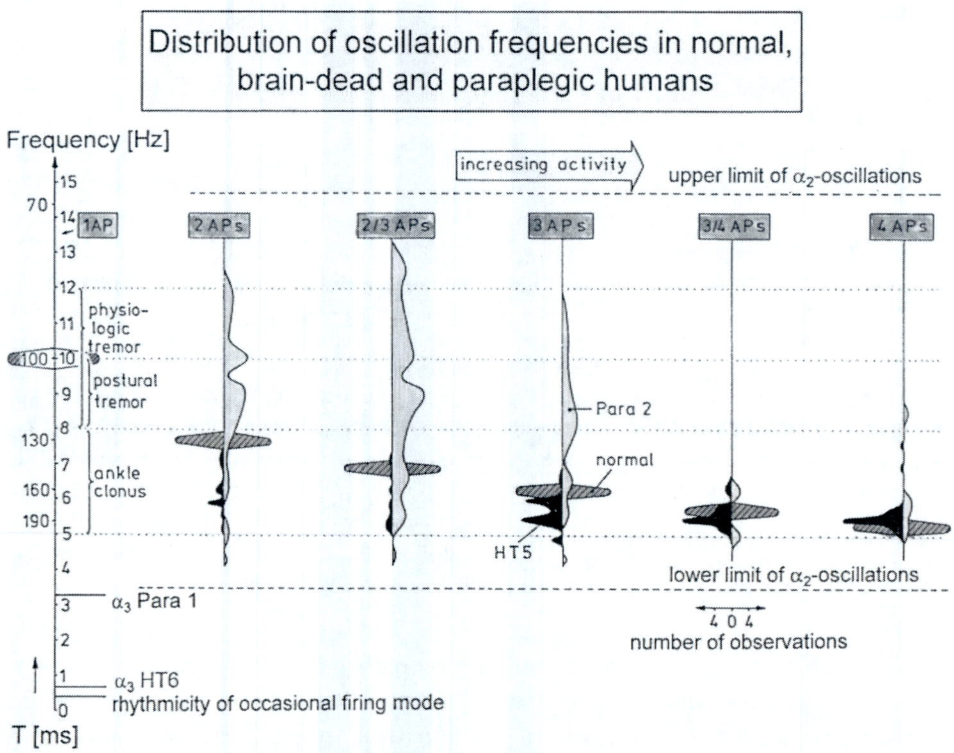

Figure 37. Frequency distributions of oscillation frequencies of continuous oscillatory firing α_2-motoneurons with an increasing number of APs per impulse train (increased activity) in paraplegic 2 (open), in brain-dead HT5 (filled), and probably normal humans (cross-hatched). Frequencies and rhythmic activity changes in the occasional and oscillatory firing mode are indicated. Ranges of physiologic tremor, postural tremor, and ankle clonus are also drawn. Note that frequencies for the brain-dead HT5 are too low, and the oscillation frequencies of the spinal cord, isolated for a long time (Para 2), are too high and too spread out as compared to the theoretically predicted frequency ranges (cross-hatched). T = oscillation frequency.

9.2. Repair of the Stability of the Pattern 'Running on Treadmill'

It is not enough to re-learn a pattern following CNS injury. It is important that the patient can maintain a given pattern as long as desired. Therefore, pattern stability also has to be repaired. When the patient with severe brain injury learned to run on a treadmill (Figure 38), the stability of the pattern for running was repaired. The stability of the repair was measured. The measure used for stability increase was how long the patient can run without losing the pattern. The longest running period per day was used to quantify repair progress. He could manage his balance problems by briefly touching the rail of the treadmill. However, when he lost the pattern, the treadmill had to be stopped and he had to concentrate for some time before starting again.

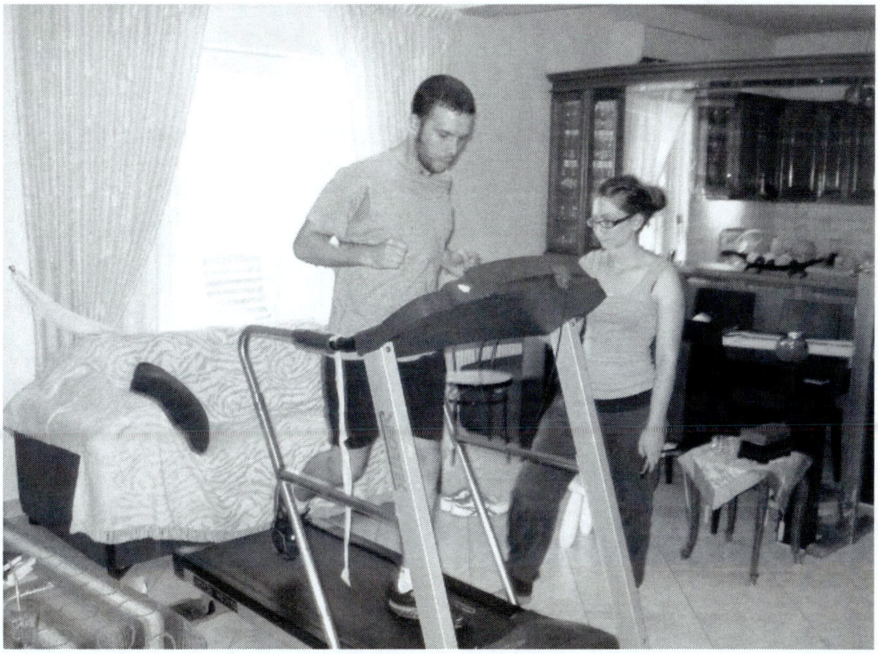

Figure 38. Patient with severe brain injury following a car accident (Sotiris) during running on treadmill.

Figure 39 shows the improvement of running at 8.5km/h on treadmill. At the beginning, he could run only 2 min before losing the pattern. Eight months later, he could run 30 min. The stability of the running pattern had increased by a factor of 15.

The increase of running pattern stability of the patient is pictured in Figure 39 by a ball moving in a potential well. For increased running times the potential well is drawn deeper and the arrows characterizing variation of phase and frequency coordination are made smaller. The question now is what contributed more to the increase of running pattern stability, the deepening of the potential well (according to the system theory of pattern formation) or the reduction of phase and frequency variation. The increase of the running stability coincided with a decrease (improvement) of the high-load CD values (therapy year 2012) [2]. The improvement of phase and frequency coordination will have therefore contributed to the improvement of the running pattern stability, but the main contribution for stability increase will probably have come from the deepening of the potential well, that is to say from a stronger establishment in the networks of the attractor running on a treadmill. The increase of the stability of the pattern running was therefore achieved by both a deepening of the potential well for running and the reduction of the variability of phase and frequency coordination, as shown in Figure 28 for jumping on springboard.

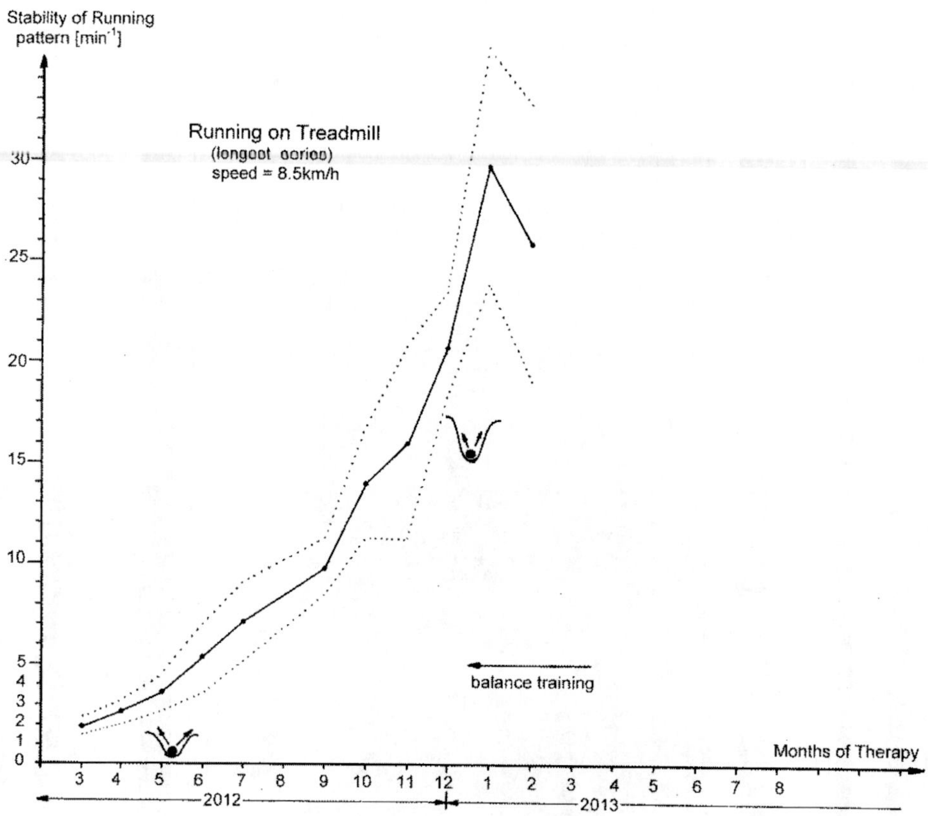

Figure 39. Stability of the movement pattern 'running on treadmill in the forward direction' is dependent on therapy time in a patient with severe brain injury. The stability of the movement pattern 'running' is quantified by the longest time in minutes the patient could stay in the running pattern with several trials and following a warm-up run. Note that the stability improved strongly. Two insets characterize the increase of the 'pattern stability' with a ball (state of the system) in a potential well. This picturing of stability has its scientific basis in the system theory of pattern formation and human neurophysiology. Within the system theory of pattern formation, the stability of a pattern is pictured by a ball (the state of the system) in an attractor layout (here the potential well). The variability of the pattern running (the 'jumping of the ball' in the potential well) is characterized by the lengths of arrows and has its scientific basis in the variability of the phase and frequency coordination of neuron firing (human neurophysiology). Note, with on-going treatment the potential well (the attractor) is getting deeper and the arrows are getting shorter (improvement of phase and frequency coordination).

Healthy humans may tire during running but they do not lose the running pattern, because the stability of the running pattern is very high. It is as if the running pattern is hard-wired, similar to crawling. This patient also occasionally lost the crawling pattern, even though crawling improved strongly during therapy. When a tennis player is serving, one can see that the pattern stability of serving with good performance can vary significantly.

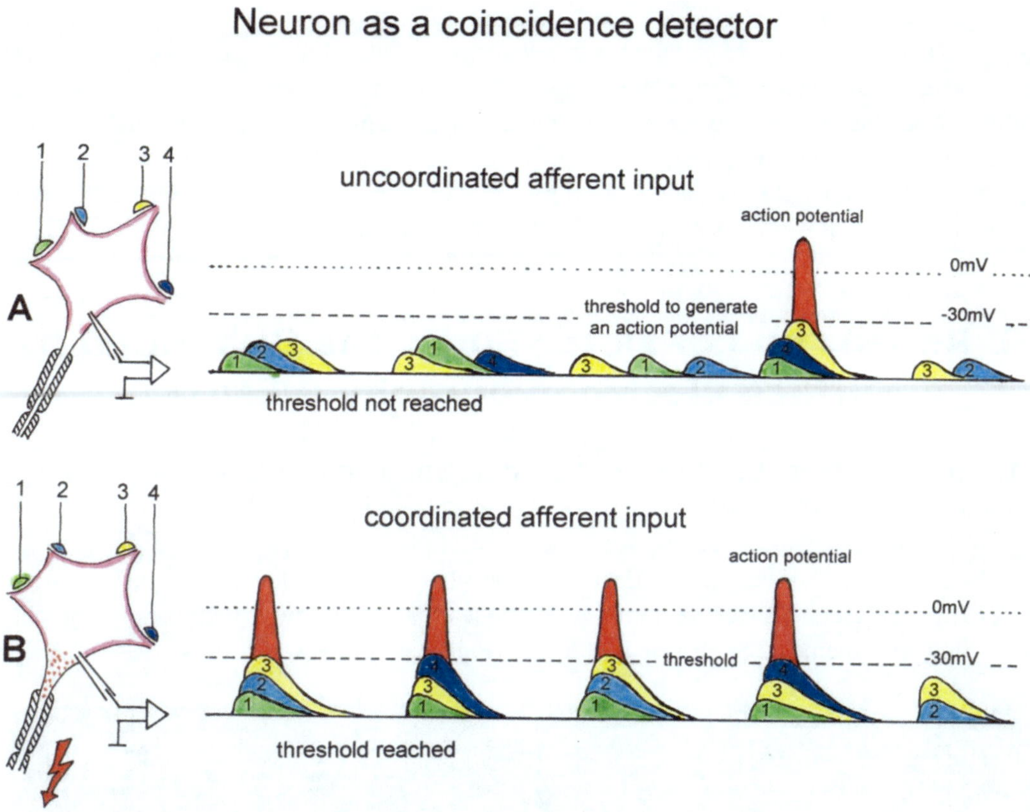

Figure 40. Neuron operating as a coincidence detector. A. Afferent input is reaching rather uncoordinated the cell soma. Only sometimes, an action potential is generated, because the threshold of action potential generation is mostly not achieved. B. The action potentials in fibers 1 through 4 are reaching time-coordinated dendrites or the cell soma. The postsynaptic potentials add up and the threshold is achieved at approximately –30mV, and action potentials are generated time-coordinated at the axon hillock. In the real CNS, many smaller postsynaptic potentials will contribute to the generation of an action potential and passive conduction from the dendrites to the cell soma must be taken into account. Coordinated afferent input may thus induce or enhance (coordinated) communication between neuronal network parts following CNS injury.

9.3. Neurons Work as Coincidence and Coordination Detectors to Improve Communication among Neurons

The impaired coordination between nerve cells and arm and leg movements following injury can be improved, especially by exercising on the special CDT device, which is very precisely manufactured. The exactness of the device guarantees that the coordinated arm and leg movement induced afferent input (like in Figure 6), which can teach the neurons of the CNS to improve their coordinated firing up to within a few milliseconds. Since the neurons work as coincidence and more widely as coordination detectors (Figure 40), this improved coordinated firing improves, for example, the communication among nerve cells and neural sub-networks (especially between networks across an injury site, for example, in a spinal cord injury) because the threshold of neuron excitation is reached earlier. The training of phase and frequency coordination via coordinated arm, leg and trunk movements, improves not only the self-organization of the corresponding sub-networks but also the functioning of the CNS neural networks in general. Some motor and other patterns re-appear upon this improved coordination at the neural and movement level. For example, a child with cerebral palsy learned to speak following exercise on the special device through learning transfer [1].

Surface Electromyography may be used to achieve correlation between neuronal network functioning and the movements necessary for movement-based learning.

10. Re-learning of Motor Functions Quantified by Surface Electromyography (sEMG)

10.1. Co-Movement: Learning in the Short and Long-term Memory from the Better Opposite Side

In Figure 34, it was shown by sEMG that the poor right side learned in the short-term memory from the good left side. In some cases, the learning by co-movement is going quickly into the long-term memory. Such case is shown in Figure 41.

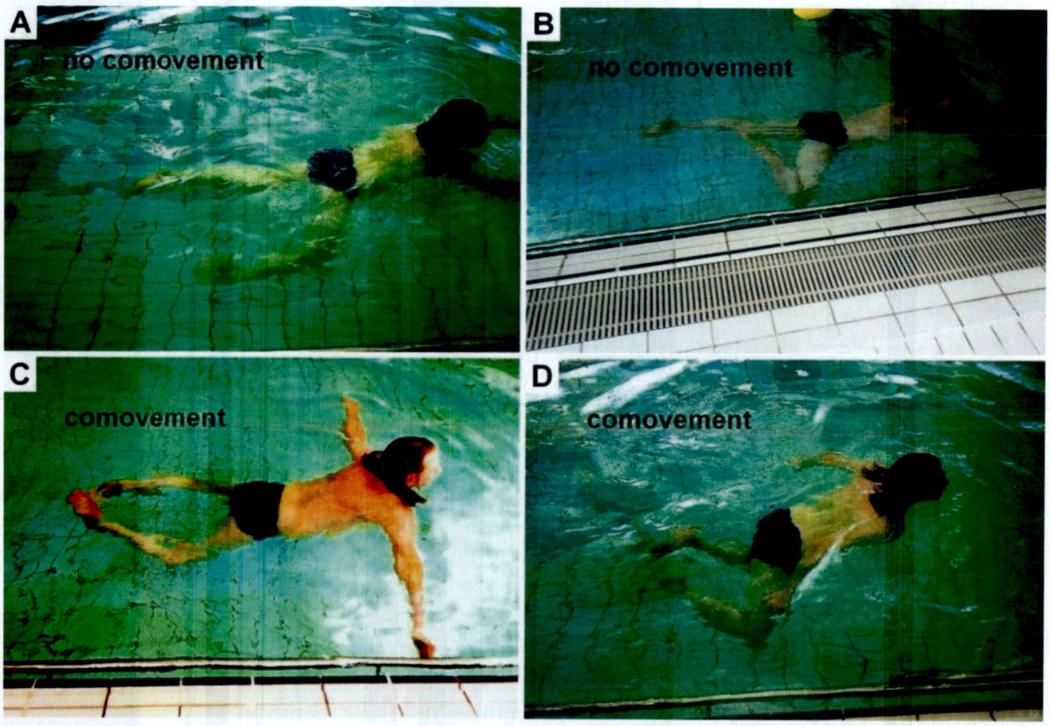

Figure 41. Co-movement between legs in a patient with an incomplete cervical spinal cord injury during swimming. When the patient was not closing the legs before flexion only one leg moved (A,B). When he closed the legs and the legs touched each other, both legs moved (flexed) (C,D). The simultaneous touch input achieved a coupling from right to left and made the not moving leg move.

The co-movement of limbs or other body parts of the body can be induced by the simultaneous input from receptors in the range of a few milliseconds. The neuron is a coincidence or coordination detector (Figure 40). With the simultaneous input from afferent fibers to a neuron the threshold to generate an action potential at the axon hillock is reached earlier than with uncoordinated input. Communication between right and left or rostral and caudal body parts can be achieved with less active connections.

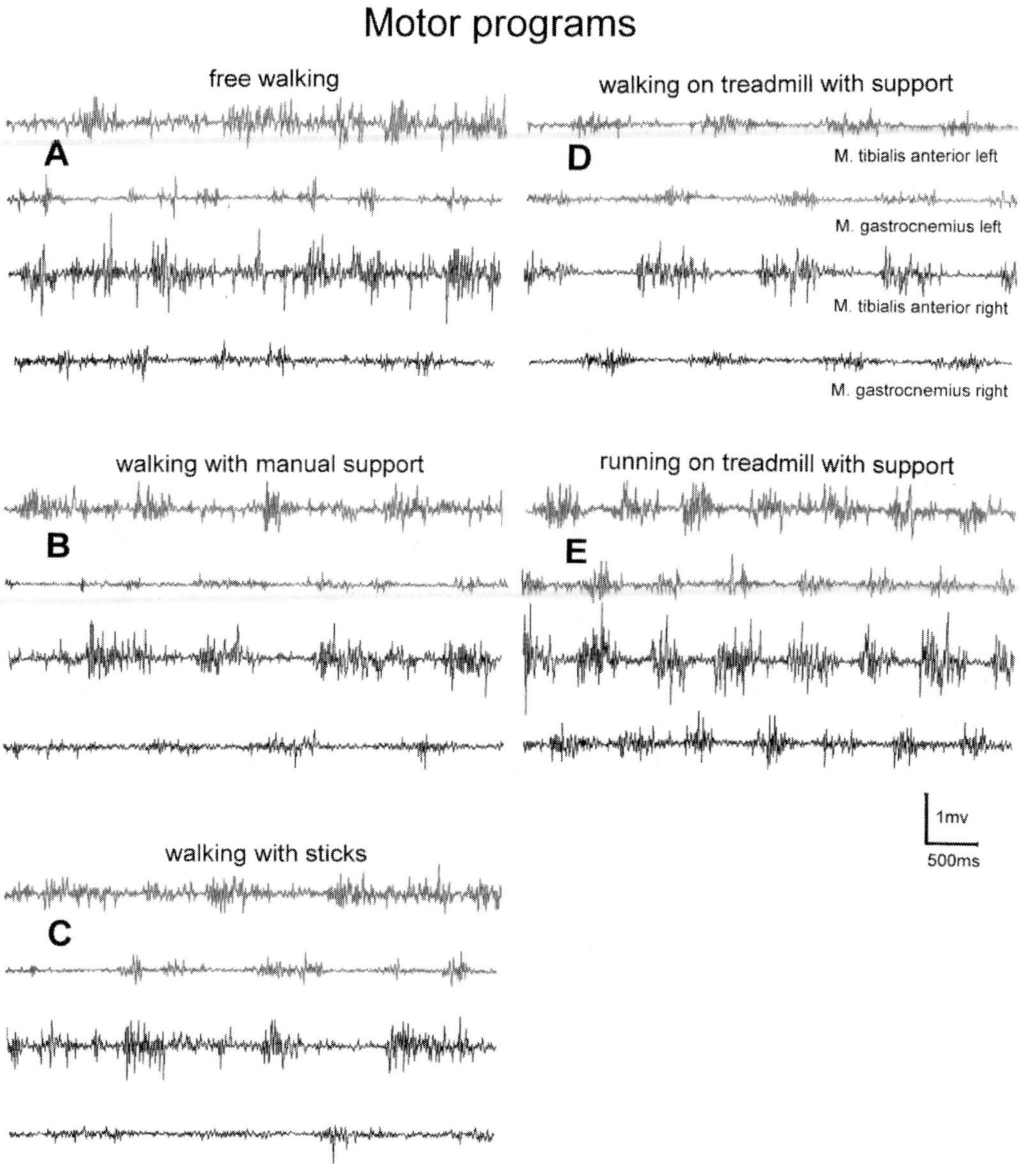

Figure 42. Surface EMG motor program of the right and left tibialis anterior and gastrocnemius muscles for free walking (A), walking with manual support (B), walking with sticks (C), walking (D) and running (E) on the treadmill with support. Note that the motor programs for walking with strong support (D,E) are much better than those for walking with little or no support (A,B,C).

The co-movement will be demonstrated here between legs. A patient with an incomplete SCI was able to swim (Figure 41). However, because he was not closing the legs before flexion, only one leg moved (A,B). But when he closed the legs and the legs touched each other, both legs moved (flexed) (C,D). The simultaneous touch input achieved a coupling from right to left and made the not moving leg move. This shows that simultaneous input will help to improve the performance of a movement.

When exercising on the special CDT device (Figure 15), co-movement will also be induced from muscle spindles and other receptors to make the 'poor' leg move better. In SCI or stroke, the knees will come into a better central position during the repeated coordinated arm and leg movements performed on the special CDT device. This co-movement works for all exact coordinated movement patterns, especially in in-phase and anti-phase movements. Exact coordinated instructions and coordinated visual input (as for example, due to interpersonal coordination [20] (Figure 26)) also enhances the earlier generation of action potentials and thus the improvement of the performance of coordinated movements.

10.2. Supported Walking Especially at High Speed Enhances Re-learning of Motor patterns

In a patient with cerebellum and cerebrum injury, changes in the walking program will be analyzed when the patient had to balance unaided. Since the vestibulocerebellum was also heavily damaged, increasing motor program deficits for increasing balance needs can be expected. Figure 42 shows sEMG motor programs for free walking (Figure 42A) and for supported walking (manual support) (Figure 42B), walking with sticks (Figure 42C), walking on treadmill (Figure 42E), and running on treadmill (Figure 42E). The motor programs for free walking (A) were worse than those for the different kinds of supported walking (B-E). With the increasing support (from B,C to D,E) the motor programs improved. Obviously, the patient had balance problems: due to the effort of attempting to maintain balance by himself, his motor programs worsened. The interpretation of this worsening of the motor programs with the increasing balance needs is that following free walking, the damaged vestibulocerebellum could no longer sufficiently coordinate balance with walking. The motor program for supported walking or running (B-E) were best for running (E), which indicates that the higher speed improves the performance and the learning as long as the patient can manage the speed.

10.3. Anti-phase Jumping on Springboard Repairs by Learning More Efficient Motor Programs than Swinging

To discover which motor patterns are best for performing rhythmic, dynamic, stereotyped movements for functional repair by learning at the neural ensemble level, sEMG was recorded for different kinds of springboard movements. Motor programs during swinging and jumping on springboards are shown in Figure 43. For swinging (Figure 43A) the tibialis anterior and gastrocnemius muscles were only slightly activated rhythmically, i.e. no real motor pattern can be seen. When jumping in in-phase, the leg muscles were more activated and motor programs can be partly identified (Figure 43B). In the right tibialis anterior muscle, synchronized oscillatory firing of motor units (pathologic) can be seen. For jumping in anti-phase with (Figure 43C), and without support by a therapist (Figure 43D), the motor programs were quite good, even though rhythmic firing of motor units can also be clearly identified during the motor bursts (right gastrocnemius in Figure 43D).

The conclusion, which may be drawn from this, is that jumping on the springboard will train the neuronal networks of the patient more effectively than swinging, since real motor programs could only be identified during jumping. The general conclusion for treatment is that sEMG can help to improve the rate and understanding of repair by learning.

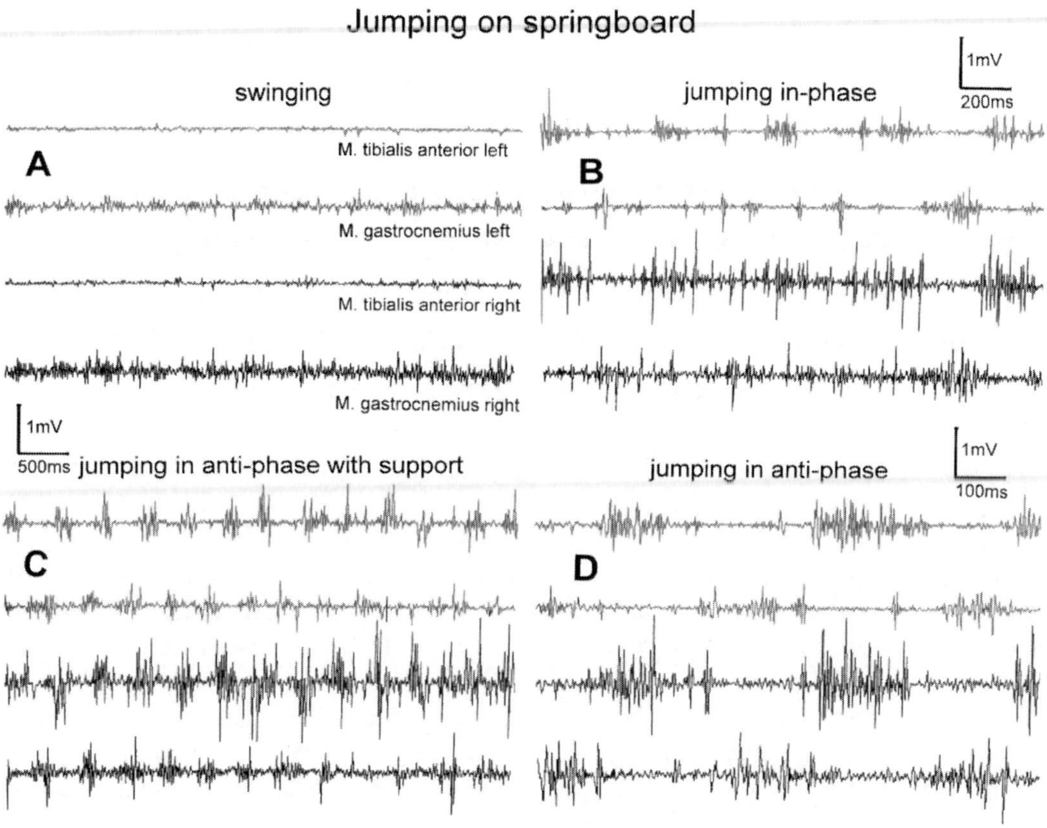

Figure 43. Surface EMG motor patterns of the right and left tibialis anterior and gastrocnemius muscles for swinging (A) and jumping on springboard in-phase (B), in anti-phase with support (C), and in anti-phase without support (D). Note that the motor programs are better for jumping in anti-phase (C,D) than for jumping in in-phase and swinging. Note further that continuous synchronous oscillatory firing of motor units can be seen on the trace taken from the right tibialis muscle for jumping in in-phase (B).

10.4. Symmetry Learning to Enhance the Efficiency of CNS Repairs

Learning transfer from one hand movement to the symmetric one has been known to occur [20]. This means that the symmetry counterpart improves without it being trained itself. It has been shown (see above, Figure 34) that the antagonicity between the tibialis anterior and gastrocnemius muscles in the poor right leg improved in the short-term memory by 36% when increasing the integrativity of exercising by changing from only leg movements to coordinated arm and leg movements. Assuming that repeated improvements in CNS function in the short-term memory will slowly manifest themselves in the long-term memory (Figure 31) would mean that there is learning transfer occurring in relation to the improvement of antagonicity from the left leg to the right leg. A particular movement can be also improved by training its symmetrical counterpart. If, for example, the walking pattern on the right side has been impaired in a stroke patient for a few years, this pathologic walking pattern has become an old-learned movement and is difficult to change. Its counterpart, namely backward walking, has however not become an old-learned movement after injury. Additional training of backward walking will therefore more efficiently improve forward walking by learning transfer. In conclusion, movements and especially automatisms have to be trained with their symmetry counterparts to enhance the efficiency of CNS repair.

11. CNS Functioning Can Be Assessed Non-invasively by Measuring the Coordination Dynamics Based on the Correlation of Measurements at the Single Neuron, sEMG and Movement Levels

11.1. Plausible Explanation of Measuring CNS Functioning

A prerequisite of repair/improvement of the human CNS by learning is that the quality of its functioning must be measured. This can be achieved when a patient or a healthy subject is exercising on a precisely calibrated device for complicated coordinated arm and leg movements. If the CNS is functioning well, the CNS can cope with the difficult coordinations, and the rhythmicity of exercising is good. If the CNS is functioning poorly, the rhythmicity of exercising is poor, because the CNS has difficulty in generating these difficult coordinated movement patterns. Such an instrument is the special Coordination Dynamic Therapy (CDT) and recording device (Figures 14,15,19,32). While performing changing coordinated arm, leg, and trunk movements, imposed by the device, CNS functioning can be measured non-invasively by the arrhythmicity of exercising with the changing coordinations between arms and legs. Further, through exercising with different combinations of arms and legs in both forward and backward directions, right-left, rostral-caudal, and forward-backward symmetry impairments of CNS movement organization can be measured, trained, and repaired by learning.

The pace and trot gait coordinations are mainly generated in the neuronal networks of the spinal cord. A newborn baby can step automatically (Figure 44). It cannot control the start and the termination of the stepping automatism. The afferent input from the heel strike is one of the stimuli, which starts the automatic stepping. With development of the infant, the neural networks of the spinal cord become more and more under the control of supraspinal centers. During early development therefore, the infant has significant difficulty in exercising on the special CDT device. This is not due to the size of the device.

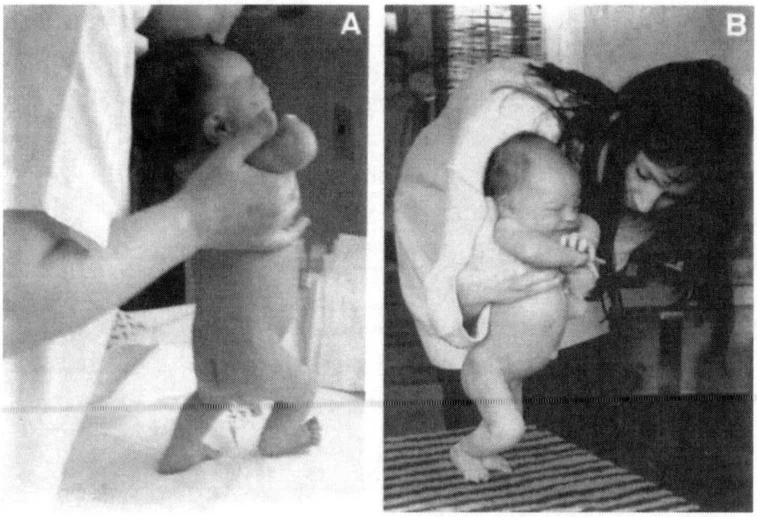

Figure 44. Automatic stepping in a newborn infant. A. A 5-day-old infant, Juliane, performing primary automatic stepping; slight backward posture. The heel of the right foot touched the ground first. B. Infant Juliane, 8-day-old, performs automatic stepping.

For performing the intermediate coordinations between pace and trot gait, the spinal cord's neuronal networks have to be varied by supra-spinal centers to accomplish these difficult movements. When the brain-injured patient is performing these movements, with the help of the device, the brain's networks are involved in the generation of these complex patterns. If the brain is functioning well, the CNS has no problem in generating the difficult coordination's and the arrhythmicity of exercising is slight. If the brain is functioning poorly, because of a brain injury, the patient struggles to generate these intermediate coordinations between pace and trot gait. The arrhythmicity of exercising is very apparent. Therefore, the measuring of the coordination dynamics is really assessing CNS functioning, including brain functioning. With respect to therapy, the training of the intermediate coordinations between pace and trot gait also repairs brain functions.

The means by which CNS functioning may be evaluated with the different measuring methods shall now be discussed (Figure 1).

11.2. Relative Phase and Frequency Coordination between the Firings of α and γ-Motoneurons and Secondary Muscle Spindle Afferents Recorded with the Single-nerve Fiber Action Potential Recording Method

With the single-nerve fiber action potential recording it was shown that the neurons of the human CNS organize themselves by phase and frequency coordination (Figures 16,17). Following injury, this organization principle is disrupted [1]. Figure 45 shows schematically a recording from a dorsal S4 root of a brain-dead human. Of the summed afferent and efferent impulse traffic the natural impulse patterns of one α_2-motoneuron, a dynamic (γ_1) and a static γ-motoneuron (γ_{21}) and two to three secondary muscle spindle afferent fibers could be extracted (for classification see Figure 8). The natural stimulations performed were pin-pricking (pain) of sacral dermatomes inside the continence automatism zone (Figure 6) and urinary bladder catheter pulling (Figures 6). It can be seen from Figure 45 that the α_2-motoneuron (α_2(O2)) fired in an oscillatory manner with 2 to 3 impulses per impulse train, and sometimes there was a break in the oscillatory firing, as it is shown in Figure 46, for the firing of the α_1-motor units (FF). The impulse train for α_1-motor units consisted of only one action potential (Figures 13,20). Phase coordinations between the α_2-motoneuron, the γ-motoneurons and the secondary muscle spindle afferent fibers are indicated by different arrows and the dotted and dashed lines. It can be seen from Figure 45 that there were many coordinations between the different neurons. The relative phase and frequency coordination seems to hold for all neurons and is an integrative mechanism for the self-organization of the neuronal networks of the human CNS.

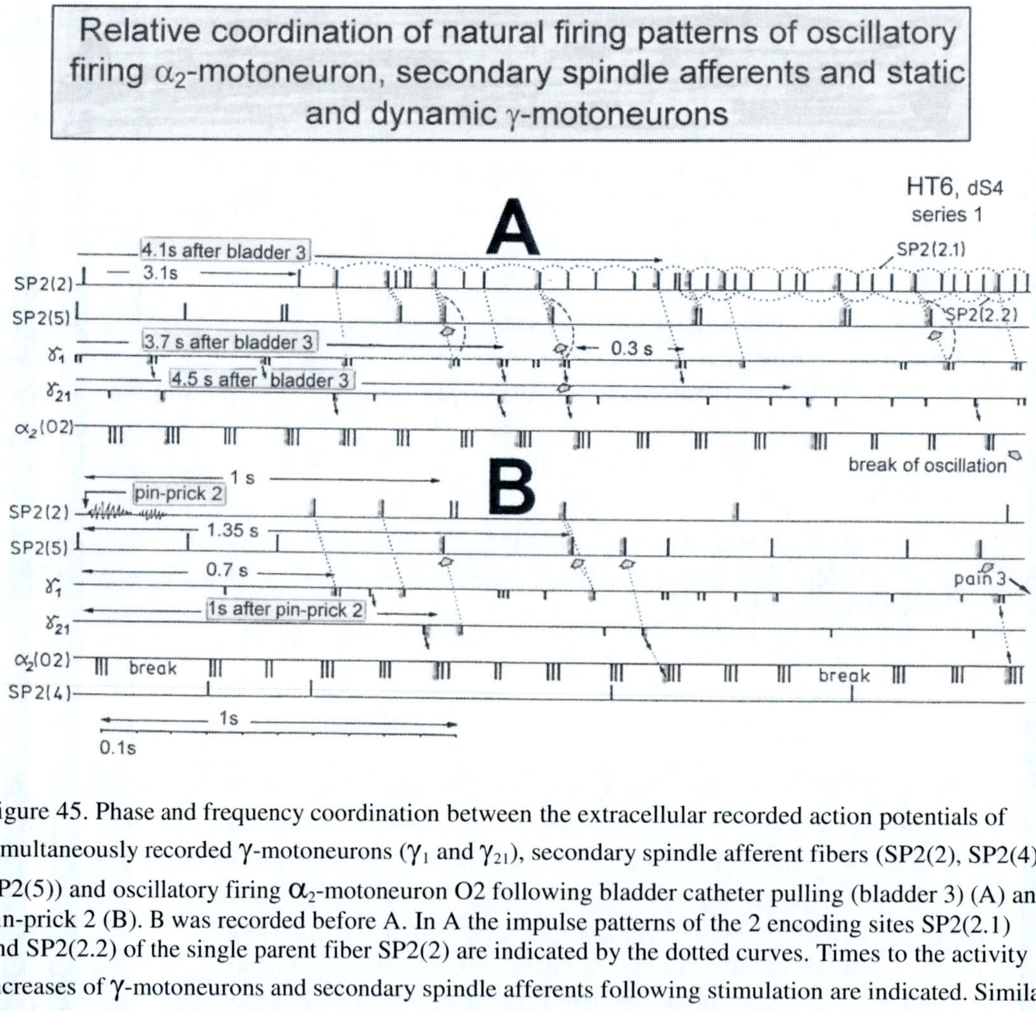

Figure 45. Phase and frequency coordination between the extracellular recorded action potentials of simultaneously recorded γ-motoneurons (γ_1 and γ_{21}), secondary spindle afferent fibers (SP2(2), SP2(4), SP2(5)) and oscillatory firing α_2-motoneuron O2 following bladder catheter pulling (bladder 3) (A) and pin-prick 2 (B). B was recorded before A. In A the impulse patterns of the 2 encoding sites SP2(2.1) and SP2(2.2) of the single parent fiber SP2(2) are indicated by the dotted curves. Times to the activity increases of γ-motoneurons and secondary spindle afferents following stimulation are indicated. Similar time intervals of the occurrence of γ-motoneuron APs and SP2(5) fiber APs (phase coordination) are indicated by the open arrows, and the similar time intervals of γ-motoneuron Aps, and α-motoneuron APs are indicated by small arrows. Similar time intervals of the APs of fibers SP2(2) and SP2(5) are indicated by the double dotted lines, those of γ_1-APs and the SP2(2) fiber APs by a dotted line, and those of γ_1-APs and the SP2(2)-SP2(5) correlation by a curved dashed line. HT6; dS4-root.

But the recording from single-nerve fibers is an invasive recording method and cannot help to quantify learning during or following therapy.

11.3. Phase and Frequency Coordination between Motor Unit Firing and Building of a Motor Program with Increasing Load Recorded with Single-motor Unit sEMG When Exercising on the Special CDT Device

Figure 46A-D shows how motoneurons coordinate their firing, and how the CNS generates a motor program during exercise on the special CDT device. Approximately 5 motor units could be activated in the right tibialis anterior muscle of a spinal cord injury patient. The motor units marked 1 and 2 had a low threshold, and the motor unit 3 a higher threshold. Their waveforms are shown time-stretched in the inserts of Figure 46A,C. For high exertion, when exercising on the special CDT device at 200N, it appears that 2 other motor units could be activated. This section shall focus on the firing of these 3 motor units, to study their coordination during the exercising on the special CDT device at the load increasing from 20 to 200N.

The patient whose record is shown in Figure 46A was exercising at 20N. Motor unit 1, with a large motor unit potential, fired in an oscillatory manner at around 10Hz and the motor unit can thus be suggested to be innervated by an α_1-motoneuron. Motor unit 2 fired at around 5Hz in between the action potentials of unit 1 to avoid synchronization, but fired in every second interspike interval due to having half of its frequency. Motor unit 2 was likely also innervated by an α_1-motoneuron, even though the frequency was low (5Hz) for the oscillatory firing of an α_1-motoneuron. There was slight inhibition acting on motor unit 1 after the 6th spike since the interspike interval greatly exceeded 100ms. When increasing the load to 50N (Figure 46B), motor unit 1 slightly increased the frequency to increase the activity, and motor unit 2 increased its frequency from to 5 to 10Hz to increase the activity. The higher threshold motor unit 3 started to fire occasionally. After the 4th motor unit 1 potential, motor unit 2 stopped firing in an oscillatory manner. It appears that motor program inhibition worked transiently on the oscillatory firing sub-network (to reduce the frequency of oscillatory firing or to stop the oscillatory firing) and gave rise to a transient break in the oscillatory firing. When exercising at 100N (Figure 46C) the way in which the motor program was generated can be seen. When an activity burst is activated, for example, in order to contract a muscle; motor units are activated to fire in an oscillatory manner for the duration of the burst. The frequency of the oscillation increased and decreased. Other motor units were also activated, but first the ones with the lowest threshold. The activated motor units avoided synchronization. When the motor program burst was supposed to be terminated, the motor unit firing stopped. They were probably inhibited. When exercising at 200N (Figure 46D), a motor program can be seen in the right tibialis anterior muscle (note the time axis change). Four weeks later, when the patient had a flu (Figure 46E,F), the motor program developed in the right tibialis anterior muscle worsened.

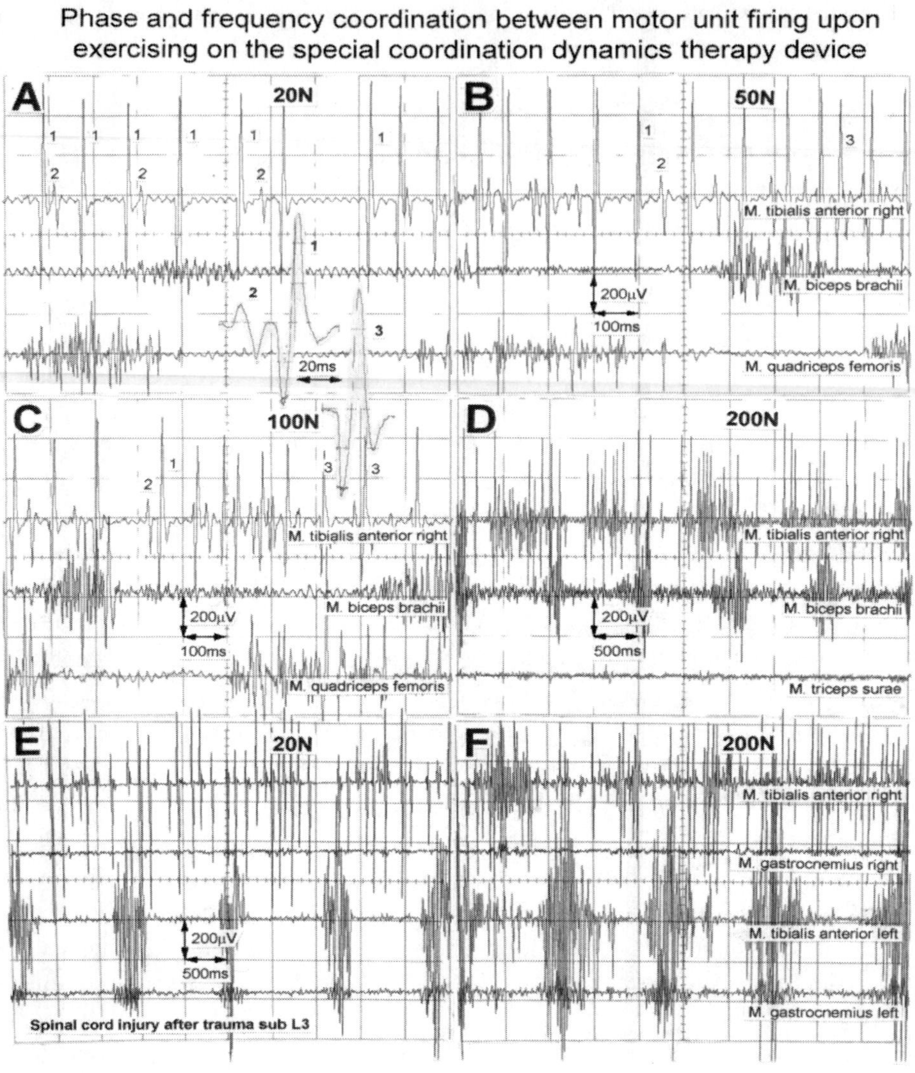

Figure 46. Phase and frequency coordination between motor unit firing (in the same muscle) during the generation of a motor program during exercise on the special coordination dynamics therapy device at loads increasing from 20 to 200Newton. The wave forms of 3 identified motor unit potentials of FF-type '1', '2', and '3' are inserted in 'A,C' and some of them are marked in 'A,B,C'. Note in 'A' that motor unit '1', firing at approx. 9Hz, coordinates its firing with motor unit '2', firing at approx. 4.2Hz, by firing twice as often. In 'B' both motor units fire non-synchronously at around 10Hz with relative phase and frequency coordination. A, B. No motor program can be seen in the right tibialis anterior muscle as in the biceps brachii and quadriceps femoris muscles during exercising at 20 and 50N. C. The oscillatory firing of motor unit 1 and 2 (and 3) was transiently interrupted (inhibited) upon exercising at 100N, thus showing the beginning of a motor program. In the motor program burst, the frequencies of the motor units are increasing and decreasing. D. Upon exercising at 200N, the motor program became more pronounced. Probably two more high-threshold motor units contributed to the motor program burst in addition to the 3 motor units identified in 'A-C'. E, F. Improvement of the motor program from 'E' to 'F' in the right tibialis anterior muscle of the same patient on another day. But due to flu the motor program of the right tibialis anterior muscle was in 'F' not as good as in 'D'. The activity in the left gastrocnemius muscle consists partly of cross-talk. Recording from a patient with a complete spinal cord injury sub L3.

The healthy motor program of the biceps brachii (spinal cord injury sub L3) looks completely different (Figure 46A,B). No coordination can be seen and analyzed. The difference is understandable. In the right tibialis anterior muscle, the CNS could only activate 5 motor units and only α_1-motor units (large motor unit potential and only one action potential per impulse train characterizes α_1-motor unit (FF) oscillatory firing (Figure 20)); no α_2 (FR) and α_3-motor units (S) were innervated. In the biceps brachii muscle, all types of motor units could be recruited, each type being in abundance. While exercising at 20N (Figure 46A), mostly only α_2 (FR) and α_3-motor units (S) (identified by the small motor unit action potential amplitude) and maybe a few α_1-motor units (FF) of small size were activated in the biceps brachii muscle. For 50N (Figure 46B), motor units of α_1-type (FF), were activated. These motor units have naturally larger amplitudes. Since the recruitment is according to the size principle in each motoneuron group and the group recruitment depends on the speed of movement, quite a complexity is present. Further, the transiently oscillatory firing of motoneurons will be shorter, especially if the patient is exercising very fast. In the partly impaired motor supply of the quadriceps femoris muscle (injury level sub L3), some motor units can be identified (Figure 46B). In the better left tibialis anterior muscle, no coordination could be extracted from the sum of the motor units activity; too many motor units were functioning to permit identification of single motor unit firing (Figure 46E,F). However, many α_1-motor units were activated in the left tibialis anterior muscle up to a size of 2mV (Figure 46F), whereas in the biceps brachii muscle they were still under 1mV (Figure 46D) for the load of 200N.

In Figure 46, relative phase and frequency coordination is shown for motor units in the same muscle, namely the right tibialis anterior muscle of a patient with a spinal cord injury sub C4/5. In Figure 23, phase and frequency coordination among motor units was shown between different muscles.

By recording single-motor units non-invasively with the sEMG phase and frequency coordination and the building up of a motor program can be identified and followed throughout a program of exercise at increasing loads. However, sEMG is not capable of quantifying the improvement of CNS functioning following therapy in human patients.

11.4. From sEMG Motor Programs to High-load Coordination Dynamics to Evaluate CNS Functioning upon Therapy

11.4.1. Pathologic Patterns of Motor Activation

Surface EMG was performed in a patient with an incomplete cervical spinal cord injury sub C5/6 while he was undergoing CDT. Figure 47 shows sEMG recordings of motor patterns while the patient was exercising on the special CDT device at the beginning of therapy (Figure 47A,B) and again after 6 months of therapy (Figure 47C,D). The motor patterns were very largely abnormal at the commencement of therapy (Figure 47A,B). The amplitude of sEMG activity was highest in muscles proximal to the injury, i.e. the biceps brachii, while significantly smaller in the muscles distal to the injury site, i.e. tibialis anterior and gastrocnemius muscles. Note the difference in calibration for displaying the amplitude. The motor patterns in the muscles distal to the injury site C5/6 show obvious improvement after 6 months of CDT (Figure 47C in comparison to Figure 47A,B).

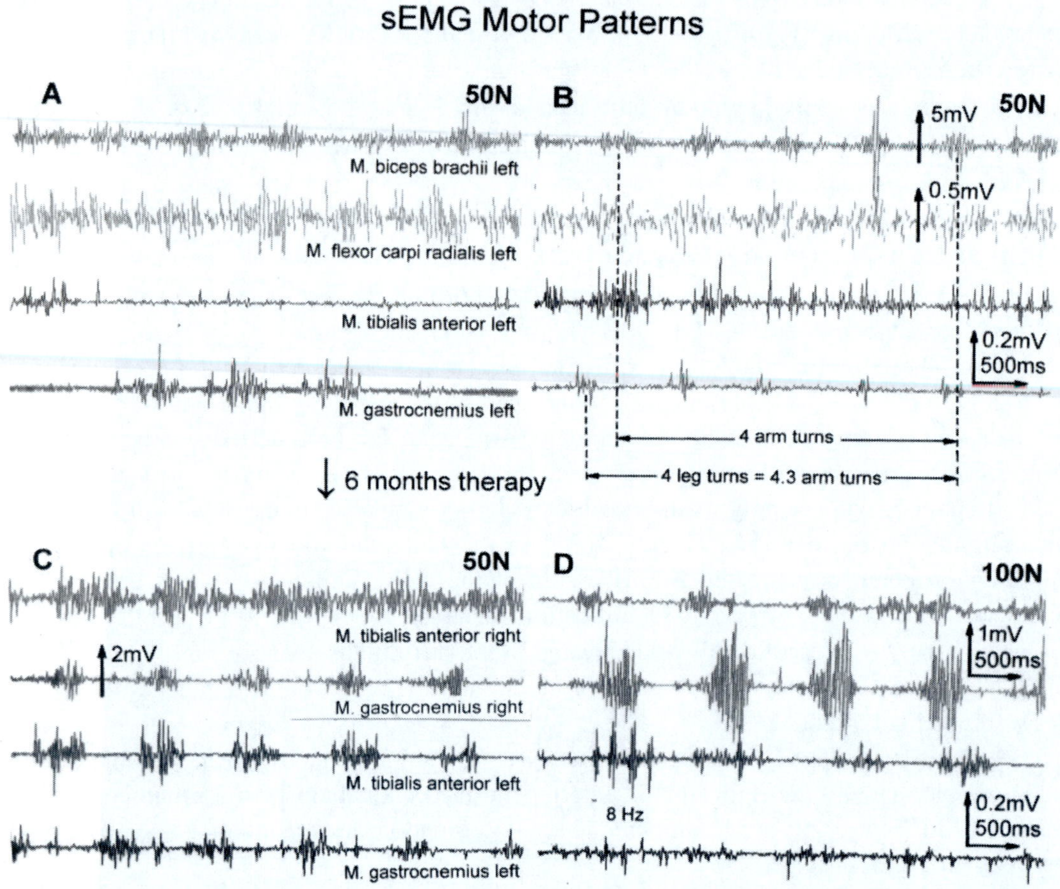

Figure 47. Motor patterns of arm and leg muscles recorded by sEMG for the patient with a 50% spinal cord injury at C5/6 levels at the beginning of the coordination dynamics therapy (A,B) and after 6 months of therapy (C,D). Note that the motor patterns of the left tibialis anterior and gastrocnemius muscles improved during these 6 months of therapy (from A,B to C); the patterns were quite varied from recording (A) to recording (B). The sEMG activity has a large amplitude for the biceps brachii muscle and a low amplitude for the tibialis anterior and gastrocnemius muscles (A,B) (note the different calibrations). The patterns were generated when the patient was exercising on the special coordination dynamics therapy device for turning. To measure stability differences of patterns (system theory of pattern formation), the arms turn at a higher frequency than the legs. The revolution relation between arm leavers and leg pedals is 19:18; this relation generates the changing coordination between arms and legs. The different turning frequencies of arms and legs can be seen in the sEMG patterns (B; 4 leg turns = 4.3 arms turns). Upon increasing the load from 50 (C) to 100N (D) during exercising on the special device for turning, the motor patterns of the left tibialis anterior and gastrocnemius muscles deteriorated (D): specific patterns were lost, antagonicity was lost, pathologic synchronization of premotor spinal α_1-oscillators at 8Hz occurred; this deterioration of motor patterns stability is in accordance with the increase of the coordination dynamics values (= increase of arrhythmicity of exercising) at higher loads.

11.4.2. Motor Pattern Stability

At the beginning of therapy, the motor patterns varied significantly while turning on the special CDT device, as can be seen in Figure 47A and 47B. This variation persisted for as long as 6 months after starting therapy. In other words, the stability of the motor pattern for exercising on the special device was very low at the beginning of therapy. This low pattern stability was also reflected in the high CDT values (high arrhythmicity of exercising = low pattern stability) at the beginning of therapy. Clinically this motor pattern variability correlated with the instability of movement performance, observed during walking and running on treadmill at the beginning of the therapy period.

11.4.3. Impairment of Reciprocal Relationship of Antagonist Muscles

The natural reciprocal relationship of agonist and antagonist muscles of the leg, namely the tibialis anterior (dorsal flexor) and gastrocnemius muscle (plantar flexor), was reasonably well preserved on the left side (Figure 47B,C) but was significantly impaired on the right side. It persisted for as many as 6 months in spite of learning therapy (Figure 47D). This impairment of antagonistic muscle action contributed to the patient's abnormal gait.

11.4.4. Measurements of Temporal Stability of Movement Patterns by Pattern Change

As explained above, the sEMG motor patterns were generated and recorded when the patient was exercising on the special CDT and recording device. The mean temporal stability of these motor patterns over one minute (= coordination dynamics value = mean arrhythmicity value) was used to quantify improvements of CNS functioning through therapy (see below). To measure differences in temporal stability of different movement patterns, such coordinated patterns, involving arm and leg movements, need to be generated with accuracy. This is achieved by the high precision mechanics of the special device. The arms turn at a slightly higher frequency than the legs. The relationship of revolution between arm leavers and legs paddles is 19:18. This difference in revolution produces the changing of coordination patterns of arm and leg movements. Such pattern change must also be seen in the sEMG motor patterns.

The way that the muscle activations changed, between the left biceps brachii muscle (arm) and left gastrocnemius muscle (leg), can be seen in Figure 47B. During 4 gastrocnemius activations (4 turns), the biceps were activated approximately 4.3 times (4.3 turns). This seemingly simple change of arm and leg movements between pace and trot gait is an extremely difficult pattern for the CNS to generate. All the muscle patterns of arms and legs have to be changed. And if the device is positioned further away from the trunk, then the trunk must perform rotating movements in coordination. Since this trunk rotation is changing with the coordination position between arms and legs, all the segmental trunk muscles change their coordination in the rostral and caudal direction. If the CNS is functioning physiologically, then the subject can turn relatively easily and smoothly. But if the CNS is injured, then such ongoing change of coordination's among arm, leg, and trunk muscles becomes a very difficult task for the CNS. Patients often sweat due to the stress, which the CNS undergoes in generating these complicated movements as the pattern keeps changing. As can be seen from Figure 47B, during the time period of the recording, the patient's CNS could manage the change of the coordination between pace and trot gait quite well. Chaotic activation was not observed on either the biceps brachii or gastrocnemius muscle traces during the movement pattern change.

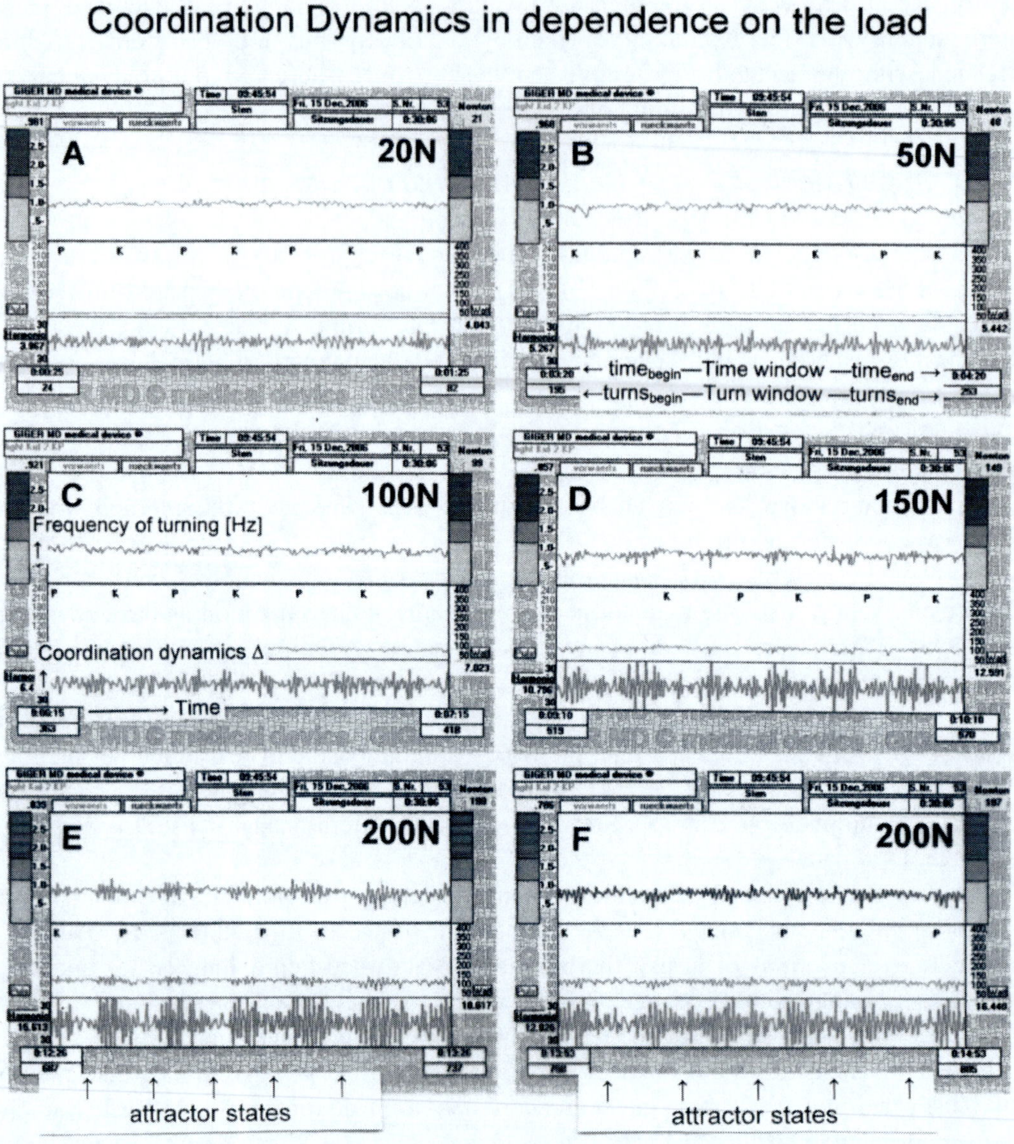

Figure 48. Coordination dynamics measurements in a patient with a 50% spinal cord injury upon exercising on the special coordination dynamics therapy and recording device for forward turning at increasing load from 20 to 200N (A-E). The recording sweeps are 1min long. F. Exercising in the backward direction at 200N. Upper trace = frequency; lower trace = coordination dynamics = arrhythmicity of exercising. P = pace gait coordination, K = trot gait coordination. "P to K" and "K to P" = coordination changes from pace to trot gait and from trot gait to pace gait. Note that for 20N (A) and 50N (B) the coordination dynamics increase is more noise like whereas from 100N (C) to 200N (E) there are rhythmic increases and decreases of the coordination dynamics with the changes of the coordination. Note further that the lowest arrhythmicity of exercising (highest pattern stability = attractor state) lies for forward exercising at 200N (E) to the right side of the pace (P) and trot gait coordination's (K) and for backward exercising (F) to the left side of the pace and trot gait coordinations.

11.4.5. Impairment of Pattern Formation Is Revealed with More Integrated CNS Activation at Higher Loads

After increasing the load on the special device from 50 (Figure 47C) to 100N (Figure 47D), the motor patterns of the left tibialis anterior and gastrocnemius muscles partly deteriorated (Figure 47D). The specific patterns and their antagonistic relationship were partly lost. Pathologic synchronization of premotor spinal α_1-oscillators occurred at 8 Hz, as can be seen from the trace of the left tibialis anterior muscle (Figure 47D). The increasing impairment of CNS self-organization with increasing load of exercising, made visible by the sEMG recordings, is in accordance with the increase of the arrhythmicity of exercising on the special CDT device for increasing load (Figure 48, increase of coordination dynamics values at higher loads).

11.4.6. Increase in Temporal Instability of Movement Patterns While Exercising against High Loads

From the above discussion, the increase of the coordination dynamics values, i.e. deterioration of coordination dynamics, when exercising against high loads (Figure 48) is understandable. It is because of the reduction in the temporal stability of movement pattern, as shown in sEMG recordings. Kinesiologically, this increased temporal instability of movement pattern formation is apparent when coordination dynamics traces are plotted for one minute against increasing load (Figure 48). As can be seen from Figure 48, the arrhythmicity of exercising increased with increasing load. Exercising at 20 (A) and 50N (B), in the recording, the arrhythmicity varied without any specific structure. At 100 (C) and 150N (D), however, structure becomes visible in the coordination dynamics traces. Note in Figure 48, that the coordination dynamics values increased with increasing load. When exercising against a load of 200N in the forward direction (E), it can be seen that there is rhythmic increase and decrease of the arrhythmicity of exercising. At times, when the arrhythmicity is small, the arm and leg movement patterns show a high degree of stability. Different patterns of coordinated arm and leg movements have different degrees of stability. Those patterns, which are more stable, have small amplitude of arrhythmicity; they are the attractor states among different movement patterns. The unstable movement patterns (with significant arrhythmicity in exercising) do not constitute attractor states. Sudden instabilities of movement patterns are perceived by the patient and healthy subjects as sudden increases in the resistance to smooth turning. These are sudden perturbations of CNS organization and are felt as a stutter, which interrupts smooth turning.

11.4.7. Exercising at High Loads Reveals Impairment in the Symmetries of CNS Organization

While the patient was exercising against a high load of 200N, rhythmic changes of the arrhythmicity of exercising were observed during forward (Figure 48E) as well as backward exercising (Figure 48F). Interestingly, the location of attractor states (patterns of high stability) were seen to be situated at different locations from the calibrated states of 'pace' (P) and 'trot gait' (K), as shown in Figure 48E,F. For forward turning, the attractor is located on right side, while for backward turning, the attractor is located on left side of the pace and trot gait patterns. This impairment of symmetry for forward-backward movements (different attractor states) only became visible with the very integrated activation of the CNS required when exercising against high loads

11.4.8. Improvement of Symmetries of CNS Organization Increased Pattern Stability

As the impairment of symmetry may be diagnosed with the highly integrated activation of the CNS, training of those movements at high loads is required to improve symmetry. As the pattern stability with respect to symmetry increases, movement performance improves (Figure 28). In the case of an asymmetrical injury and attendant asymmetric CNS organization, Figure 28B,C), the movement pattern 'jumping-in-anti-phase' can be so unstable that the patient is unable to perform the movement. In terms of system theory, the potential well for 'jumping-in-anti-phase' in the attractor layout is very shallow or does not exist anymore (Figure 28C). In a patient with incomplete spinal cord injury C5/6, initially, only in-phase jumping was possible without support. The anti-phase jumping needed assistance and effort from the patient. With training, the patient re-learned to jump in anti-phase; the attractor basin for 'jumping-in-anti-phase' became deeper and the phase and frequency variability became smaller so that the movement pattern became stable and thus the movement could be performed independently. At the end of therapy, the patient was even able to jump unassisted with a skipping rope.

11.5. Conclusion of Measuring CNS Organization by the Relationship between Single-nerve Fiber Action Potential Patterns, sEMG Patterns and Coordination Dynamics Patterns

Within the System Theory of Pattern formation, CNS functioning was quantified by pattern stability during pattern change while exercising on the special CDT device. Pattern stability was given by the deepness of the potential well and the variability of phase and frequency coordination among the neurons of the neuronal networks as demonstrated in Figure 28 (and Figure 39) by the moving (indicated by arrows) of a ball (the pattern state) in a potential well. The change of the coordinated movement patterns is generated, when the subject is exercising on the special coordination dynamics recording device, where the coordination between arms and legs, imposed by the device, changes continuously between pace and trot gait. The change of the coordination's can be seen when the therapist is looking at the position of arms and legs. The change of the arm and leg coordination's can also be made visible by the change of motor patterns (sEMG), as shown in Figure 47B.

The temporal stability of the intrinsic coordination tendencies is measured by the deviations in differential stability during the performance of these rhythmic movements. The plotting of the differential stability over time of the frequency of exercising generates an attractor layout for this special movement (Figure 33) and the mean stability per minute can be measured by the arrhythmicity of exercising (df/dt, f = frequency). This value, the so-called coordination dynamics value (Δ = df/dt:f), quantifies CNS functioning objectively, integratively and non-invasively.

But as shown in Figure 46, the generated motor pattern, and therefore the CNS organization, is conditioned to some extent by the load under which the patient exercises. For low-load exercising at 20N (46A), only few motor units were activated (tibialis anterior right trace). With increasing load, more and more motor units were activated until they fused (46A to D). Also, the coordination dynamics values depend strongly on the load (Figure 48) as one might expect. With increasing load the coordination dynamics values Δ are higher (worse) and the integrativity of CNS organization increases. As will be shown below, the repair of the integrated functions, i.e. the ability coordinate various parts of the body together, need more time. But the repair of the integrated functions is the most meaningful real repair, which includes substantial learning transfer, i.e., functions such as continence, which cannot be specifically trained can be repaired by learning transfer. The CNS of the patient is routinely assessed by Coordination Dynamics values for low load (20N) and at higher loads. The sum of the coordination dynamics values is then calculated. This is the so-called 'high-load' test.

Impairment of pattern stability and pattern performance, which can be seen in sEMG recordings, correlate with the impairment of the coordination dynamics (increased coordination dynamics values). The quality and stability of motor pattern organization correlates, on the other hand, with the quality and stability of neuronal network organization, measured with the single-nerve fiber action potential recording method. CNS functioning can therefore be effectively assessed by measuring the coordination dynamics.

Movement-based learning, accomplished by CDT, is successful in repairing the CNS through functional reorganization. However, in severe brain and spinal cord injuries (SCI), structural repair is also required, including the building of new neurons from stem cells. It was shown that new motoneurons could be built in SCI, but only to a very limited extent [14]. Cell physiology and genetics, especially epigenetics, have to be correlated to movement-based learning to further enhance the functional and structural repair of the human CNS.

12. Learning and Communication via External Loops of Oscillators As a Principle of Interlacing Brain Parts for Cooperation

12.1. The Caudal Spinal Cord As a Suitable Place to Study Learning in the Human CNS

I now tried to go deeper into the complexity of neuronal network self-organization changes of the human nervous system to understand more about neuronal network learning. The learning for repair following CNS injury is one way to learn about neuronal network learning in the healthy CNS, because regulations of CNS functioning are impaired and now allow insight into their functioning. For studying learning, the urinary bladder repair is used because firstly, single-nerve fiber recording method can be used, and secondly, because the somatic and the vegetative nervous system divisions are involved in the learning process. The communication between different brain parts can be studied.

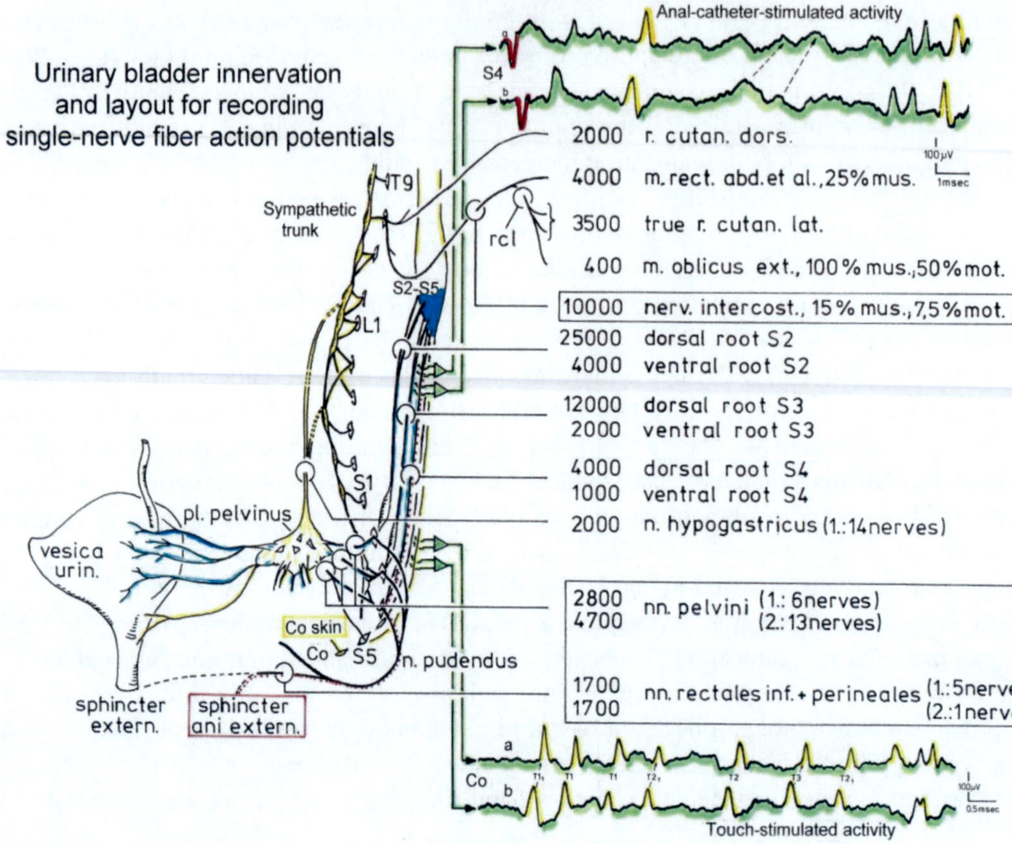

Figure 49. Urinary bladder innervation (anatomy; structure) and recordings of single-nerve fiber action potentials from a S4 and a coccygeal root (electrophysiology; function). Number of myelinated nerve fibers of the nervi rectales inferiores and perinales, the nervi pelvini and the nervus hypogastricus. T = thoracal; L = lumbal; S = Sacral; rcl = ramus cutaneus lateralis, no pure skin nerve; %mus = % of nerve fibers leading to muscles; %mot = % of nerve fibers which are motoric; 1 and 2 = Nerve fiber counts from cadavers 1 and 2. Number of nerves in the bracket gives the number sub-nerves of which the nerve consisted. The innervation pathway of the external bladder sphincter is unclear. The S4 root recording informs about single-nerve fiber activity running into the spinal cord (from the bladder receptors) and running out of the cord to the urinary bladder (bladder efferents) upon natural stimulation. Functions of the urinary bladder function and the sacral micturition center in the spinal, as a part of the human CNS, can therefore be analyzed. The Co root (no efferents; something like a skin nerve) recording informs about skin receptor activity in the coccygeal dermatome.

12.2. Human Neurophysiology for a Deeper Understanding of Bladder Repair by Learning

Since learning and learning transfer during movement-based learning may need a few years, it is of great importance to use such learning treatment (behavioral information), which can really repair the urinary bladder. The proper behavioral information for learning transfer was developed by using the single-nerve fiber action potential recording method. By analyzing natural impulse patterns in sacral nerve roots from the somatic and parasympathetic nervous system upon bladder, bowel and skin stimulations, intravesical stimulation, and movements, it will be found that jumping on a springboard (Figures 29,30) and exercising on the special CDT device (Figure 15) are especially efficient to induce learning transfer from movements to bladder functions. The free jumping or jumping on springboard are specifically designed to repair patterns, including transfer of learning from somatic to autonomic functions, discussed in detail below, and the exercising on the special CDT device is primarily used to reduce the variability of phase and frequency coordination and to thus increase the stability of network states by decreasing pattern fluctuation (Figure 28). Movements are exercised to generate a physiologic attractor layout of patterns again and the coordinated firing of neurons and neural sub-networks is trained to increase pattern stability by reducing the fluctuation of pattern states (Figures 28).

The innervation of the bladder is schematically shown in Figure 49. Nerve fiber counts of myelinated nerve fibers indicate the thickness of nerve roots and nerves. Even though single-nerve fiber action potentials can also be recorded from peripheral nerves [1], high-quality recordings can only be obtained from thin and long sacral caudal nerve roots. The very thin coccygeal nerve root contains mainly skin afferents to the coccygeal dermatome and is ideally suitable to analyze skin receptors [1]. A recording is shown in Figure 6. To avoid adding up single-nerve fiber action potentials, the summed activity has to be low to be able to split up the multi-unit recording into several natural single-nerve fiber impulse patterns (Figures 5,6). The lower sacral nerve roots are suitable for that. An original recording from a S4 root upon anal catheter pulling to stimulate the somatic and parasympathetic nervous systems is also shown in Figure 152. Afferent (main amplitude upwards) and efferent single-nerve fiber action potentials (amplitude downwards) can be seen. The ventral S4 root is an ideal root to record single-nerve fiber action potentials from efferent (running out of the spinal cord) and afferent fibers (leading into the cord) (Figure 3). In caudal sacral nerve roots, there are also afferent fibers in ventral roots and efferent fibers (motor fibers) in dorsal roots. The innervation of the bladder from the sacral micturition center is leading through the nerve roots S2 till S5.

Since the repair of bladder function in patients with SCI probably also requires the reorganization of pathways outside the spinal cord, the connection of the plexus pelvinus to rostral plexuses is shown in Figure 50. The picture allows understanding of just how complex and distributed the autonomic nervous system is. The bladder is mainly innervated by the somatic and parasympathetic nervous system and probably a little by the sympathetic nervous system (internal bladder sphincter). In Figure 51, the innervation of the human organs is shown via the sympathetic and parasympathetic divisions. By recording single-nerve fiber action potentials from caudal sacral nerve roots, only natural impulse patterns in somatic and parasympathetic nerve fibers can be expected to be obtained because it was recorded from the S3 to S5 dorsal and ventral nerve roots.

The complexity of the autonomic nervous system is tremendous (Figure 50). The thin ventral S4 root (motor root) in itself shows quite a complexity (Figure 7). The thick myelinated nerve fibers are prominent. By enlarging the scale, the non-myelinated nerve fibers become visible (Figure 155B of [2]). Their number is in the range of 5-times higher than those of the myelinated fibers. The diameter of the thinnest myelinated fibers, from which it is possible so far to record single-nerve action potentials, is in the range of 3 to 4μm (Figure 8). To record from these thinly myelinated fibers is difficult but important. To repair the urinary bladder by learning, we must be able to record also from parasympathetic nerve fibers and distinguish their activity from the activity of the different groups of γ-motoneurons to understand bladder function.

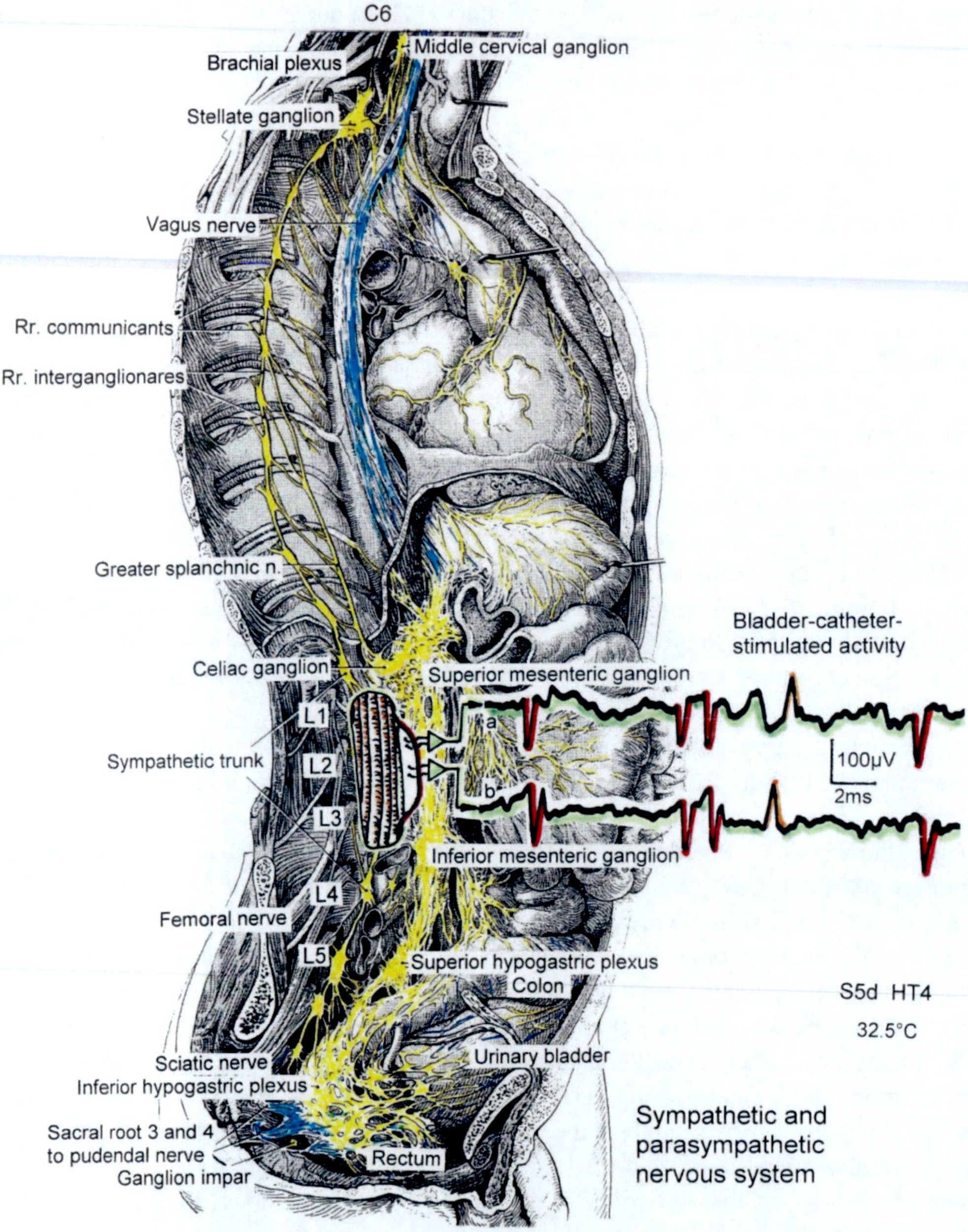

Figure 50. The picture illustrates the complexity of the autonomic nervous system. Connections of the different plexus bypass the spinal cord and offer the structure for a functional repair of vegetative functions (and may be of somatic functions) like urinary bladder control, cardio-vascular performance, and breathing. The recording of single-nerve fiber action potentials from human nerve roots on the other hand show that in this complexity of the human nervous system structure it is possible to record activity from several single neurons under rather natural conditions.

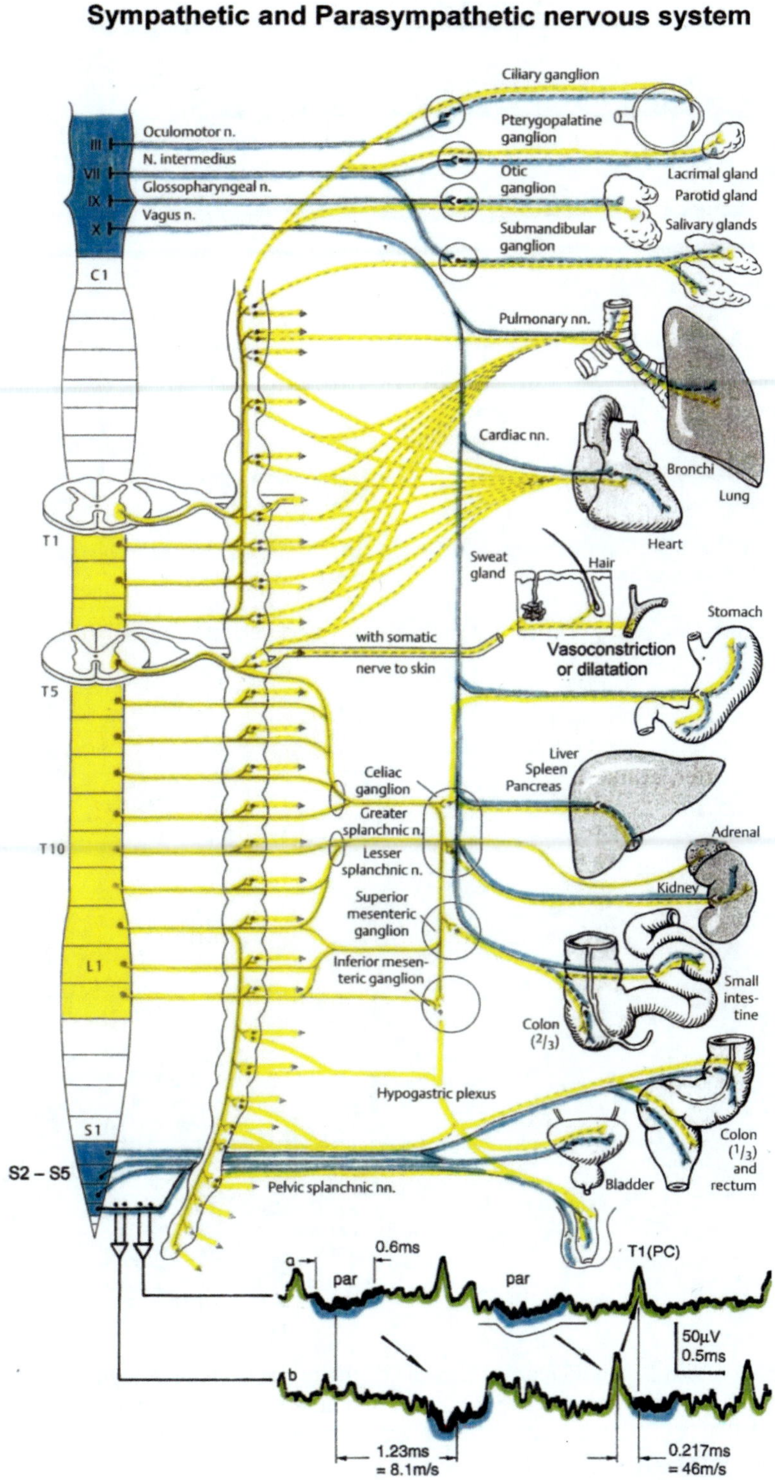

Figure 51. Schematic diagram of the sympathetic and parasympathetic nervous system. Yellow = sympathetic, blue = parasympathetic. The recording from a sacral root shows single action potentials of preganglionic neurons (par) and a skin afferent fiber.

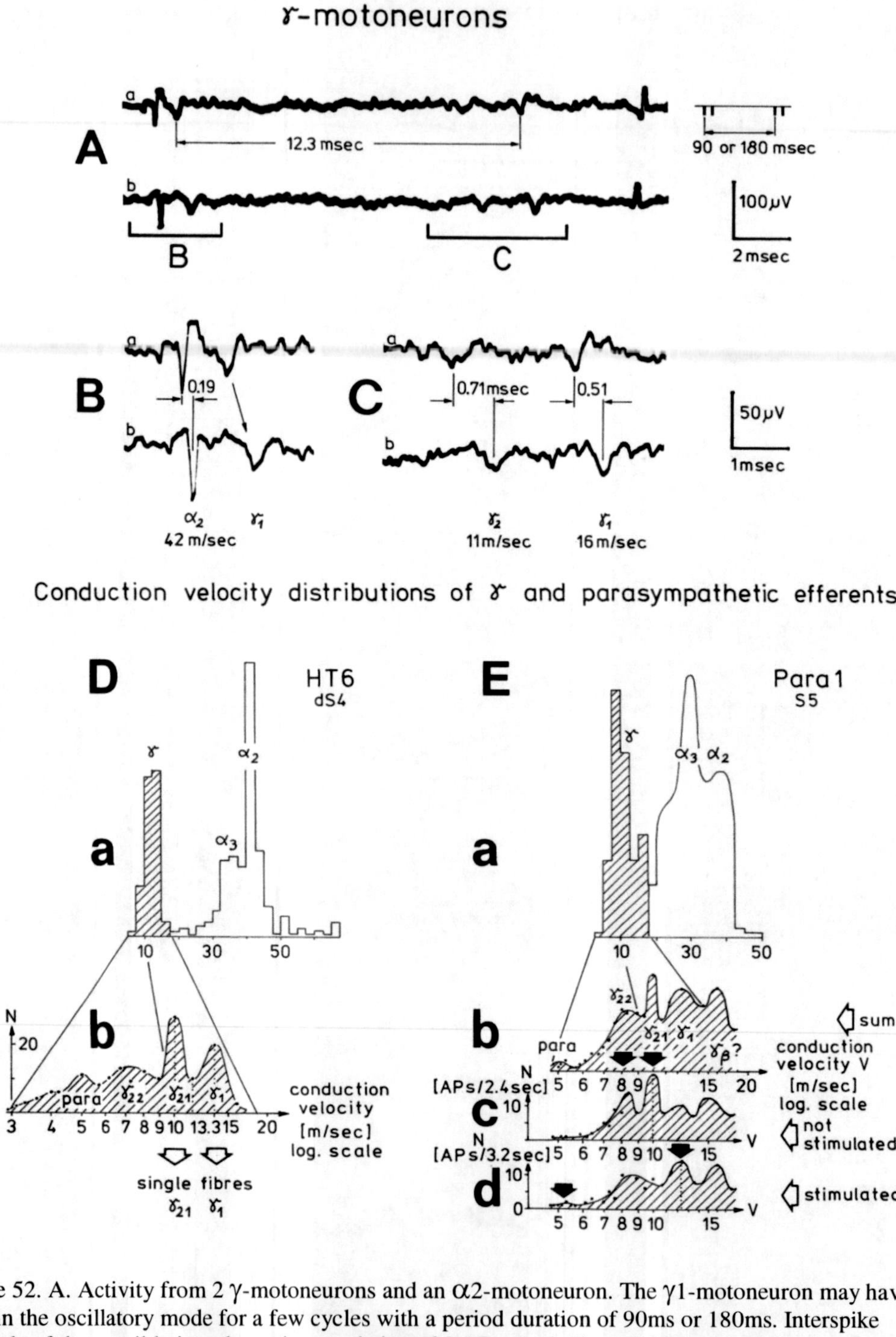

Figure 52. A. Activity from 2 γ-motoneurons and an α2-motoneuron. The γ1-motoneuron may have fired in the oscillatory mode for a few cycles with a period duration of 90ms or 180ms. Interspike intervals of the possible impulse train, consisting of 2 APs, are indicated (12.3ms). The insert shows the possible oscillation cycle period. HT6; dS4.

B,C. Time and amplitude expansions of A. Conduction velocities are indicated. Note that with the increasing conduction velocity, the AP amplitude increases and the duration decreases.

D,E. Conduction velocity distributions of motoneurons for the brain-dead human HT6 (D) and paraplegic 1

(E). To make intrafusal motoneuron and parasympathetic peaks visible, the main γ-peak of Da and Ea was plotted on a log scale in Db and Eb. The distribution peaks are labeled with the groups they most likely represent. In E, the distribution Eb is split into the distribution upon no additional stimulation (Ec) and upon additional stimulation (Ed). Note that in the non-stimulated distribution (Ec) the static γ-motoneuron peaks (γ_{22}, γ_{21}) are highest, whereas under stimulation (Ed) the parasympathetic (Para) and the dynamic γ-motoneuron peaks (γ_1) are highest. When plotting the velocities in Db and Eb logarithmically, the conduction times were first grouped by a conduction time histogram and the column values were then used (conduction distance = 8 mm) to construct conduction velocity distribution curves.

Since parasympathetic fibers are thin, their single-nerve fiber action potentials are small and the identification of their action potentials in the summed impulse traffic is difficult. Also, the γ-motoneurons are thin and its different groups are difficult to separate, especially when the recording temperature is low. In the conduction velocity distribution histogram of Figure 7 (upper right) the parasympathetic fiber groups cannot be distinguished from those of the γ-motoneurons, making it difficult to identify impulse patterns of single γ-motoneurons and parasympathetic efferent fibers.

12.3. Identification of Peaks of γ-Motoneurons and Parasympathetic Fibers in Conduction Velocity Distributions on Log scale

On a linear scale the conduction velocity peaks may fuse and can be separated by their different functions only partly (Figure 52Da,Ea). By first constructing histogram classes for conduction times and plotting them on a log scale from the column values, the velocity peaks separated. Figure 52D,E shows that the fused γ-peaks of the linear plots (Da, Ea) split up into different peaks on the log scale (Db, Eabc). Since first histogram classes of conduction times were constructed using a linear scale, it was also possible to study the dynamics of peaks or peak values upon stimulation. The group identification of single-nerve fiber action potentials is necessary to explore human CNS self-organization and learning.

Figures 52Db and 52Eb show several distribution peaks in which the γ_1 and γ_{21}-peaks could be identified from earlier functional considerations. By comparing the velocity distributions obtained upon stimulation with those without stimulation (Figure 52E), and reasonably assuming that the dynamic intrafusal motoneuron peaks are higher upon stimulation (Figure 52Ed) and the static ones are higher with no stimulation (Figure 52Ec), a second static γ_2-peak (γ_{22}) can be identified and at least one further rather dynamic motoneuron peak (para) can be seen. This peak distribution observed in the paraplegic patient could also be found in the measurements in HT6 (Figure 52Db). That the "Para" peak contains activity from parasympathetic fibres will be shown later. The identification of parasympathetic efferents contributes to the understanding of urinary bladder functions under physiologic and pathophysiologic conditions and its repair by CDT. The nerve fibre groups obtained from conduction velocity distributions for single fibres are in accordance with the sub-compound action potentials of compound action potentials [27].

In the recording from a paraplegic patient in Figure 53 the different groups of parasympathetic fibres and γ-motoneurons do not fuse, probably because of the higher recording temperature during an operation to implant an electrical bladder stimulator. Such a procedure is now obsolete because the bladder can be repaired by CDT.

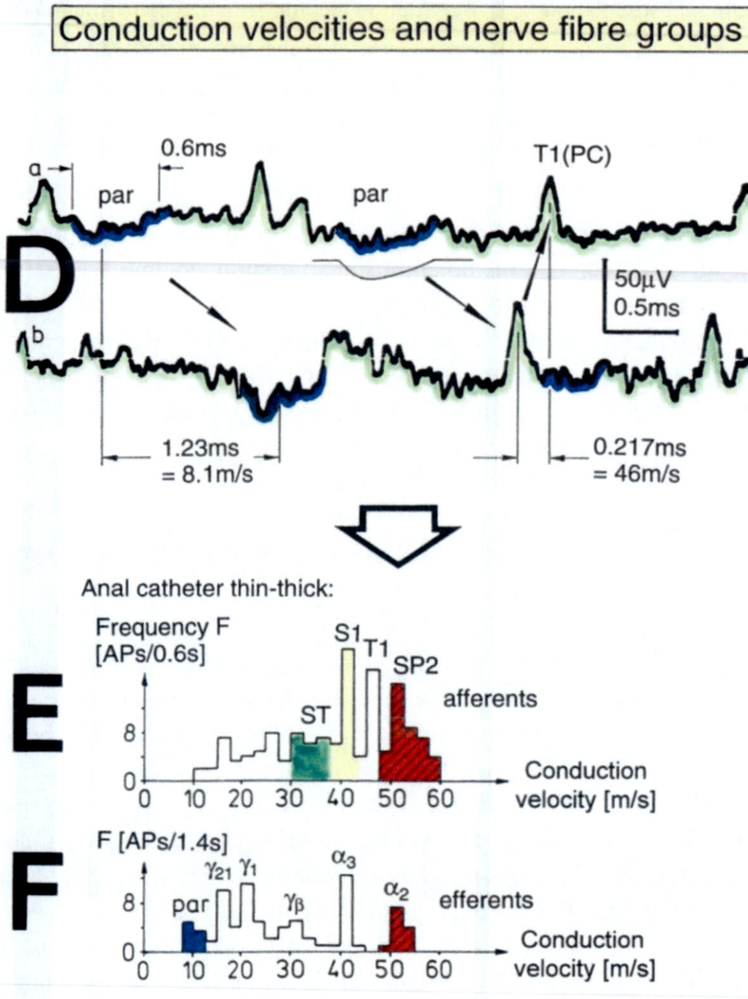

Figure 53. Sweep piece of recording (D) and conduction velocity distributions (E,F) taken from time intervals following a change of a thin anal catheter (∅ = 12 mm) for a thick one (∅ = 20 mm). Note the manifestation of the parasympathetic peak.

12.4. Location and Stimulation of Receptors for Continence

The receptors of skin, urinary bladder and anal mucosa, the muscle spindles, which were innervated by secondary spindle afferents, and the muscles, which most likely were innervated by α-motoneurons in paraplegics 7 and 9 and in the brain-dead human HT6, are marked in Figure 54 in the lower pelvis. The (rhythmic) activation of these receptors by anal and bladder catheter pulling and skin stimulation of sacral dermatomes, to stimulate simultaneously the somatic and parasympathetic divisions to induce learning transfer, can be simulated and achieved during therapy, if a patient with severe cervical spinal cord injury jumps rhythmically on springboard with no weight support (Figure 29).

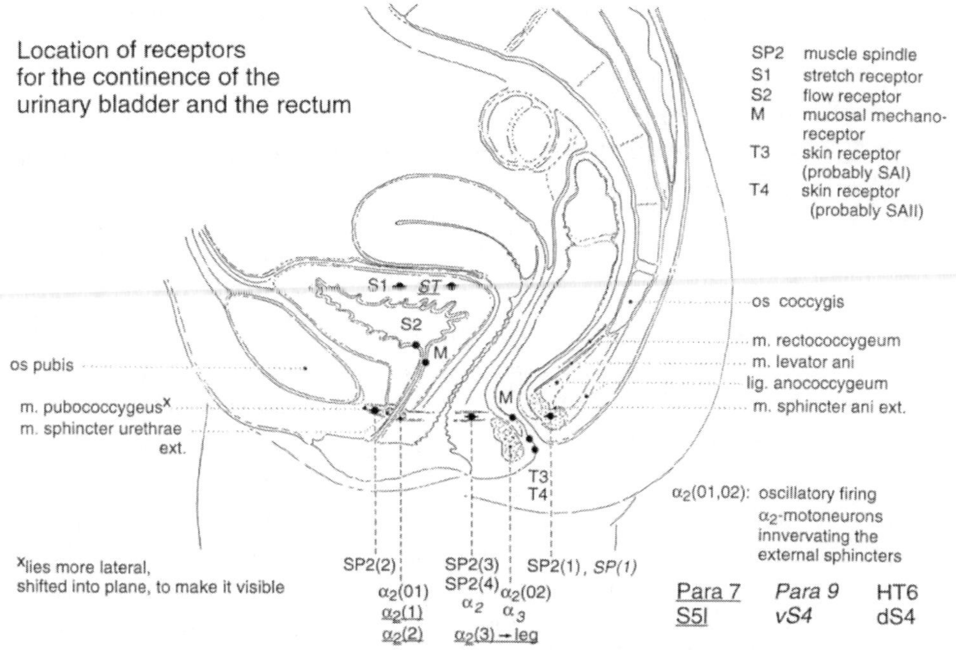

Figure 54. Location of receptors and muscles for the continence of the urinary bladder and the rectum, innervated by motoneurons the activities of which were recorded, in the brain-dead human HT6 (dS4 root), paraplegic 9 (vS4 root) and paraplegic 7 (nerve root S5).

Following retrograde bladder filling urinary bladder stretch (S1) and tension receptor afferents (ST) can be stimulated and measured. In Figure 12 the conduction velocity distribution histogram shows the activation of S1, ST and secondary muscle spindle afferents (SP2) for a bladder filling of 750ml. Jumping on springboard with a full bladder or colon and rectum activates the somatic and parasympathetic divisions much more because more receptors are activated. Before exercising on the special device or jumping on the spring-board, patients often first empty their bladder and rectum to stay continent. To understand urinary function it is important to study bladder filling at the neuron level.

12.5. Bladder Functioning at the Neuron Level

With the single-nerve fiber action potential recording method it has so far been possible to record single-nerve fiber action potentials from nerve fibers down to a diameter of approximately 3.5 μm in undissected thin long nerve root fascicles. It is therefore possible to record natural impulse patterns from parasympathetic efferents (par), urinary bladder stretch and tension receptor afferents (S1, ST), mucosa afferents from mechanoreceptors of the bladder, the urethra and the anal canal (M), from afferents responding to fluid movement (S2), and from α_2, α_3 and γ-motoneurons and muscle spindle afferents innervating the external striated urinary bladder and anal sphincters (or functionally associated pelvic floor muscles), and to analyze regulatory and organizational mechanisms of parasympathetic neurons and motoneurons in the human CNS.

During a laminectomy, natural impulse traffic to and from the CNS can be recorded from sacral nerve roots. The summed impulse traffic can be split into the natural impulse patterns of single afferent and efferent fibers. Following retrograde bladder filling (as in the clinical diagnostic [1,2]) and the identification of the neuron type, with the use of the classification scheme for human nerve fibers (Figures 7,8), the natural impulse patterns of identified afferent and efferent fibers can be obtained and analyzed. The obtained natural impulse patterns answered an age-old question: Is it the firing rate of a neuron that codes the information transmitted and processes it, or does the precise timing of cell discharge code information? As the different pictures show, the information is coded by specific impulse patterns, including the precise timing and the firing rate. *Neuronal network learning therefore includes changes of specific impulse patterns, precise timing, and firing rates among neurons* apart from structural changes. Learned functions are those ones which have pattern stability in the multi-dimensional space.

In response to retrograde bladder filling of a brain-dead human, the self-organization of a premotor spinal α_2-oscillator innervating the external striated bladder sphincter is shown in Figure 12. Because the axon of motoneuron O1 had a recurrent fiber (Figure 12A) at the recording site, each single AP of this motoneuron could be identified safely by the AP of the recurrent fiber. The function of the motoneuron was to secure bladder continence. The activity from urinary bladder receptors, i.e. the activity of bladder stretch (S1), tension (ST), and flow receptor afferents (S2) (Figure 12E), was an adequate afferent input to the motoneuron. Phase relations between the firing patterns of bladder afferent fiber S1(1) and oscillatory firing urinary bladder sphincter motoneuron O1 can be seen in the schematized firing patterns in Figure 12B. They were an indicator that the natural impulse pattern of the S1(1) fiber was an adequate drive of the sphincter motoneuron O1. For retrograde bladder filling up to 550 ml, motoneuron O1 only fired occasionally (Figure 12D,F). This was the storage phase of the bladder, during which the intravesical bladder pressure increased only little. For higher bladder filling volumes, the motoneuron switched via the transient oscillatory firing mode to the continuous oscillatory firing mode (Figure 12D) to generate a higher activity (Figure 12F) for a stronger drive of the urethral sphincter to more strongly secure continence when the storage phase was nearly passed and the bladder pressure increased. For bladder filling volumes higher than 800 ml, the activity of the spinal oscillator decreased again; probably, the oscillator became inhibited (Figure 12D). Pain fibers (not shown in Figure 12G) may have inhibited the oscillatory firing to protect the bladder from mechanical damage. The overflow mechanism was initiated. It is likely that fluid entered the trigonum vesicae to activate flow receptors, so that the flow receptor activity (S2) increased strongly (Figure 12E). The premotor spinal oscillator O1, of which motoneuron O1 is probably a part, was organized by the adequate afferent input (S1, ST, S2, ...) induced by bladder filling. For the oscillator to be formed in the premotor neuronal network in the spinal cord, consisting of motoneurons and interneurons, it needs a certain preformation of the networks (connectivity, synapse efficacies, membrane properties of neurits etc.) and adequate afferent input patterns. In modeling such networks, the organization of the premotor spinal oscillators in the spinal cord neuronal networks cannot be separated from its space-time distributed adequate afferent input patterns, giving rise to self-organization. From Figure 12B it can be seen that the stretch receptor afferent APs (S1(1)) show a relative phase correlation to the impulse trains of the oscillatory firing motoneuron O1. The successive interspike intervals (IIs) of the S1(1) afferents have, on the average, a value similar to that of the oscillation period of oscillator O1 (Figure 12C).

Other premotor spinal oscillators, activating the external striated bladder sphincter not shown here, will also have self-organized themselves by the same or similar afferent input. However, these oscillators sub-serving the same function will be correlated in their firing in the way that they do not fire in synchrony but are distributed in their phases with respect to one another (probably by relative inhibition with respect to each other) to secure that the sphincter muscle does not show rhythmic movements (tremor). Such non-synchrony correlated oscillatory firing of several oscillators has in a small time window been measured (Figure 24B). In the case of a non-injured CNS, this premotor spinal oscillator O1 could also be activated volitionally from supraspinal centers. It would be very interesting to see how the afferent input patterns from supraspinal centers to the spinal premotor network would look.

By recording from the dorsal S4 root of the brain-dead human HT6, impulse trains from another oscillatory firing motoneuron and its driving afferents were measured, which served quite a different function (Figure 12F). The α_2-oscillator O2 (Figure 12A,B) innervating the striated external anal sphincter was activated by secondary muscle spindle afferent activity, induced by the anal catheter-stretched muscle spindles, probably located in the anal sphincter or functionally associated pelvic floor muscles. Also, mucosal and skin receptors within the anal reflex area will induce self-organization of premotor oscillators activating the external anal sphincter to secure anal continence. It is evident from the impulse patterns shown schematically in Figure 12B that the impulse trains of this oscillator O2 show a phase relation and an interspike interval relation to its driving spindle afferent APs. But no synchronized firing can be seen between the sphincter α_2-motoneurons, innervating the external bladder sphincter (O1) and anal sphincter (O2), the CNS networks probably tried to avoid an increased physiologic tremor (macroscopic rhythmic activation of muscles, due to sub-optimal regulation). Important for the application of human neurophysiology to neurotherapy by learning is the duality of the functions of the sphincter motoneurons and secondary muscle spindle afferents, sub-serving somatosensory and autonomic (parasympathetic) functions. In animals, it was also found that sympathetic fibers innervate muscle spindles [28]. These measurements indicate that parasympathetic fibers also innervate muscle spindles in the parasympathetic innervation area (S2-S5).

Motoneurons innervate the external sphincters of the bladder and the anal canal sub-serve somatic functions (contraction of the sphincters on volition or for protection reaction) and parasympathetic functions for the coordination of the detrusor function (parasympathetic) and the external sphincter function. This duality of the sphincter motoneurons (and spindle afferents in the parasympathetic domain) makes the pattern change from continence to protection reaction understandable and makes learning transfer between somatic and parasympathetic patterns likely. The motoneurons build up two phase relations per oscillation cycle with other motoneurons and secondary spindle afferents for somatic activation (Figures 56,58), and buildup 3 phase relations per oscillation cycle when the parasympathetic division is also activated (Figures 58, see below). The neuronal networks of the somatic and the parasympathetic nervous systems are interlaced and interact with each other. It should therefore be possible to improve parasympathetic functions when improving somatic functions by coordination dynamics therapy, especially as there is indication that parasympathetic efferents also fire rhythmically.

Repair of bladder function through the reorganization of networks seems to be difficult because there is false neuronal network organization in the parasympathetic nervous system (overactive (spastic) detrusor) and the somatic division (spastic external (striated) bladder sphincter), and there is false interaction of the interlacing somatic and parasympathetic networks (detrusor-sphincter dyssynergy: when the detrusor contracts, the external sphincter is also contracting instead of opening (relaxing)). However, repair is possible through CDT.

12.6. Parasympathetic Activation of the Detrusor Can Be Assessed by Parasympathetically Induced Muscle Spindle Afferent Activity

It was shown that the activity in parasympathetic efferents could be measured, identified and distinguished from the activity of γ-motoneurons in conduction velocity distribution histograms (Figures 52,53). Since the action potential amplitudes of parasympathetic efferents is small (Figure 53), it would still be very difficult to analyze the organization of the parasympathetic nervous system and its coordinated functioning with the somatic nervous system, as in the control of the urinary bladder. But if some secondary muscle spindles in the parasympathetic range are also innervated by parasympathetic efferents, besides somatic efferents (γ-motoneurons), then the activation of the parasympathetic nervous system can also be measured by the activity of secondary muscle spindle afferents. Since the action potentials of the secondary spindle afferents are comparably large (thick fibers), the activation of the parasympathetic nervous system could easily be indirectly assessed. From the similarity of changes of spindle afferent activity and detrusor pressure changes it can be concluded that some muscle spindles in the domain of the sacral parasympathetic nucleus are partly controlled by the parasympathetic division and that the muscle spindle and the detrusor activation have similar time courses (Figure 162 of [1,2]).

12.7. Relative Phase and Frequency Coordination between the APs of α and γ-Motoneurons and Secondary Muscle Spindle Afferents with No Additional Stimulation and Upon Touch, Pin-prick, and Bladder Catheter Pulling

Above it was shown that natural impulse patterns can explain pattern change, which is to say that the generation of integrative patterns can partly be understood at the single neuron level. Natural impulse patterns were related to integrative CNS organizations. How can this aid in understanding why jumping on springboards contribute substantially to the repair of bladder function? A better understanding can be achieved to go deeper into the complexity of the cooperative and competitive interplay among neurons, which means going deeper into the complexity of phase and frequency coordination of CNS self-organization. The phase and frequency coordination of coordinated firing will now be analyzed in more detail.

Time correlations of afferent and efferent impulse patterns are easy to detect in the brain-dead individual since the oscillatory firing of α_2-motoneuron (O2) was regular, like an inner clock (Figure 12). The phases of fusimotors and spindle afferent APs could be defined with respect to the impulses of that inner clock. In the paraplegic, the rhythmic firing was rather irregular. The motoneuron firing therefore can no longer be used as a time reference basis. More phases between the extracellular action potentials of the different fibers are necessary to fully describe the correlation between the simultaneous impulse patterns. In Figure 55Cg, the mutual phases between the APs of the different fibers are defined. Figure 55Ch,i shows the corresponding phase distribution histograms. Since too few phases occurred in a sweep piece of 0.8 s duration, phases occurring in certain time intervals were pooled and plotted in Figures 56 and 57.

In this special pathologic case, the α_2 and α_3-motoneurons fired rhythmically with impulse trains consisting of one action potential (AP), in contrast to the physiologic firing patterns, in which α_2 and α_3-motoneurons fire with impulse trains consisting of more than one AP. The identification of motoneurons by conduction velocity is not absolutely safe, since group conduction velocity ranges overlap. It is very unlikely, nevertheless, that one of the motoneurons was an α_1-motoneuron (FF) (missing high conduction velocity and high AP amplitude and no correlation to primary spindle afferent fibers), even though they fire physiologically with impulse trains consisting of one AP. It has been shown that oscillatory firing α_1-motoneurons are mainly driven by time locked primary spindle afferent fibers (Figure 66C of [2]). The firing patterns of the α_2 and α_3-motoneurons are strongly pathologic with respect to the length of the oscillation period and the impulse train length so that it is impossible in this paraplegic to identify the kind of motoneurons by their discharge patterns of oscillatory firing; this would be possible were the neuronal network driving the motoneurons to fire in a physiologic manner.

In Figure 56, the interspike intervals (IIs) and the phases are shown for similar time intervals given in Figure 55. Before stimulation, within the time interval 1-6 s, the α_3-motoneuron fired every 100ms, the γ_1-motoneuron every 100 to 130ms, and the SP2(1) fiber every 80 to 150ms (Figure 56Aa). The α_2-motoneuron mostly fired every 300ms and the SP2(2) fiber every 250ms. At that particular time interval, similar phases (phase relation of broad peak type) occurred twice per α_3-oscillation period between the APs of the α_3 and γ_1 axons, between the γ_1 and the SP2(1) fibers, and between the α_3 and the SP2(1) fibers (Figure 56Ba). One phase relation occurred between the impulses of the α_3 and α_2-motoneurons, and two between the α_3 and the SP2(2) fibers. The broad phase relations between discharge patterns are interpreted as interactions between populations of neurons. Following different stimulations, interspike interval (II) distributions and phase relations changed with time.

Upon touching sites 1 – 5 (Figure 55B), the IIs of the almost oscillatory firing γ_1-motoneuron reduced in size to be more similar to those of the oscillatory firing α_3-motoneuron (Figure 56Ab). The changing of the different phase relations indicates changes in the interactions between neuronal sub-networks (Figure 56Bb). Upon touching sites 6 to 10 (especially sites 6, and 7 (inside anal reflex area)), the IIs of the almost oscillatory firing γ_1-motoneuron increased again (Figure 56Ac). A transient partial synchronization occurred between the different nerve fibers (Figures 56Bc,59c). Upon pin-pricking sites 1 - 5, the IIs of the almost oscillatory firing γ_1-motoneuron reduced again to have a similar II distribution as the α_3-motoneuron. The α_3-motoneuron even slightly increased its IIs (decrease of activity), so that the II distribution of the oscillatory firing α_3-motoneuron and the now oscillatory firing γ_1-motoneuron became very similar (Figure 56Ad).

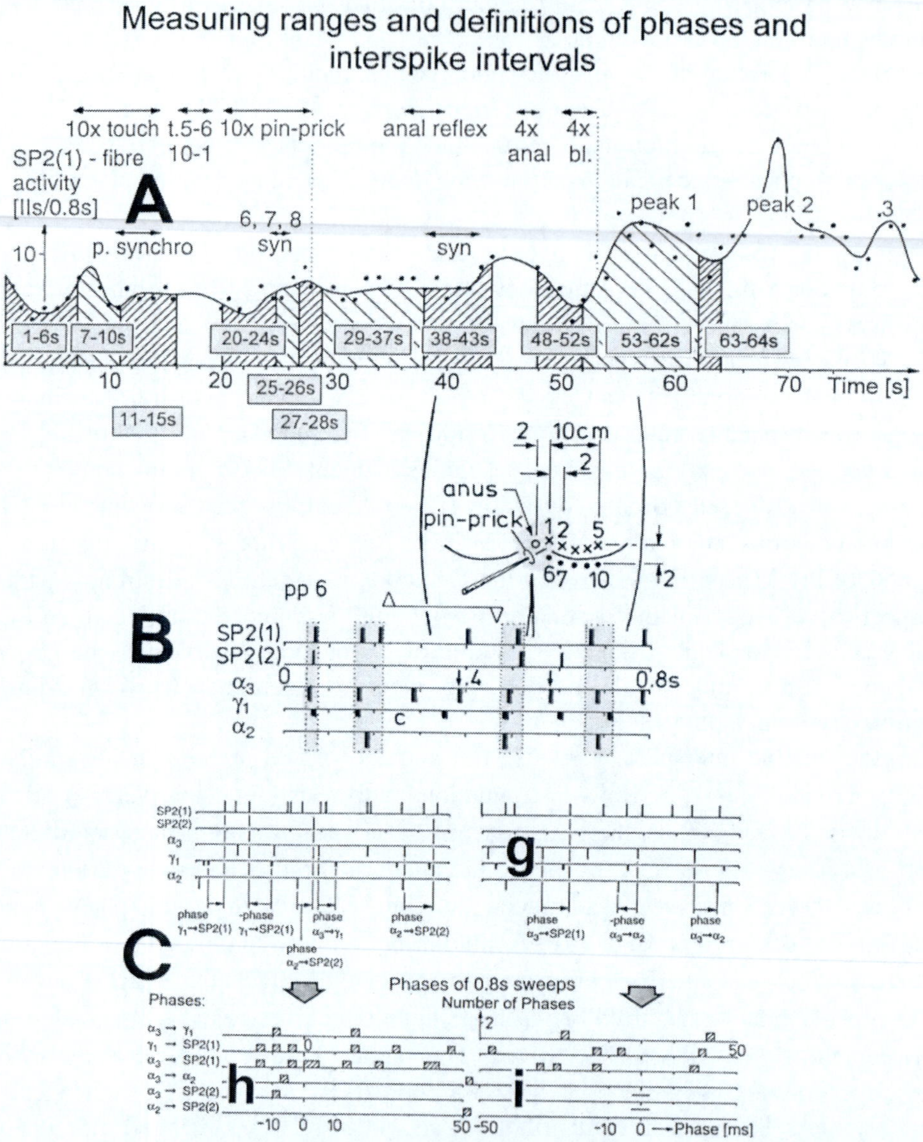

Figure 55. Measurement ranges and definitions of phases for the analysis of phase and frequency coordinations between motoneurons and spindle afferents. Coordination (synchronization) between firing patterns can be directly be in B.

(A) Activity level of secondary muscle spindle afferent fiber SP2(1) in dependence on time. 10x touch = touching sites 1 to 10 shown in B; t. 5-6 = touching alongside the skin from site 5 to site 6; 10x pin-prick = pin-pricking sites 1 to 10; anal reflex = anal reflex stimulation; 4x anal = fourfold anal catheter pulling; 4x bladder = fourfold bladder catheter pulling; peak 1 - peak 2 - 3 = first, second and third peak of spindle afferent activity due to parasympathetic activation; p. synchro = partial synchronization; syn = synchronization of α and γ-motoneurons and secondary muscle spindle afferents. Note the synchronization of the firing patterns following pin-prick 6 inside the anal reflex area.

(B) A set of single impulse patterns of secondary muscle spindle afferents (SP2(1,2)) and $\alpha$and γ (intrafusal)-motoneurons and sites of stimulation. The small arrows in the impulse pattern of α_3-motoneuron (S) point to a shortening of the oscillation period following pin-prick 6 (pp6). The triangles indicate the beginning and the end of pin-pricking.
(C)(g,h,i) Definitions of the phases between the different motoneurons and spindle afferents in 2 sets of impulse patterns (g), and the corresponding sets of phase relation distributions (h,i). Para 9; vS4.

Upon pin-pricking sites 6 and 7 (inside the anal reflex area), α and γ-motoneurons and secondary muscle spindle afferents showed similar II distributions (Figure 56Ae). Only one phase relation was organized per oscillation cycle between the different nerve fibers (Figures 56Be, Figure 59e). A synchronization between the APs of the different nerve fibers occurred, as can be seen from the direct impulse patterns (Figure 55B). The occurrence of similar II distributions of, and transient constant phases between, the APs of the α_3, γ_1 and SP2(1) fibers is interpreted in the way that, in its oscillatory firing the oscillatory firing α_3-motoneuron built up an external loop to the muscle spindle innervated by the γ_1 and SP2(1) fibers. The γ-loop became integrated into the oscillatory firing of the α_3-motoneuronal network. Before pin-pricking, the γ-loop, consisting partly of the γ_1 and SP2(1) fibers, also contributed to the oscillatory firing, since on average there existed phase relations. With the pin-pricking, however, the II distributions also assimilated, so that this γ-loop was directly included into the oscillatory firing of the α_3-network rather than only contributing to the drive of it. The build-up of an external loop to the periphery by spinal oscillators is commonly used when a patient with a spinal cord injury is jumping on a springboard (see Figure 62, below). Upon pin-pricking sites 8, 9 and 10 (outside of the anal reflex area) and following pin-pricking of site 10, the II distribution of the SP2(1) fiber shifted away from those of the α_3 and γ_1 axons. The oscillatory firing α_3-motoneuronal network had abolished its external loop, even though it was still getting drive from it. Upon anal reflex stimulation and catheter pulling, the external loop was not built up again.

Following touch, pin-prick and anal reflex stimulation, but not painful catheter pulling, mostly two phase relations existed in paraplegic 9 between the activity of the α_3, γ_1 and SP2(1) fibers per oscillation period (100-140ms) of the α_3-motoneuron (Figures 56,57), but the phase relations changed with ongoing time. With the activation of the parasympathetic division, upon bladder catheter pulling (Figure 55A), three phase relations occurred per α_3-motoneuron oscillation period (Figure 57Bc,e). At the peaks of parasympathetic activation, three phase relations occurred (Figure 57Bc,e), and only two phase relations were present with little parasympathetic activation (times between the peaks 1 and 2) (Figures 55A,57Bd). Even though the functional units, consisting of fusimotor and α-motoneuron neuronal networks and spindle afferents fibers, were rather unstable in paraplegic 9, in comparison to the brain dead human, an important difference between skin (somatic) and bladder (parasympathetic) stimulation occurred. Another (third) phase relation per α_3-motoneuron oscillation period occurred with the activation of the parasympathetic division. The activated parasympathetic neuronal network of the sacral micturition and defecation center seems to have channeled input to the oscillatory firing somatic neuronal network.

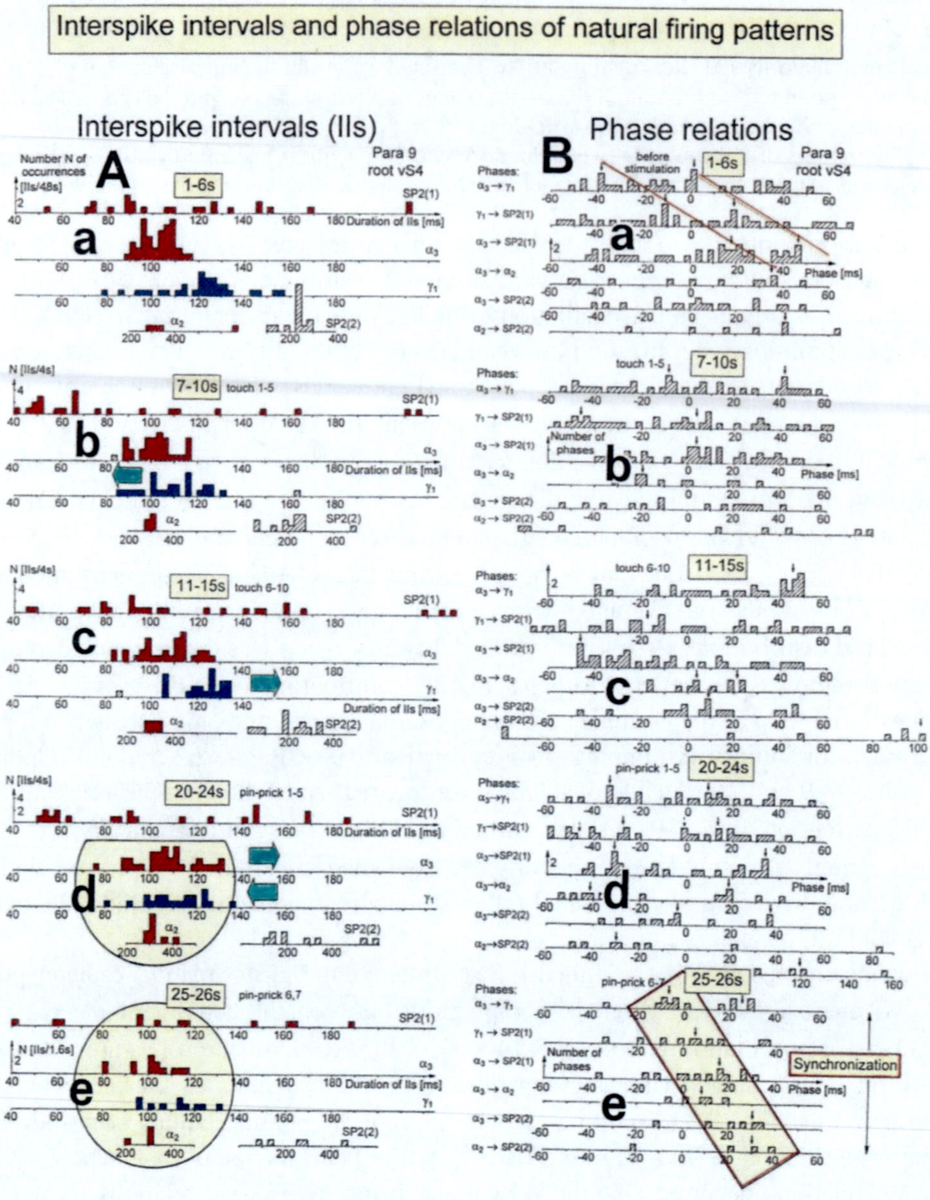

Figure 56. Relative phase and frequency coordination between α and γ-motoneurons and secondary muscle spindle afferents due to touching and pin-pricking sacral dermatomes as in Figure 5G. (A) Interspike interval distribution of spindle afferents SP2(1) and SP2(2), α_2 (FR) and α_3-motoneurons (S) and the dynamic fusimotor γ_1 for different time intervals upon touch, pin-prick and anal catheter pulling. Interspike intervals (IIs) were collected from several sweeps of 0.8 s duration per second. External loop generation and frequency coordination of α and γ-motoneurons and secondary muscle spindle afferents are marked by the semi-circle and the full circle. The large arrows point to the increase and decrease of the mean II of the distribution. Unsafe identification of α_2 and α_3-motoneurons (or vice versa) because of loss of specific oscillator properties. (B) Histograms of the phases between afferent and efferent fibers for the time intervals indicated, upon different stimulation. Phases were collected from several sweeps of 0.8 s duration per second. The small arrows indicate phase relations. Phase coordination is indicated in a,e. Para 9; vS4.

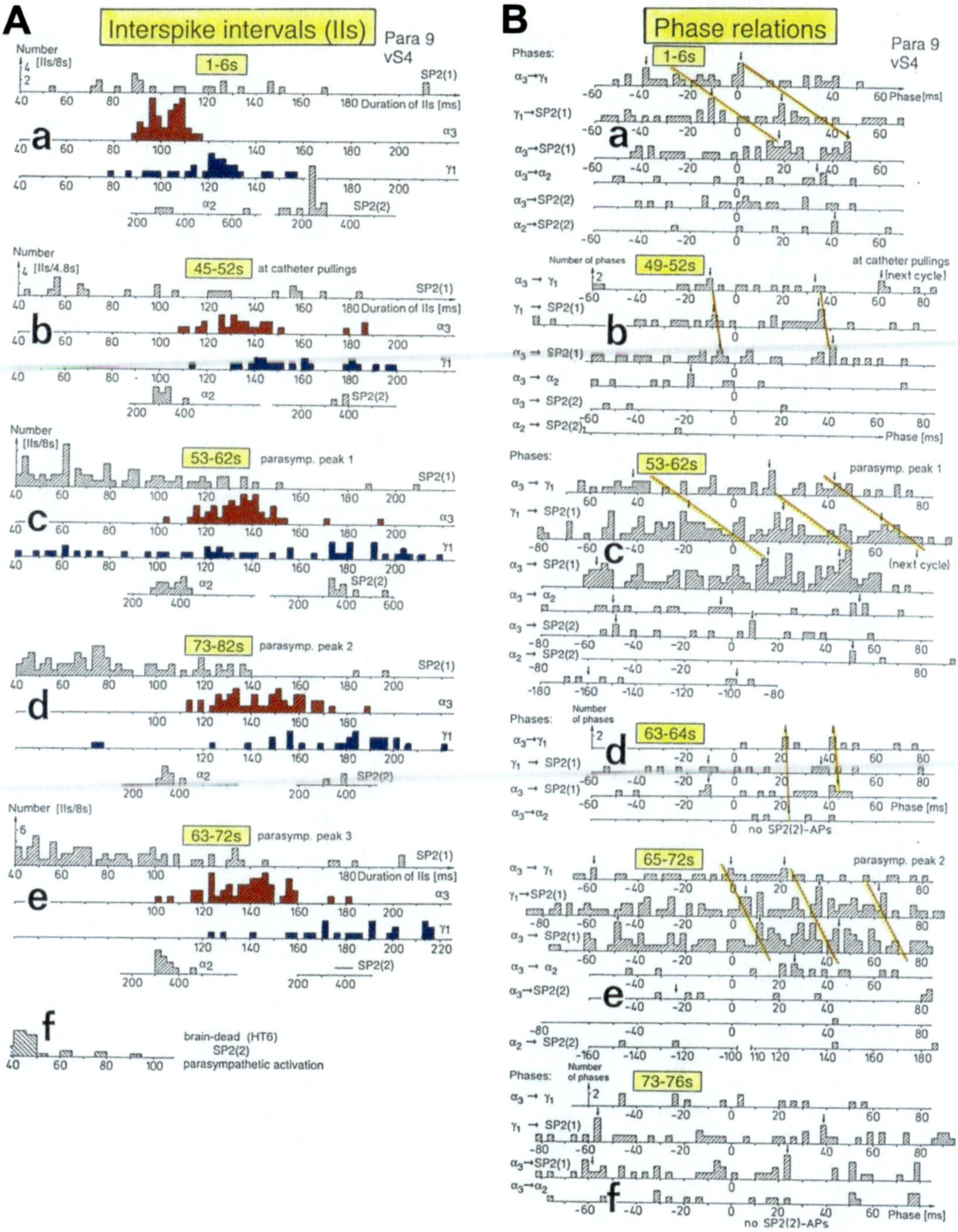

Figure 57. Interspike intervals (IIs) and phase relations for time intervals indicated in Figure 162. For legend, see Figure 56.

12.8. Phase Relation Changes between the Action Potentials of the α and γ-Motoneurons and Secondary Muscle Spindle Afferents in Paraplegic 9 upon Somatic and Parasympathetic Activation of the Sacral Micturition Center

As shown in Figures 56 and 57, the number and the values of phase relations changed between the firings of the different nerve fibers upon different stimulations. In the brain-dead human HT6, two phase relations were found between the α_2-motoneuron and the secondary muscle spindle afferent fiber SP2(2), the α_2 and the γ_1-motoneuron [29], and also in the paraplegic two phase relations that often exists between the firings of the different nerve fibers. It may be that a third phase relation occurs when the activated parasympathetic division channels an additional input to the oscillatory firing somatic neuronal network. It is therefore worthwhile to further analyze the number of occurring phase relations per oscillation cycle upon different somatic and parasympathetic stimulations.

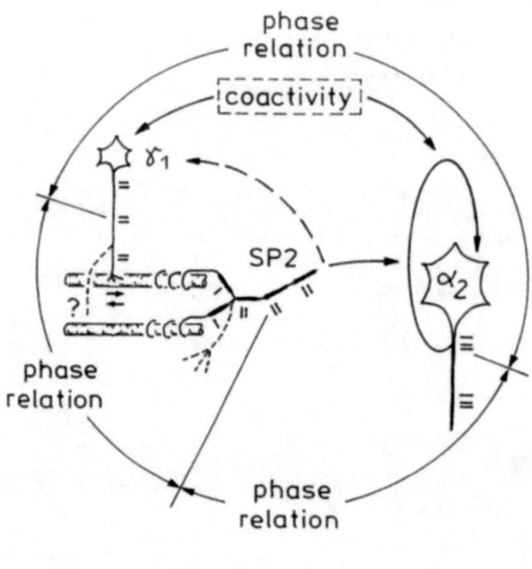

Figure 58. Schematized existing phase relation between α_2 and γ_1-motoneurons and a secondary muscle spindle afferent fiber (SP2). Parallel existing phase relations between other parent afferents and the α_2-motoneuron and between parent secondary spindle afferents are not shown. Phase relation means, the increased occurrence of phases in ms in a certain phase range between the action potentials (APs) of the two compared nerve fibers. The complex afferent and efferent muscle spindle innervation was not apparent. Small arrows at intrafusal muscle fiber indicate local contraction, which is in nuclear chain fibers readily transmitted to the place of afferent innervation. A possible reason for the doublet firing of the SP2 fiber is pictured to occur from single APs (schematized by bars) of two myelinated endings, not necessarily from pacemaker switching. More endings of the parent SP2 fiber and γ_1-motoneurons are indicated by dashed line branches. "Coactivity" indicates a correlation between γ and α-motoneuron spinal cord circuitries for higher activations.

Since two phase relations occurred per oscillation cycle between the α_3 and γ_1-motoneurons and the SP2(1) fiber (Figure 56Ba) in paraplegic 9, and also their IIs were rather similar, it is concluded that the neuronal networks of the α_3 and γ_1-motoneurons formed together with the spindle afferent fiber SP2(1) a part of a functional unit. The neural ensemble is built by weights of synapses and projections between the convergence of several fusimotors on one muscle spindle and by the divergence of muscle spindle projections onto several rhythmically firing populations of neurons driving α and γ-motoneurons. Such a functional unit is partly pictured in Figure 58 and schematized drawn by 3 circles in Figure 59. The α_2-motoneuron and the SP2(2) fiber belonged to another functional unit (another ensemble) (longer IIs and the existence of only 1 phase relation). The two functional units (ensembles) are characterized in Figure 59 by two sets of three circles each. The two functional units interacted with each other since a phase relation existed between the α_2 and α_3-motoneurons (Figure 59).

Before stimulation (but with the anal and bladder catheters positioned), there were two phase relations in unit 1 (Figure 59a). When touching sites 1 to 5 (Figure 55), only slight changes occurred in the two units with respect to the number of phase relations (Figure 59b). But when touching sites 6 to 10, a partial synchronization occurred (Figure 59c) and functional unit 1 reduced the number of phase relations to one. When pin-pricking sites 1 to 5, two phase relations occurred again in unit 1 (Figure 59d). Upon pin-pricking sites 6 and 7, the number of phase relations between all the components of the two units dropped to one (Figure 59e), and synchronization occurred between the firing patterns (Figure 55B). Since in the brain-dead human two phase relations per oscillation cycle were observed in the functional units [29], it is possible that synchronization and the existence of only one phase relation for 2 to 3 seconds reflected a slight pathologic organization of the networks.

Even though upon touching sites 6 to 10 (Figure 59c), or upon pin-pricking sites 6 to 7 (Figure 59e), only one phase relation existed in unit 1, and synchronization occurred with both stimulations, it is shown in Figure 60 (and Figure 151 of [2]) that the touch afferent input organized a different functional state of unit 1 than pin-pricking. The response time till the shortening of the oscillation period was longer than the oscillation period ($\approx$ 100ms) for pin-prick and shorter for touch. It was shown in Figure 60 that repetitive touch stimulation (most effective inside the anal reflex area) reinforced the sustained stretch reflex of the anal sphincter (continence pattern), and repetitive pin-prick stimulation replaced the continence pattern by the protection reaction of the anal sphincter. The number of phase relations alone therefore only provides limited information on the functional state of the organization of the neuronal networks of the human spinal cord. Measurements of a number of parameters are necessary to yield a rather complete description of the functional state of neuronal networks.

Following pin-prick 8 and 10 and with no stimulation two phase relations existed again in functional unit 1 (Figure 59f,g), in some similarity to pre-stimulation status (Figure 59a). Following two anal reflex stimulations, partial synchronization occurred in the components of the two units, and mainly two-phase relations existed (Figure 59h). But the organizational state was still not very similar to the pre- (Figure 59a) or post-stimulation state in unit 1 (Figure 59g), since the parasympathetic division was slightly activated following anal reflex stimulation as was measured by the impulse pattern (increase of doublet activity) of the secondary muscle spindle afferent fiber SP2(1). Therefore, probably one phase relation was due to the somatic activation in similarity to Figure 59c,e and the other phase relation was due to the activation by the parasympathetic division. During bladder catheter pulling (Figure 59i) and with no stimulation (Figure 59k), the number of phase relations and possibly the functional organization was again similar to the pre stimulation state (Figure 59a).

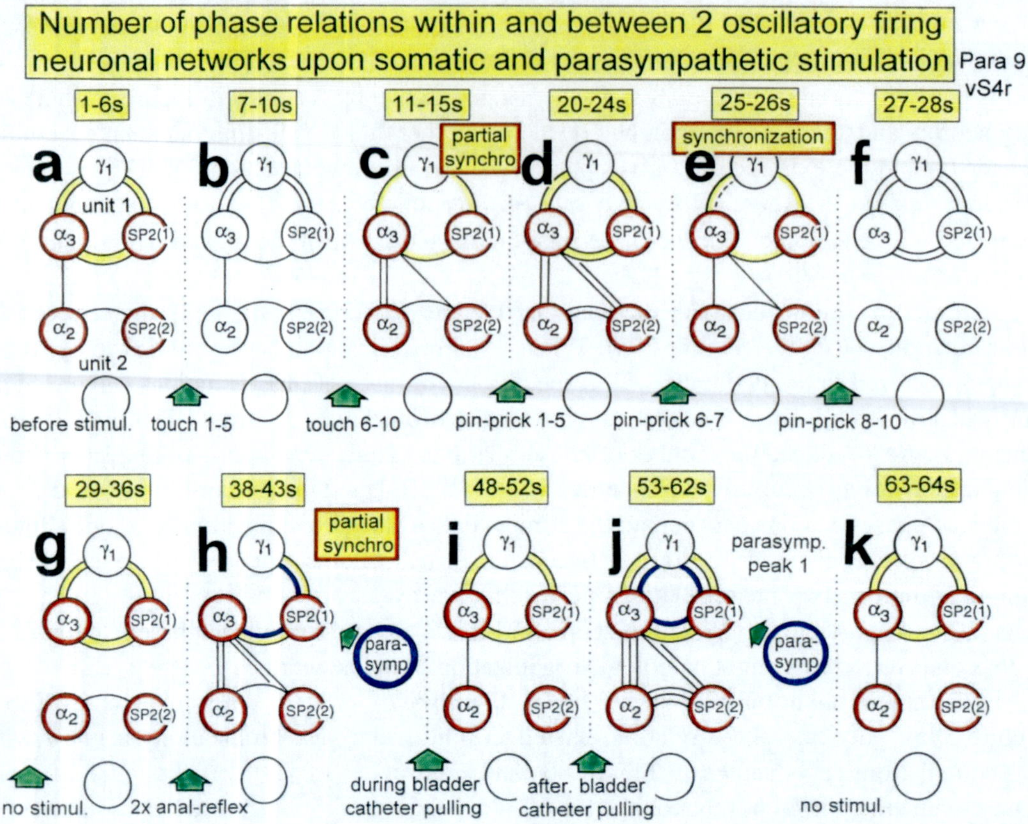

Figure 59. Number of phase relations within and between the two functional units α_3/γ_1/SP2(1) and α_2/-/SP2(2). Time intervals are those of Figure 2A. Note that in 'a' the functional unit 1 is with two phase relations per oscillation period in a stage similar to those seen in the brain-dead individual; with synchronization only 1 phase relation occurred (e) and the parasympathetic division channeled an extra phase relation to interact with the somatic division (j).

Following strong (painful) bladder catheter pulling with a strong activation of the parasympathetic division (time interval 53-62s (Figure 59j)), measured by the increased doublet firing (for doublet firing see page 514 of [1]) of the SP2(1) fiber, the functional organization of the sacral micturition center of the disconnected spinal cord changed completely. Functional unit 1 was now correlated by three phase relations per α_3-oscillation cycle. The functional unit 2 also showed 3 phase relations per α_2-oscillation cycle, and interacted with the functional unit 1 by 3 phase relations as well (between the α_3 and α_2-motoneurons; Figure 59j (53-62s)).

Between the first and second parasympathetic peak (Figure 55A) at the time interval 63-64s (Figure 59k), the organization form of the two functional units was similar to that before the first parasympathetic activation (49-52s) (Figures 59), only the values of the phase relations changed (Figure 57Bd). With the second strong activation of the parasympathetic division (parasympathetic peak 2, time interval 65-72s) the functional unit 1 was bound together again by 3 phase relations (Figure 55A), in similarity to the first strong activation of the parasympathetic division, measured by the burst firing of the secondary muscle spindle afferent fiber SP2(1) and the increased doublet firing of the SP2(1) fiber [14]. The functional unit 2 was disorganized, but phase relations still occurred between the α_3 and the α_2-motoneurons and the SP2(2) fiber. The α_2-neuronal network and the fusimotor networks, driving the SP2(2) spindle afferent fiber, were integrated differently. After the second strong parasympathetic activation, in the time interval 73-76s (Figure 55A), the functional organization of the two functional units in the spinal cord was similar to that before the activation of the parasympathetic division. Functional unit 2 was slightly disorganized as the SP2(2) fiber strongly reduced its firing.

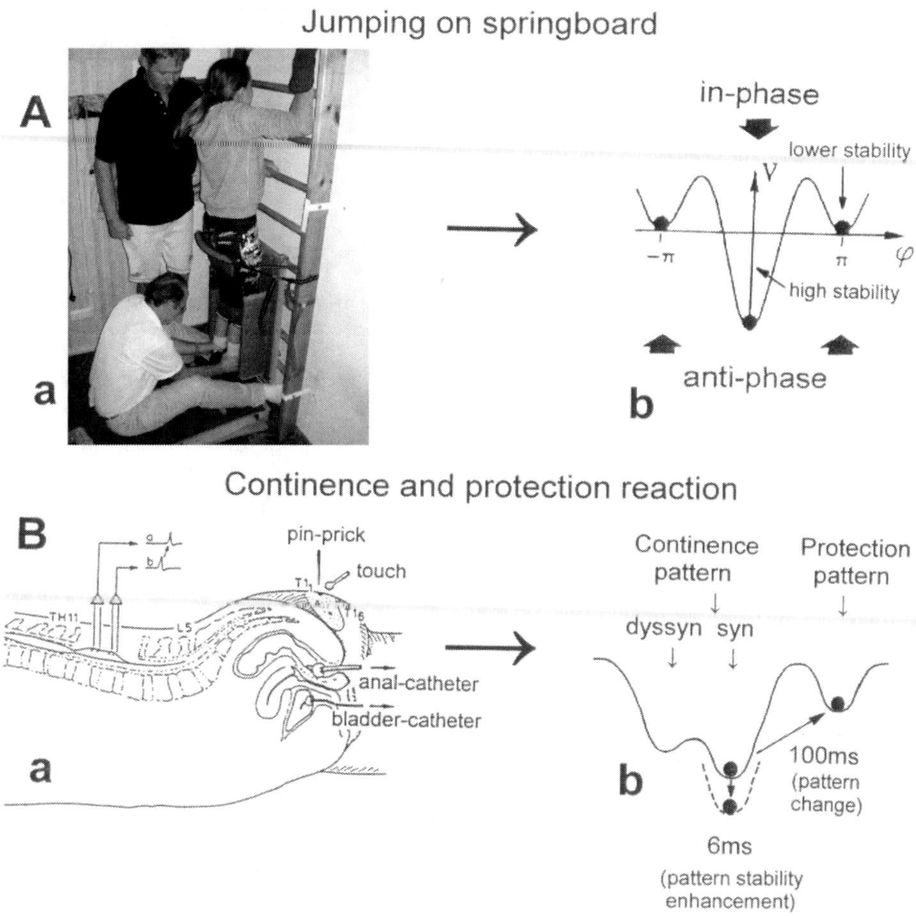

Figure 60. Attractor layout for jumping on springboard in comparison to that of continence and protection.

12.9. The Need to Improve the Stability of Phase and Frequency Coordination to Allow Specific Pattern Formation and Learning Transfer

A young mother, with stress incontinence after giving birth to her first child, could strongly improve her continence status by jumping on the springboard in addition to other training. Her CNS is not injured. It was the periphery that required repair by means of changing the CNS.

In severe cervical spinal cord injury, however, solely jumping on springboard (Figure 60A) is not sufficient for bladder repair. First, of course, the patient has to regain movement functions back (especially the trunk stability) to be able to perform the jumping on springboard. Further, the self-organization of CNS networks by phase and frequency coordination must be improved to make learning transfer from movements to bladder functions possible, since in every CNS injury the phase and frequency coordination is impaired. Large instabilities in phase and frequency coordination will not allow specific pattern formation as a basis for learning transfer. However, the stability of phase and frequency coordination can be improved when the patient is exercising on special coordination dynamics therapy devices, especially the one shown in Figure 15. In the following, measurements are presented to understand the necessity of improving the stability of phase and frequency coordination for functional repair.

The importance of stable phase and frequency coordination to allow specific pattern formation and consequential learning transfer to other patterns can be understood at the collective variable level and the neuron level. The behavioral information F_{inf} of the coordination pattern dynamics, characterized by equations of motion of collective variables, $dX/dt = F_{intr}(X) + \sum c_{inf}F_{inf}(X,t)$, affect the whole coordination pattern dynamics, including stability, rather than only certain coordination patterns. If the behavioral information includes the exercising of extremely coordinated, integrative movements, like exercising on the special CDT device for turning, then the quality of the CNS self-organization can be enhanced by improving the exactness of self-organization, namely the precision of phase and frequency coordination between neuron and neural assembly firing. By improving the precision of organization of the intrinsic dynamics $F_{intr}(X)$, that is, the specific variability of the injured networks, certain patterns do then already re-appear (Figure 28). Neurons often serve more than one network pattern at the same time by time-sharing of neuron firing, which in this way give rise to learning transfer among the activated patterns. If sub-networks are improved in the organization of one pattern, the organization of the other pattern will also improve. Neurons involved in the organization of breathing and activating intercostal muscles, for example, are also involved in the organization of trunk stability. By reducing spasticity of the trunk (in patients with Parkinson's) breathing will also improve. Similarly, sphincter motoneurons are involved in continence and pelvic floor weight bearing. If the pelvic floor is not trained during pregnancy, incontinence sometimes occurs and continues into the antenatal period.

12.10. Phase and Frequency Coordination between Oscillatory Firing α_2-Motoneurons and their Adequate Afferent Drive in Brain-Dead Human

The relative phase and frequency coordination between the APs of the oscillatory firing α_2-motoneuron O2 and the secondary muscle spindle afferent fibre SP2(1) has partly been shown in Figure 25, and can be clearly seen in the original recordings in Figure 9 taken from a brain-dead human. The firing of the oscillator and the sweep pieces, which are shown time-expanded, are indicated at the summary trace (Figure 16A). Figure 16B,C shows the action potential (AP) impulse train of oscillator O2 in connection with one of its driving spindle afferent AP. Because of the duration of the phase relation of around zero milliseconds between the firing of the driving SP2(1)-fibre and the impulse train of the oscillatory firing motoneuron O2, the SP2(1)-fibre AP (every second AP of the SP2(1)-fibre) appeared at a similar time as the impulse train. Because the AP of the spindle afferent fibre had a characteristic waveform, it was easy to extract its impulse pattern from the summed impulse traffic of this S4 dorsal root. During touch-induced skin afferent activity (Figure 16A), the activities of the motoneuron and the spindle afferent fibre were covered by the skin afferent activity. After the cessation of the skin afferent activity the afferent and efferent APs were found again at their expected time positions of the regular firings. The phase coordination between the firings of the oscillatory firing motoneuron O2 and the secondary muscle spindle afferent fibre SP2(1), at the time when records B,C were taken, was 1.6ms (3ms - 1.4ms, Figure 16B,C). In Figure 16D, the relative frequency coordination between the firings of the SP2(1)-fibre and the impulse train of the oscillator is indicated. For the time period evaluated, the correlation between the firing of the motoneuron and the spindle afferent fibre was in the range of between 3 and 5ms (Figure 16D).

12.11. Relative Frequency Coordination

In Figure 17, considerations concerning the relative frequency coordination are extended to the activity of further afferent fibers and γ-motoneurons of the same root. Figure 17G shows sweep pieces of the original recordings; A through F show the interspike interval distributions of spindle afferents and γ-motoneurons. It can be seen from the overlapping of the oscillator frequency distribution ranges (and the half of it), and from the interspike interval distributions of the afferents that, from the viewpoint of frequency coordination, fiber SP2(1) contributed strongly to the drive of oscillator O2, whereas there was a weaker contribution from other afferents (less overlapping between the distributions of the afferents and the range of the basic frequency or the first harmonic of the oscillator). Also, γ-motoneurons showed only little frequency correlation at that time period.

Figure 61 shows the interspike interval distributions of more afferents (including the afferents for bladder filling; stretch receptor afferents S1(1), S1(2)) of another root, together with the oscillation period range (and the half of it) of a second α_2-oscillator (O1). By comparing the oscillation periods (and their halves) and their ranges with the interspike interval distributions of the afferents, it can be suggested which afferents made a (frequency coordination) contribution to the drive of what oscillator at that time interval. For example, the S1(1) urinary bladder stretch afferent fiber activity contributed to the drive of oscillator O1 (activating the external bladder sphincter) because its interspike interval distribution overlaps strongly with the range of the oscillation periods of O1. But the S1(1) distribution does not overlap with the range of the oscillation periods of oscillator O2, or with their halves or quarters. The S1(1) afferent fiber will therefore not have made a substantial contribution to the drive of oscillator O2. On the other hand, the secondary muscle spindle afferent fiber SP2(12) activated oscillator O2 innervating the external anal sphincter, since its interspike interval distribution overlaps with the range of O2 oscillation periods. But the secondary muscle spindle afferent fiber SP2(12) did not activate oscillator O1, as its interspike interval distribution does not overlap with its oscillation period range or half of it (Figure 61).

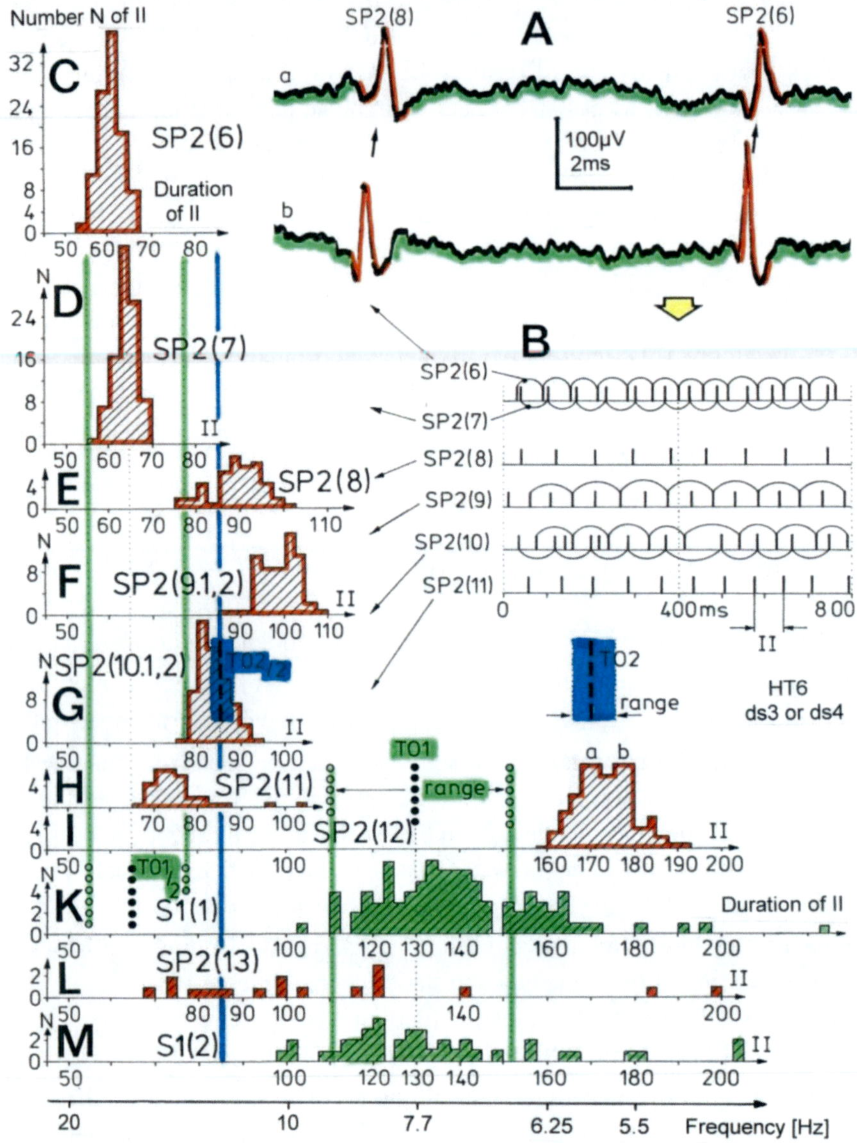

Figure 61. Measurements from brain-dead human HT6 from different spinal cord segments after retrograde bladder filling (700 to 800 ml), with the exception of "I", which was obtained before filling. A. Sweep piece of a recording from a dorsal S3 or S2 root filament. It can be seen that the secondary muscle spindle afferent SP2(6) AP can be distinguished by the wave-form on the two traces from the secondary spindle afferent fiber SP2(8) AP (different amplitude of the three phases of the triphasic APs). B. Simultaneously recorded impulse patterns of the six parent secondary spindle afferents SP2(6) through SP2(11) obtained from dS3 or dS2 root recordings. The impulse patterns of SP2(6) and SP2(7) fibers are not separated to show the similarity of the patterns. The impulse patterns of the parent spindle afferents SP2(9) and SP2(10) are split into patterns of the single endings (single ending activity partly connected by circle lines) with the assumption that single endings of parent secondary muscle spindle afferents should have interspike intervals of duration longer than 50 ms. C to H. Interspike interval distributions of six simultaneously recorded single secondary spindle afferent endings. F, G. Interspike interval distributions of parent fibers, which are the sums of the distributions from the two activated endings. I. Interspike interval distributions of a secondary spindle afferent fiber (SP2(12)) of a

coccygeal root. K, L, M. Interspike interval distributions of single-fiber afferent activity from a lower sacral dorsal root. In L, most likely the activity from a secondary spindle afferent fiber is shown. In K and M most likely the interspike intervals from afferents (S1(1) and S1(2)), innervating stretch receptors of the urinary bladder wall, are shown. In G, H and K, the durations of the oscillation periods (mean and range) of the oscillatory firing α_2-motoneurons are indicated by thick dashed and dotted lines; the motoneurons innervate the external anal sphincter (TO2) and the external bladder sphincter (TO1). The sites of innervation of the oscillatory firing motoneurons are identified (and distinguished from each other) by anal reflex stimulation, bladder filling and catheter pulling. Note that the TO1 and TO2 ranges and their halves overlap with the interspike interval distributions of the secondary spindle and stretch receptor afferents (relative frequency coordination).

By comparing interspike interval distributions of afferent fibers with oscillation period distributions it can be estimated what afferents made a (frequency coordination) contribution to the drive of the spinal oscillators. These considerations need no knowledge of the connectivity of the neuronal networks. In the frequency coordination between the firings of afferents and oscillators and among oscillators, entrainment or coordination may occur sub- or super-harmonically. The energy transfer, and therefore the coupling strength will be smaller if the APs coincide in their firing less often. As indicated by my measurements, the coupling and the relative coordination during the self-organization of the neuronal networks of the human spinal cord are of an enormous complexity; this *self-organization is induced by sets of mutual impulse patterns from stimulated receptors, which are ordered, in time and space (for skin receptors see Figure 6), so as to reflect, in the spinal cord and higher centers, the interplay of the body with the external and internal world. The learning method to improve and repair the human nervous system is "pattern-based learning" in which movement-based learning is a large part of it.*

12.12. Impaired Organization of Premotor Spinal Oscillators Following Spinal Cord Injury as an Indicator for Pathologic Network Organization

Following a spinal cord injury, the oscillatory firing networks lose specific properties. The Eigen-frequencies of the premotor spinal oscillators change from a narrow to a broad frequency band (Figure 37). Self-organized α_2-oscillators fire physiologically at an Eigen-frequency (varying within a small frequency band as probably indicated with the hatched distributions in Figure 37) with impulse trains consisting of 2 to 3 action potentials. Following brain death, this Eigenfrequency band enlarges (black area in Figure 37). Following a spinal cord injury, the Eigen-frequency band enlarges strongly and includes in this case the frequencies between 4 and 14Hz for firing with 2 or 3 action potentials per impulse train (Figure 37). The premotor spinal oscillators have lost their specific properties and could now be excited at frequencies at which they physiologically would not be excited. By generalization, if rhythmically firing sub-networks of the CNS lose their specific properties the self-organization of the CNS becomes rather chaotic. The tremendous specific complex and exact organization is lost.

12.13. Explanation for a Spastic External Bladder Sphincter

In the brain-dead human HT6 of Figure 61, the premotor spinal oscillator TO1 could, according to the overlap of the narrow frequency band harmonically (for simplicity only this case of activation is considered), only be driven by the activity of the bladder stretch receptor afferents S1(1) and S1(2). This oscillatory firing α_2-motoneuron, innervating the FR-type muscle fibres of the external bladder sphincter, becomes activated with the activity of the stretch receptor afferents, when the bladder is filled to secure continence. All the secondary muscle spindle afferents cannot activate the external bladder sphincter harmonically, because their frequency bands did not overlap (frequency band overlap in the frequency range 6.5 till 9Hz). Only the stretch receptor bladder afferents can activate the external bladder sphincter to secure continence. Following a spinal cord injury, the Eigen-frequency band broadens from 6.5-9Hz to 4-14Hz. According to Figure 61, also the secondary muscle spindle afferents SP2(8), SP2(9.1,2), SP2(10.1,2), and SP2(12) could activate the motoneuron innervating the external bladder sphincter. The striated external bladder sphincter can be activated by secondary muscle (and probably other) afferents, not involved in continence. This broadening of the frequency band could be one reason for spasticity (pathologic continuous contraction) of the external bladder sphincter. The patient in Figure 29, with a severe cervical spinal cord injury, had a spastic external bladder sphincter when therapy was commenced. She was continent, but could not empty the bladder – the emptying of the bladder was achieved by a suprapubic catheter.

12.14. Reduction of Spasticity of the External Bladder Sphincter

If one could improve the functioning of the sacral micturition center and reduce the frequency band of the oscillatory firing premotor spinal oscillators, innervating the external bladder sphincter, then this kind of spasticity of the external bladder sphincter could be reduced. I named this training 'oscillator formation training'. It was shown (Chapter VII of [2]) that this spasticity reduction could be achieved by learning and learning transfer. But, as shown in Figure 62, the reduction of the spasticity of the external bladder sphincter is only the first step to repairing bladder functions. The synergia of the urinary bladder (Figure 60Bb) needs to be re-learned. Following CNS injury (including spinal cord injury) not only the frequency coordination, but also the phase coordination among neuron firing becomes impaired.

12.15. Stable Phase Coordination in the Brain-Dead Individual

Existing relative phase relations among α and γ-motoneurons and secondary muscle spindle afferents in a paraplegic are indicated in Figures 56 and 57 by small arrows. In Figure 59 the natural impulse patterns were used to show the interaction between the somatic and parasympathetic nervous system divisions in the sacral micturition center. Next, the representation of the phase relations (shown in Figures 56,57) is changed to reveal phase stability, which makes it possible to find reasons for the pathologic organization of neuronal networks in spinal cord injury. We shall start with a physiologic case, specifically a brain dead individual.

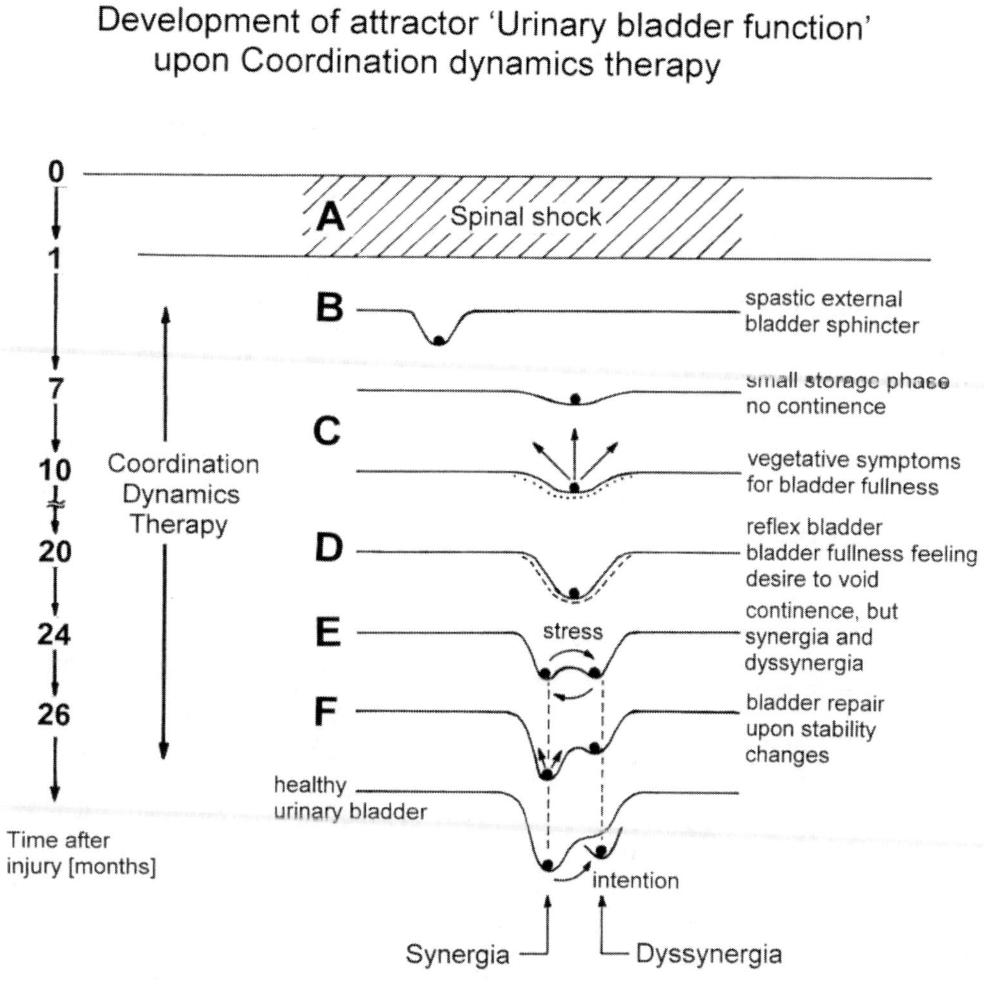

Figure 62. Evolution of the attractor layout of bladder functioning induced by learning transfer from movements to bladder functions upon CDT. The region around each local minimum of the potential landscape acts like a well that weekly traps the system into a coordinated state. Black balls correspond to stable minima of the potential. With learning the pattern, 'spasticity of the external bladder sphincter' vanishes, and the patterns for bladder functioning ('synergia' and dyssynergia') appear anew and gain their physiologic stability (physiologic deepness of each basin of attraction). The corresponding attractor layout for physiologic bladder functioning is given. Fluctuation of pattern state (the black ball) (C), and their decrease (F), due to the impairment of phase and frequency coordination of neuron firing, is pictured in 'C' and 'F' by long and short arrows. Dotted and dashed lines indicate the re-occurrence of bladder sensation. Note that more than two years of optimal CDT were needed for bladder repair.

To make the phase relation changes better recognizable with time, a representation of phase relations is used, which comes from the measuring of the speed (frequency) of rotation. The speed of rotation of a turning cylinder with a spot on its surface can be measured with a stroboscope. If the stroboscope flashes light with the same frequency as the cylinder is turning, the spot on the circumference seems to stand still. There is a constant phase between the two frequencies (frequencies are the same or multiples of each other). If the phase relation changes, the spot will move. If no phase relation exists between the turning of the cylinder and the flashing of the light, no spot will be seen. In similarity to stroboscopic measurement of frequencies of turning cylinders, the phase relation between two oscillatory firing spinal oscillators is pictured in Figure 35A. A time axis is introduced on the horizontal line, to make phase relation changes visible in dependence on time.

In Figure 35Aa, the loop excitation is pictured for this oscillator model. In Figure 35Ab, the phase relation between the SP2 fiber activity incidence and the oscillatory firing is pictured on the circumference of the oscillation period cylinder of the oscillator. Figure 35Ac,d,e shows different phase relations, namely a constant phase relation (c), a changing phase relation (d), and no phase relation (e). In Figure 35B, phase relation changes are plotted between an α_2-motoneuron and the activity of a secondary muscle spindle afferent fiber and between an α_2 and a γ_1-motoneuron. The data were taken from Figures 4 and 5 of [29] of a brain-dead individual (probably normal with respect to the number of phases per oscillation cycle and with respect to phase changes). It can be seen that there were two phase relations per α_2-oscillation cycle, and that the phase relation changed only little with time. The phase coordination between the firings of the α_2 and a γ_1-motoneuron and the secondary muscle spindle afferent fiber was stable.

12.16. Unstable Phase Coordination in the Patient with a Spinal Cord Injury

In Figure 36A,B different phase relation changes are plotted from Figures 56 and 57 with respect to the α_3-oscillation cycle (A) and the α_2-oscillation cycle (B). It can be seen that the different phase relations changed strongly in value over time (upon different stimulation), and that also the number of phase relations per oscillation cycle changed. The phase stability of the cooperative and competitive interplay among neurons became impaired. Whether the change of the number of phase relations from 2 to 3 following the activation of the parasympathetic nervous system in the sacral micturition center (Figures 57,59) is physiologic is not clear.

12.17. Impaired Neural Network Functioning and Learning Because of Impaired Phase Stability Following SCI

The most obvious difference of the phase relation changes between the above mentioned brain dead human and the paraplegic (SCI) was that in the paraplegic the phase relations varied very much, whereas they changed only a little in the brain dead human. The strong phase relation changes in the paraplegic can be interpreted as instability in the organization of neuronal networks. The correlation of neuronal sub-networks was unstable in relation to those of the brain-dead human. Assuming that the neuronal network organization and functioning was rather physiologic in the brain dead human with respect to the firing patterns of the premotor spinal oscillators, the functioning of the networks became unstable following spinal cord injury (SCI). The more frequent occurrences of changes of phase relations between the different nerve fibres, in combination with the changing number of phase relations per oscillation, may mean that sub-networks reacted and interacted more quickly and easily with others according to the afferent input. Especially because the oscillatory firing networks lost specific properties, their resonance frequencies changed from a narrow to a broad oscillator frequency band (Figure 37), which means that the oscillators were not excited at a certain frequency any more, but by a broad frequency band. They were now being excited at frequencies at which the physiologically functioning oscillator would not. Over-activation and mass effects could be the result. On the other hand, certain networks could escape from driving afferent influence by changing their phase-by-phase escape to avoid interaction. Functionally distant networks are no longer reached, which also would result in a loss of specific properties. Therefore, because of the loss of specific properties, some interactions could have occurred more easily and others not at all.

12.18. Change of the Neuronal Network Organization Following Spinal Cord Injury - Pathologic Network Organization

Following spinal cord injury the spinal cord neuronal networks have been observed to change their organization, and this can be quantified by six changes of organization.

1. Following spinal cord injury, the spinal oscillators strongly widen their oscillation frequency band (Figure 37), which means that their firing lacks rhythm. An increase of the more irregular oscillator firing can also be seen in the broadened distributions of the interspike intervals of the impulse trains of oscillatory firing α_2 and α_3-motoneurons [30]. The loss of specificity of oscillatory properties will at least partly be due to the loss of supraspinal inhibition because muscles below the paretic spinal cord injury are over-activated.
2. Following the injury, the phase relations among the oscillators and between oscillators and their driving afferents (for example, secondary muscle spindle afferents) become very instable (Figure 36).
3. Because of the widening of the frequency bands (Figure 37) and the instability in phase coupling (Figure 36), the oscillators partly lose their rhythmicity and their coordination and cooperation properties following CNS injury. The rhythmicity of movements and their coordination and coupling of arms and legs is reduced. The loss of rhythmicity and the loss of coordination can be observed easily when measuring the coordination dynamics or when letting patients walk or crawl.
4. The spinal oscillators are not under full volitional control by the patient any more (Figure 48 of [31]). Sometimes, paraparetic or tetraparetic patients can switch on motoneurons to fire in an oscillatory manner, but cannot switch them off again.
5. Following natural stimulation, the recruitment of motoneurons in the occasional firing mode (low activity mode of motoneuron firing; see Figure 12D for bladder filling up to 550ml) according to the size principle in each nerve fibre group changes [32,33]. The level of motoneuron activation increases following spinal cord injury (probably due to loss of inhibition) and the slowly conducting α_3-motoneurons (S) are recruited before the faster conducting α_2-motoneurons (FR), which is pathologic.

6. The coupling of α and γ-motoneurons changes in strength following natural stimulation [32].

The loss of rhythmic oscillator properties and also the change of motoneuron firing in the occasional firing mode (when the oscillators have not been self-organized because of low activation) indicate deterioration in the self-organization of the spinal cord neuronal networks below the injury and can at least partly explain the pathologic CNS network organizations of spasticity and increased clonus.

The impaired phase and frequency coordination's, measured among α and γ-motoneurons and secondary muscle spindle afferents by an increased variability of coordination, was introduced in a first approximation in the equation of motion at the collective variable level of description for jumping on springboard ($d\varphi/dt = - dV(\varphi)/d\varphi + (Q\xi_t)^{1/2}$) by a variability increase of strength Q (see above). This increased variability was pictured by the increased fluctuation of pattern states (pictured by a ball) in an attractor basin (Figure 28B).

12.19. Re-Learning of Phase and Frequency Coordination

This partly impaired phase and frequency coordination between the firings of single neurons and neuron assemblies (Figures 37,36) can be expected to have consequences in the coordination between arm and leg movements, because motoneurons innervate muscle fibres, and rhythmic coordinated firing of single motor units has been measured electromyographically with surface electrodes [34-36] (Figures 22-24). Indeed, the coordination between arm and leg movements is partly or fully lost following CNS (brain and spinal cord) injury and is often not taking place at all in the malfunctioning CNS.

This partly impaired phase and frequency coordination at the single neuron level, the assembly level and the macroscopic level can be measured macroscopically when the patient is exercising on a special coordination dynamic therapy device (Figure 19) on which arms and legs turn with a slightly different frequency (transmission 19 (arms) : 18 (legs)). The phase coordination between arms and legs is imposed by the device. The loss of phase and frequency coordination between arm and leg movements becomes visible and measurable by the arrhythmicity of turning when the patient is exercising on the device. During a turning cycle and during the change of coordination between arms and legs the turning frequency increases and decreases. This frequency variation (df/dt:f; f = frequency) can be recorded, quantified and displayed on a computer screen (Figure 19).

During the functional reorganization of the injured CNS of patients, the relative phase and frequency coordination of neuron firing has to be entrained as exactly as possible by the movement induced afferent impulse patterns from the receptors (learning through feedback information) to restore coordination in the range between 3 and 5 milliseconds (Figure 16). The device has therefore to impose the exercising patient coordination in the millisecond range for the different coordination's of arms and leg movements between pace gait and trot gait. The simple pace and trot gait coordinations can often be performed easily by the patient, but not the difficult intermediate coordinations (pattern change). Therefore, the continuous change from the easy to the difficult coordinations, and backwards diagnoses the extent to which the patient's nervous system can produce both the easy and the difficult organizational states. If the neuronal networks of the CNS easily generate the movement state, then the frequency variation of turning is small during the turning cycle, and if the movement state is difficult to organize by the CNS, then the frequency variation is large.

Because it is sometimes the case that in the pathologically functioning CNS, intermediate coordinations are often easier than pace and trot gait are, and a coordination coordinate is needed to judge the coordination dynamics. The pace and trot gait coordinations are used for the calibration, since both coordinations between arm and leg movements occur naturally during rhythmic coordinated (automatic) movements like commando-crawling, crawling, walking and running. The quality of CNS functioning can be measured by pattern change, namely the coordination dynamics.

The imposed coordination of arm, leg and trunk movements performed while exercising on the special coordination dynamic therapy and measuring device, is in accordance with the coupling possibilities of premotor α_1 (8-12Hz), α_2 (6-9Hz) and α_3-oscillators (0.4 (may be down to 0.1) -4Hz), even though the frequencies are only a relative coordination parameter, whereas phase is an absolute coordination parameter.

When the hand levers are turned at between $\approx$ 0.4-1.5 Hz, the resulting frequency difference in turning between arms and legs is approx. 8.5 Hz (low α_1-oscillator frequency or high α_2-frequency) for low hand turning frequency of 0.5Hz (low α_3-frequency). A slower turning of the hand levers would directly train more the premotor α_2-oscillators ($f < 8.5$Hz). Faster turning of the hand levers (higher α_3-frequency) would directly train the α_1-oscillators in the higher frequency range ($f > 8.5$Hz). Therefore, similar frequencies appear with respect to the frequency of turning on the device for measuring CNS organization and reorganization as has been measured for premotor spinal oscillators.

12.20. Learning to Improve the Recruitment of Motoneurons in the Occasional and Oscillatory Firing Mode

When turning the levers steadily with medium or high strength the premotor spinal oscillators in the patient's CNS are entrained because premotor spinal oscillators self-organize themselves for high activation (Figure 12D). The turning of the levers is therefore an oscillator (or assembly) formation therapy. The members of premotor oscillator assemblies (most likely the motoneuron and interneurons) are entrained to improve activation and inhibition by adjusting, for example, the efficacies of the corresponding activated synapses. By turning the levers with little strength at approx. 0.4Hz (releasing the power (load)-setting knob), the motoneurons get only partly organized into premotor spinal oscillators. The motoneurons are firing mainly in the occasional firing mode and are trained for a better recruitment according to the size principle (rhythmicity of repeated recruitment $\approx$ 0.4Hz (Figure 15 of [31])). Between the motoneuron firings in the occasional and oscillatory firing modes, a better coordination of both firing modes is entrained (Figures 15,16 of [31]).

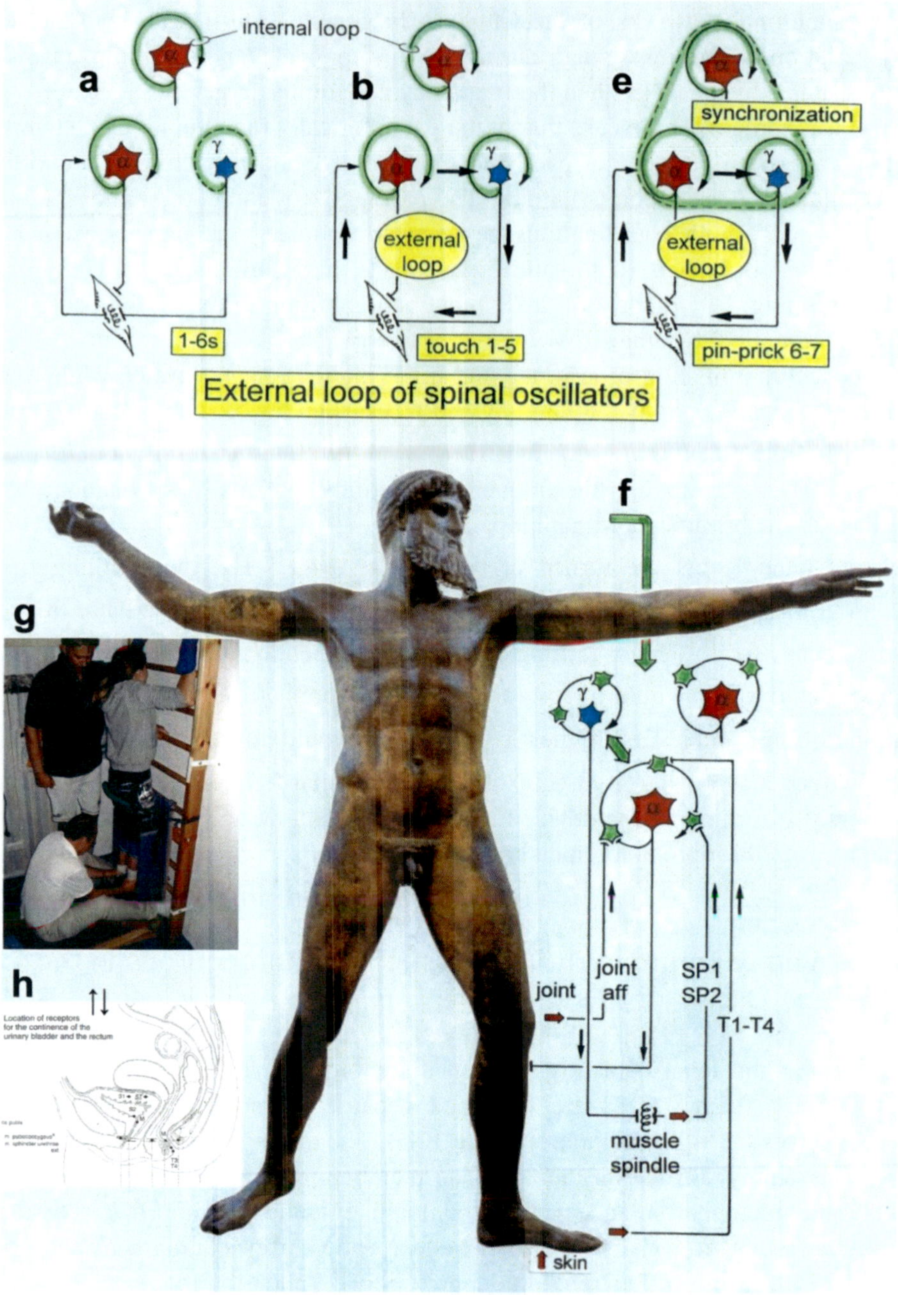

Figure 63. Spreading of oscillatory firing from α-motoneuron neuronal network to include muscle spindles (periphery) and synchronization of different α and γ-motoneuron neuronal networks caused by touch and pin-prick stimulation.

(a) α-motoneuron neuronal networks fired oscillatory (solid line loop), γ-motoneuron neuronal network did not or did only partly (dashed line loop), upon no additional stimulation.

(b) Oscillatory firing α and γ-motoneuron neuronal networks built up a phase relation with muscle spindle afferents and efferents (external loop to the periphery, indicated by thick arrows) upon touch 1-5.

(e) Oscillatory firing α (internal circuitry loop) and γ-motoneuron neuronal networks (external loop) synchronized (broad peak phase relation) upon pin-pricks 6-7. The dashed line loop represents

synchronization.
(f) Oscillatory firing α (internal circuitry loop) and γ-motoneuron neuronal networks (external loop) are extended by analogy from the continence muscles to muscles for locomotion. The open arrows indicate that it may be possible to synchronize spinal oscillators by rhythmic afferent input, generated by rhythmic movements (such as jumping on a springboard or running), and to re-preformat the neuronal circuitry by synapse remodeling to fire more physiologically oscillatory to reduce spasticity and improve locomotion. Extensive pathologic movement like tremor may entrain neuronal circuitry to increase tremor movement. The Greek god is a bronze statue of Zeus found close to the cape of Artemision 460 BC.
(g) Supported jumping on the springboard in anti-phase. The patient with the severe cervical spinal cord injury is supported by the author.
(h) Location of receptors for the continence of urinary bladder and rectum stimulated upon jumping on springboard.

12.21. Building up of External Loops to the Periphery by Premotor Spinal Oscillators

It will be shown now that with the building up of simultaneous phase relations between α, γ and SP2 fibers and the assimilation of interspike interval (II) distributions (coordination's of rhythms), an external loop of premotor spinal oscillators is built up to the periphery, which makes it possible to directly influence the firing of spinal oscillators by rhythm training. The somatic and parasympathetic pattern organizations in the sacral micturition center can simultaneously be entrained by jumping on springboard (Figure 63g) (including the stimulation of movement (Figure 63f) and bladder receptors (Figure 639h)) to allow movement-based learning in the continence and movement patterns and to induce learning transfer from movements to urinary bladder functions. A repair of neuronal network patterns of the functionally disconnected sacral micturition center in spinal cord injury seems possible.

In Figure 57Ba it can be seen that there existed two phase relations between the firings of the α_3, γ_1 and SP2(1) fibres, which means that the γ-loop, including the γ_1 and SP2(1) fibres, contributed to the drive of the α_3-oscillator (probably of α_2-type). However, since the II distributions are different (Figure 56Aa), the γ-loop was not a part of this spinal oscillator; it was only contributing to the drive of it, as pictured in Figure 63a. Since the II distribution of the fusimotor γ_1 was often rather broad (for example Figures 56Aa,57Ac), its driving network was not or only almost oscillatory firing (the dashed line loop in Figure 63a). Upon touching sites 1 to 5 (Figure 55) approximately every second, phase relations occurred between the α_3, γ_1 and SP2(1) fibres, even though reaching a different value (Figure 56Bb,c) and the II distributions of the α_3 and γ fibres (Figure 56Ab) and the SP2(1) fibre (Figure 56Ac) became assimilated. The γ-loop became directly connected to the oscillatory firing network: the premotor spinal oscillator (the network driving the α_3-motoneuron; the α_3-motoneuron is a part of the network) built up an external loop to the periphery (Figure 63b). Upon pin-pricking sites 6 and 7, one phase relations occurred between the firings of the α_3, γ_1 and SP2(1) fibres (Figure 56Be) and also the II distributions assimilated (Figure 56Ae). The spinal α_3-oscillator had built up a full external loop to include the γ-loop in its oscillatory firing (Figure 63e). Since there was transient synchronous firing of unit 1 with unit 2 (including the α_2 and SP2(2) fibres (Figures 56Be,59e)), probably the α_2-oscillator also built up an external loop to the periphery.

12.22. Extension of the External Loop Generation of Spinal Oscillators to Non-continence Muscles

If one extends the integration of the γ-loop in the oscillatory firing of spinal oscillators innervating striated continence muscles to muscles activated for locomotion (Figure 63f), then for example, while jumping without weight support (Figure 63g), sky-walking without weight support, running (Figure 38) or other coordinated, rhythmic, stereotyped, dynamic movements, at least oscillatory firing α_2 or α_3-oscillators build up external loops to the periphery. It seems partly possible during jumping on springboard (Figures 29,30) to synchronize spinal oscillators with the jumping rhythm, mainly given by the properties (Eigen-frequency) of the springboard. Especially the α_3-motoneuron networks can be entrained efficiently by jumping on springboard, since the 'Eigen frequency' of the springboard and that of the α_3-oscillator network are both in the range of 1Hz (for entrainment, see below). The extension of premotor network organizations serving bladder functions to premotor network organizations serving movement functions is most likely justified, since rhythmic firing of single FF and FR-type motor units, which are innervated by α_1 and α_2-motoneurons, has been recorded by surface electromyography (sEMG) during rhythmic, coordinated movements (Eigen-frequency around 1 Hz) in patients with spinal cord injury (Figures 22-24). Upon exercising on the special CDT device (Figures 15,32), single FF-type motor units in arm and leg muscles coordinated their rhythmic firing with respect to phase and frequency. These single-motor unit recordings performed with sEMG support the recordings from single-motoneuron axons (Figure 20).

12.23. External Loop of Premotor Spinal Oscillators and Rhythmic, Dynamic Stimulation of Motor and Bladder Functions

While jumping on a springboard (Figure 63g) (and other rhythmic movements like sky-walking or running) premotor spinal oscillators organize themselves to fire transiently in an oscillatory manner according to the motor pattern and build up an external loop to the periphery (Figure 63). If the frequency of the rhythmic movement has an integer relationship to the Eigen-frequencies of the premotor networks and more rostral networks, these premotor networks get entrained for more specific self-organization (see below). When jumping on a springboard (Figure 63g) not only the motor networks get activated; also the external sphincteric motoneurons, innervating the external bladder and anal sphincter, as a part of the pelvic floor, are rhythmically activated to counteract the rhythmic weight changes of the intestine. Further, the rhythmic, dynamic, stereotyped up and down movements stimulate stretch, tension, flow, and mechanical receptors of the bladder (detrusor and proximal urethra) (Figures 54,63h). This rhythmic movement-related sensory input with ≈1Hz bears similarity to the sensory input stimulated by bladder and anal catheter pulling with ≈1Hz during the measurements (Figures 55,57). Repetitive phase relation changes in and among neural ensembles will occur in some similarity to the changes following catheter pulling (Figure 57B). Since the neurons involved in the generation of movement and continence (and micturition) patterns (especially if the neurons serve both functions at the same time) are synchronously, rhythmically activated, the pathologic bladder patterns get entrained from the rather physiologic jumping movement, in some similarity to co-movement. The synchronized activation of the somatic and parasympathetic networks allows efficient learning transfer, since the neurons work coincidentally, or more generally, as coordination detectors (Figure 40). If there is fluid in the bladder and material in the bowel and rectum, the continence stimulus is stronger. Also, walking and running will stimulate and change the intrinsic dynamics of the micturition and defecation centres, but not as strong as the jumping on springboard. The walking and running on a treadmill in severe cervical spinal cord injury is performed under weight support, whereas no weight support is needed during jumping on springboard (Figure 63g). The sympathetic nervous system division, probably innervating the internal urinary bladder sphincter (smooth muscle), will also be activated synchronously. Since the frequency of jumping is around 1 Hz, similar to the 'Eigen-frequency' of the α_3-motoneuron oscillators, these oscillator networks should be entrained most efficiently (see below).

12.24. Entrainment of Premotor Spinal Oscillator Networks by Rhythmic Movement-induced Afferent Input and Inputs from Supraspinal Centers

If one approximates for high activation spinal neuronal networks into premotor spinal oscillators (driving the motoneuron) and propriospinal oscillators (generating by coupling with one another movement patterns), then premotor spinal oscillators can be handled in a first approximation as single linear oscillators. The premotor spinal oscillators and the spinal pattern generating networks are self-organized and driven by afferent and supraspinal inputs. When training dynamic, rhythmic, stereotyped movements, the premotor spinal oscillators approximated as linear oscillators are driven by movement-induced afferent input from the periphery (mainly the legs) and surrounding pattern generating networks, and possibly supraspinal inputs. These spinal oscillators and most likely their neuronal network can be entrained at least by use of the external loop for better self-organization. But the functional repair by learning only becomes physiologic if volitional control (by ascending and descending tracts) is intertwined.

If one assumes that loop circuits do not only exist between the premotor spinal oscillators and the periphery, but are a general structure within the CNS, then motor learning involves the formation of loop circuits (or better loop network circuits) between the cortex and the periphery involving the sensory cortex and the thalamus. When a linear oscillatory system is driven by an external periodic input, its response contains both frequency components. This is also, in general, true with nonlinear oscillators. However, in this case, if the external frequency is close to the Eigenfrequency of the oscillator itself, then it is possible to have a response at the external frequency only. This phenomenon is known as entrainment or synchronization. It is of paramount importance with respect to biological oscillators because it allows them to 'latch on' to the environment. Thus a rhythm with a free-running period of 24.7 hours may be synchronized to 24 hours when exposed to the natural sequence of day and night.

An oscillator with one degree of freedom can be described by the equation:

$d^2x/dt^2 + f(x, dx/dt) + \omega_0 x = E \cos\omega_1 t$

(x = variable, t = time, f = nonlinear term, $\omega_0 = 2\pi f_0$ = frequency of the oscillator, ω_1 = entrainment frequency, E = amplitude)

There exists a set of paired values of amplitude E and the absolute difference in frequencies $\Delta\omega = |\omega_0 - \omega_1|$, such that the output of the system only contains the frequency ω_1. Figure 64 shows a typical example. Entrainment occurs in the shaded part of the plane. If E and $\Delta\omega$ are below the curve (outside the shaded area), the frequencies ω_0 and ω_1 are usually present. In the case that $\Delta\omega = 0$ but E is very small, the phase of oscillations is not influenced by the input. For further details, including the van der Pol oscillator, see [37].

Upon jumping on springboard, the entrainment frequency ω_1 (the jumping frequency) is 0.9 to 1Hz, close to the Eigen-frequency of the premotor spinal α_3-oscillators ($\approx$ 1Hz). The α_3-oscillators and the networks they are integrated in are entrained directly. For the entrainment of the α_2-oscillators (Eigenfrequency $\approx$ 6Hz) and α_1-oscillators (Eigenfrequency $\approx$ 10Hz) sub-harmonic and super-harmonic entrainment must be considered. When a nonlinear oscillatory system is driven by an external periodic input z_k, the entrainment can be harmonic (z_k itself has the oscillation period T of the oscillator; case of the α_3-oscillators), sub-harmonic (z_k has a period which is an integer multiple of T, mT; case of the α_2 and α_1-oscillators) or super-harmonic (z_k has a period which is an integer fraction of T, T/m) [38]. With the increasing order of sub-harmonic entrainment, the entrainment strength reduces for the same coupling strength or entrainment amplitude E. For the α_2 and α_1-oscillators the entrainment is sub-harmonic and the entrainment strength is therefore reduced. In α_2-oscillators, two entrainment phases per oscillation period were commonly observed, which correspond to 'in-phase' and 'anti-phase' coordination of arms and legs, which enhances entrainment. Also, the change of the number of phase relations between the neural assemblies in the sacral micturition centre may indicate changing entrainment or coupling communication between the somatic and parasympathetic nervous system divisions. However, oscillator models are still far away from human network properties. The jumping on springboard is very rhythmic and stereotyped (frequency of jumping $\approx$1Hz). But firstly, the movement-induced afferent input enters the network at different levels (premotor neuronal network, propriospinal oscillatory network, brainstem network and higher centres); secondly oscillators can be driven from different sources; and thirdly, often the rhythmic input patterns consist of impulse trains with increasing interspike intervals and with delays between the responses from different receptors and receptor types. For natural sets of simultaneous impulse patterns of numerous skin receptors induced by touch or pin-prick, see Figure 6, and for natural impulse patterns of secondary muscle spindle afferents, see Figure 45. In simulations of networks consisting of populations of interacting oscillators the known natural afferent input patterns have to be used, and oscillator network structures have to be used, which give the measured output patterns under both physiologic and pathophysiologic conditions. Even though the oscillator models used are far away from human self-organization network forms, their results are still very useful for the interpretation, under certain conditions, of measured data and to better understand the jumping on springboard training (as a part of the coordination dynamics therapy), which contributes substantially to the repair of urinary bladder function. For jumping on the springboard in in-phase or anti-phase and for sky-walking, the amplitude/stride length has to be large to be efficient for repair by movement-based learning.

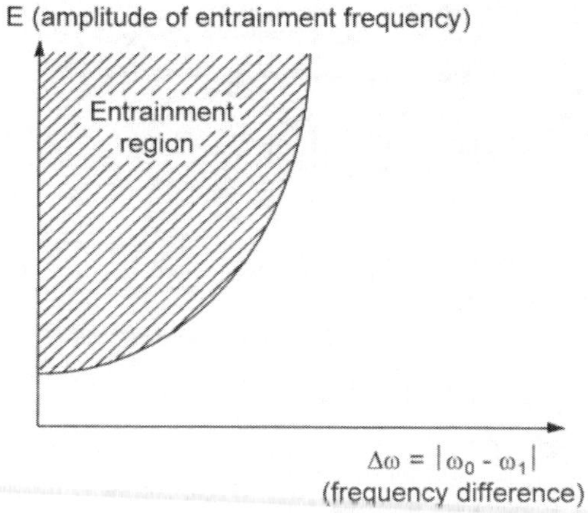

Figure 64. Illustration of the relation between the amplitude E of the external input and the difference in frequencies $\Delta\omega$, which produces entrainment of a nonlinear oscillator.

12.25. Stimulation of the Parasympathetic and Somatic Division via their Receptors of the Pelvic Floor and Intestine to Induce Learning Transfer from Movements to Urinary Bladder Functions for Cure

It was shown that the parasympathetic division can be activated, as well as the somatic one, by anal and bladder catheter pulling (Figures 55,57). The activated parasympathetic neuronal network channeled input to the rhythmically firing somatic neuronal network via an additional phase relation (Figure 59j). This joint stimulation of the somatic and parasympathetic division is used in coordination dynamics therapy to use movements which substantially activate the somatic and the vegetative nervous system to induce learning transfer from movement to bladder patterns to cure urinary bladder function. Upon jumping on springboard, without weight support, in severe cervical spinal cord injury (Figure 29), the pelvic floor (including the sphincters) is mechanically stimulated (stretch of muscle spindles) under the weight of the intestine. Also, the stretch and tension receptors of the bladder and rectum are activated upon the rhythmic up and down movements. The desire or urge to void is also strongly stimulated when the patient is exercising on the special coordination dynamics therapy device, especially in the lying position (Figure 32). But this stimulation of the parasympathetic and somatic divisions is related to the improvements in the short-term memory of the phase and frequency coordination of the neuronal networks of the sacral micturition center, including the functional units pictured in Figure 59. Both, the rhythmic mechanical stimulation of the continence organs (Figure 29) and the improved CNS organization in the short-term memory upon exercising on the special device (Figure 15) lead to a strong activation of the sacral micturition centre, which may in turn lead to strong increases of the heart rate and blood pressure, which are vegetative symptoms to inform supra-spinal centres about bladder fullness in complete spinal cord injury, if the bladder is filled and if there is plenty of material in the colon and rectum.

13. Stability of Premotor Spinal Network Oscillators and their Phase and Frequency Coordination

13.1. The Study of Impaired Coordination among Neurons as a Tool to Understand CNS Self-organization and Neural Network Learning

To understand learning strategies of human neuronal networks, network organization of the injures or degenerating human CNS must be studied, because in the healthy CNS regulations are so nicely working that network organizations and network learning is difficult to observe. From patients with spinal cord injury (SCI), single neuron and single-motor unit firing patterns can be nicely recorded and studied. To understand the coordinated firing of network oscillators, their stabilities and their organization and repair by learning, the degenerative disease Parkinson is used. From the learning in the injured nervous system, we learn how neural network learning is achieved in the healthy nervous system.

In the CNS we encounter a specific and an unspecific degeneration. This unspecific impairment of phase and frequency coordination can be treated by CDT. The specific impairment of phase and frequency coordination, namely the impaired inhibition among nerve cells (Parkinson) can only partly be treated. By performing CDT at the limit, crucial new nerve cells can be built. Unfortunately, however, it is likely that more nerve cells are dying than can be built anew. This compounds the fact that older patients with Parkinson's are unable or unwilling to train at limits.

13.2. Oscillatory Firing of Motoneurons and Motor Units

Motoneurons can fire occasionally, transiently in an oscillatory manner and continuously in an oscillatory manner (Figure 12). In everyday life, they largely fire occasionally and transiently in an oscillatory manner, depending on the strength and speed of muscle activation. Transient oscillatory firing means that the motoneuron fires for a few oscillations, stops firing and fires for some further oscillations (Figures 21,46). The oscillatory firing pattern differs according to the motoneuron type (Figure 8). In Figure 20 the different patterns are shown schematically and on an original record taken with the single-nerve fiber action potential recording method and by surface electromyography (sEMG) from single motor units. α_1-Motoneurons innervate FF-type muscle fibers and fire rhythmically with impulse trains consisting of 1 action potential in the range of 10Hz. α_2-Motoneurons innervate FR-type muscle fibers and fire rhythmically with impulse trains consisting of 2 to 5 action potentials in the range of 4 to 7 Hz. The extracellular action potential amplitude of the α_2-motoneurons is smaller than that of the α_1-motoneurons and the FR-type motor units have much smaller amplitudes than the motor unit potentials of FF-type muscle fibers. The α_3-motoneurons innervate S-type muscle fibers and fire in an oscillatory manner at a frequency of around 1 Hz with long impulse trains (up to 50 action potentials per impulse train). The motor unit firing of single S-type muscle fiber motor units could not been safely identified by sEMG because their amplitudes are still smaller than those of FR-type motor units and are thus difficult to identify. The impulse patterns of oscillatory firing motoneurons obtained with sEMG are similar or the same as those obtained with the single nerve-fiber action potential recording method (Figure 20). Since sEMG is a non-invasive recording method, oscillatory firing can be recorded easily when using appropriate patients.

It was shown that the oscillatory firing of α_1-motoneurons depends very much on the primary muscle spindle afferent drive and that α_1-motoneuron oscillators are not very stable [19]. The α_2-motoneuronal network oscillators obtain drive from several secondary muscle spindle afferents. The α_2-motoneuron oscillators are more stable since they do not depend so much on the afferent drive. The α_3-motoneuron oscillators receive polymodal input; this means they get input from several kinds of afferent fibers. They seem to be very stable. The problem is that they are difficult to record since the motoneuron axons are thin and the motor unit potentials are small. From the theoretical standpoint, α_2-motoneuron oscillators are the most interesting because they are quite stable, and they can be measured with the single nerve-fiber action potential recording method and sEMG.

13.3. Continuous Synchronization of Network Oscillators Is Pathologic

Above α_1, α_2, and α_3-motoneurons were shown to fire in an oscillatory manner for high activation and FF, FR, and S-type muscle fiber motor units, they also innervate to fire in an oscillatory manner with the same impulse patterns. It was further shown that there was a relative phase and frequency coordination between the oscillatory firing motoneurons and also between the oscillatory firing of the motor units. The coordination was such that the motoneurons or the motor units did not fire synchronously unless exposed to strong and sustained stimulation. Such non-synchronous coordination seems to make sense, since synchronously oscillatory firing motor units could give rise to rhythmic muscle movements and tremor. It will be shown that in Parkinson's disease, patients with motor units upstream, the motoneurons do synchronize their firing. A mechanism, which muscle generates movement and tremor by the pathologically functioning CNS will be explained. Since during tremor, α_1-motor units do synchronize their firing with oscillatory firing α_2-motor units to generate tremor, differences in the stability between α_1 and α_2-motoneuron oscillators and the building up of an external loop to the periphery by the premotor spinal oscillators will be clarified further, through measurements.

13.4. The Triggering Mechanisms of Parkinsonian Tremor and Large-scale Coordination

In Figure 65A,B two mechanisms of tremor induction are shown. The patient was positioned on a chair at the special coordination dynamics therapy device but was not exercising. In Figure 65A, one FR-type motor unit in the gastrocnemius muscle (innervated by an *α_2-motoneuron) started to oscillate (first triggering mechanism)*, i.e. fired rhythmically with an impulse train consisting of 3 motor unit potentials. Most likely, the gastrocnemius muscle was not moving. Then more FR-type motor units and maybe some S-type motor units started to fire synchronously (Figure 65C). The gastrocnemius muscle may have moved a bit with a frequency of 4.1Hz. The triceps brachii muscle moved, because a movement artifact was recorded from it (Figure 65C). With the start of FF-type motor unit firing, synchronized with the FR-type motor unit firing (Figure 66A,B and Figure 65E,F, tibialis anterior r.), the muscle moved rhythmically.

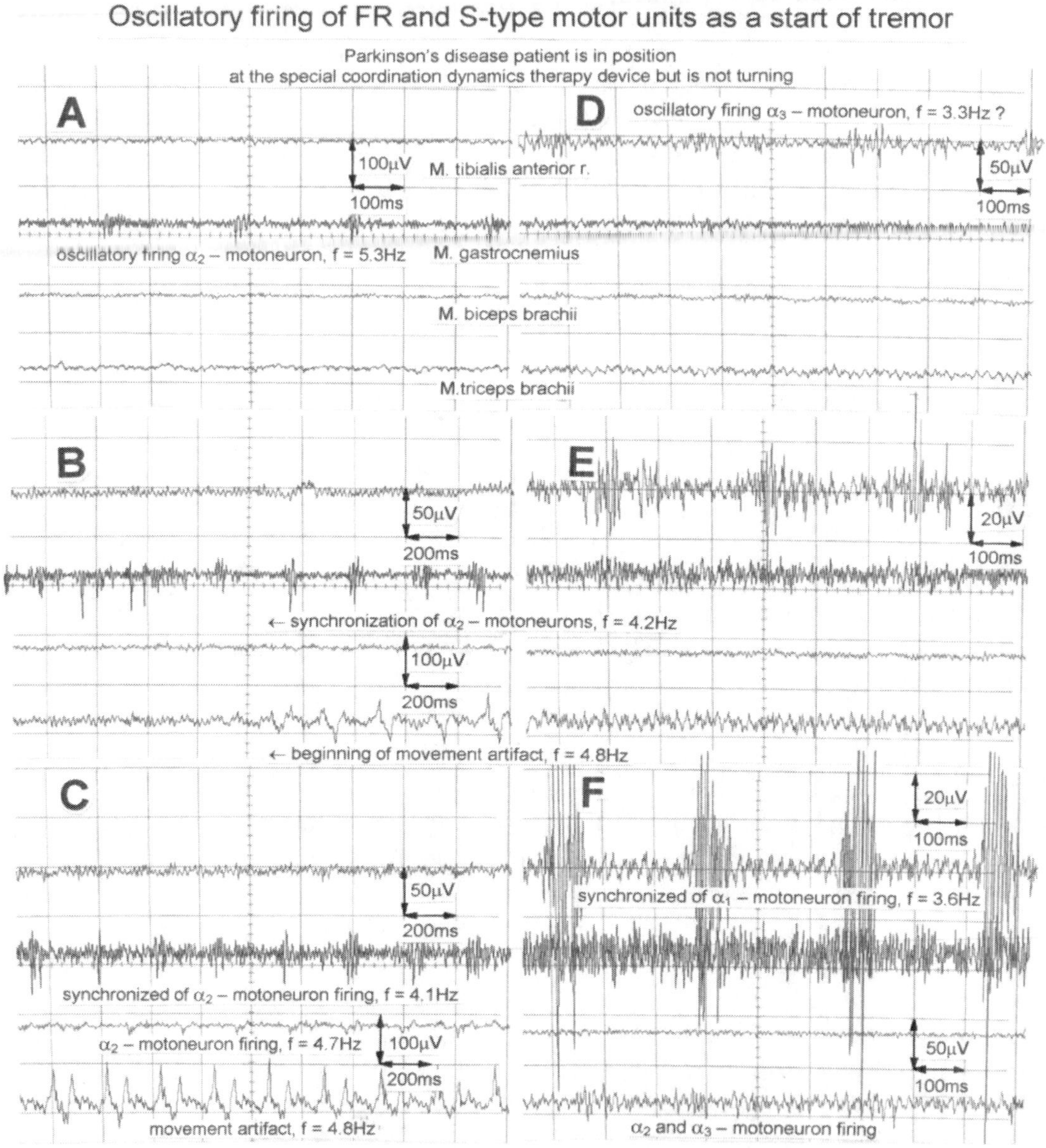

Figure 65. Oscillatory firing of α_2 – motoneurons and the synchronization of oscillatory firing α_2 and α_1 – motoneurons. Recordings were done with sEMG. A. Oscillatory firing of a single α_2 – motoneuron motor unit (f = 5.3Hz; amplitude ≈ 35μV, triplet-firing) in M. gastrocnemius. B. Synchronization of mainly oscillatory firing α_2 – motoneuron motor units (FR, f = 4.2Hz) in gastrocnemius muscle. With the synchronization of the oscillatory firing α_2 – motoneurons a movement artifact was induced in the electrodes on M. triceps brachii. C. Synchronized oscillatory firing of α_2 – motoneuron motor units, mixed with a few α_1 – motoneuron motor units of small amplitude in gastrocnemius muscle.

In musculus biceps brachii an α_2 – motoneuron motor unit fires at with 4.7Hz not synchronizedly with the α_2 – motoneuron firings in gastrocnemius muscle. Rhythmic movement artifact in triceps brachii muscle (f = 4.8Hz), indicating an arm tremor of 4.8Hz. D. Possible oscillatory firing of an α_3 – motoneuron motor unit (f = 3.3Hz; amplitude ≈ 10μV; long impulse train) in tibialis anterior muscle; a few α_2 – motoneuron motor unit potentials (larger amplitude) fire synchronizedly. E. Synchronized oscillatory firing of mainly α_2 and α_3 – motoneurons (f ≈ 3.3Hz). F. Synchronized firing of α_1 – motoneuron unit potentials (FF-type) of large amplitude (up to 200μV) with the oscillatory firing α_2 and α_3 – motoneuron motor unit potentials (FR and S-type) in gastrocnemius muscle a few minutes later (resulting frequency = 3.6Hz). Rhythmic firing of α_2 and α_3 – motor unit potentials in the triceps brachii muscle, in some coordination to the rhythmic firing in the tibialis muscle, give rise to substantial rhythmic movements of tibialis anterior muscle. D, E, and F are consecutive recordings. Recordings from a 70-year-old patient with Parkinson's disease for 6 years (U.H.); tremor on both sides.

A *second tremor or muscle movement triggering mechanism concerned FR-type (and may be S-type) motor units synchronizing their oscillatory firing.* In the left part of Figure 65B the FR-type motor units oscillated but not in synchronization; no movement artifact at the electrodes of the triceps brachii muscle was recorded. When the FR-type motor units started to oscillate in a synchronized manner at a frequency of 4.2Hz (right part of Figure 65B), then a rhythmic movement artifact (f = 4.8Hz) was generated at the surface electrodes of the triceps brachii muscle. Since the triceps brachii muscle is sited in the arm and the gastrocnemius muscle in the leg, the oscillatory firing networks of the arm (network assemblies in the intumescentia cervicalis) and the leg (assemblies in the intumescentia lumbosacralis) did communicate with each other. There were therefore large-scale synchronization and coordination at least between the intumescentia cervicalis and lumbosacralis, which is a distance of approximately 300mm.

The full muscle movement or tremor was generated when the FF-type motor units (innervated by α_1-motoneurons) synchronized their firing with the firing of the FR-type motor units (Figure 65E,F) (third triggering mechanism). With the synchronization of the FF-type motor units, the muscle movement or tremor frequency changed to match frequency coordination between the driving oscillatory firing of α_1 and α_2-motoneurons. In Figure 65D, the common frequency of the synchronized oscillatory firing α_2-motoneurons (and maybe α_3-motoneurons) was 3.3Hz. With the synchronization of the α_1-motoneurons, the common synchronized frequency increased to 3.6Hz (Figure 65F). The α_1-motoneurons could have fired in an oscillatory manner at a frequency of 7.2Hz, so that a single FF-type motor unit potential could have fired in coordination with every second muscle activation burst of the tremor. If another α_1-motoneuron oscillated at 10.8Hz, the innervated FF-type motor unit potential could have fired in coordination with every third activity burst.

In conclusion, in this patient the Parkinsonian tremor started with the synchronized oscillatory firing of FR-type motor units (innervated by α_2-motoneurons). With the synchronized firing of FF-type motor units (innervated by α_1-motoneurons) the full muscle movement or tremor was generated.

The reason that the FF-type motor units synchronized their rhythmic firing and that the FF-type motor units synchronized their firing with those of the FR-type motor units and not vice versa will be partly analyzed below in the course of explaining oscillator stability differences between α_1 and α_2-motoneuron oscillators. α_2-oscillators have a high and α_1-oscillators a low stability.

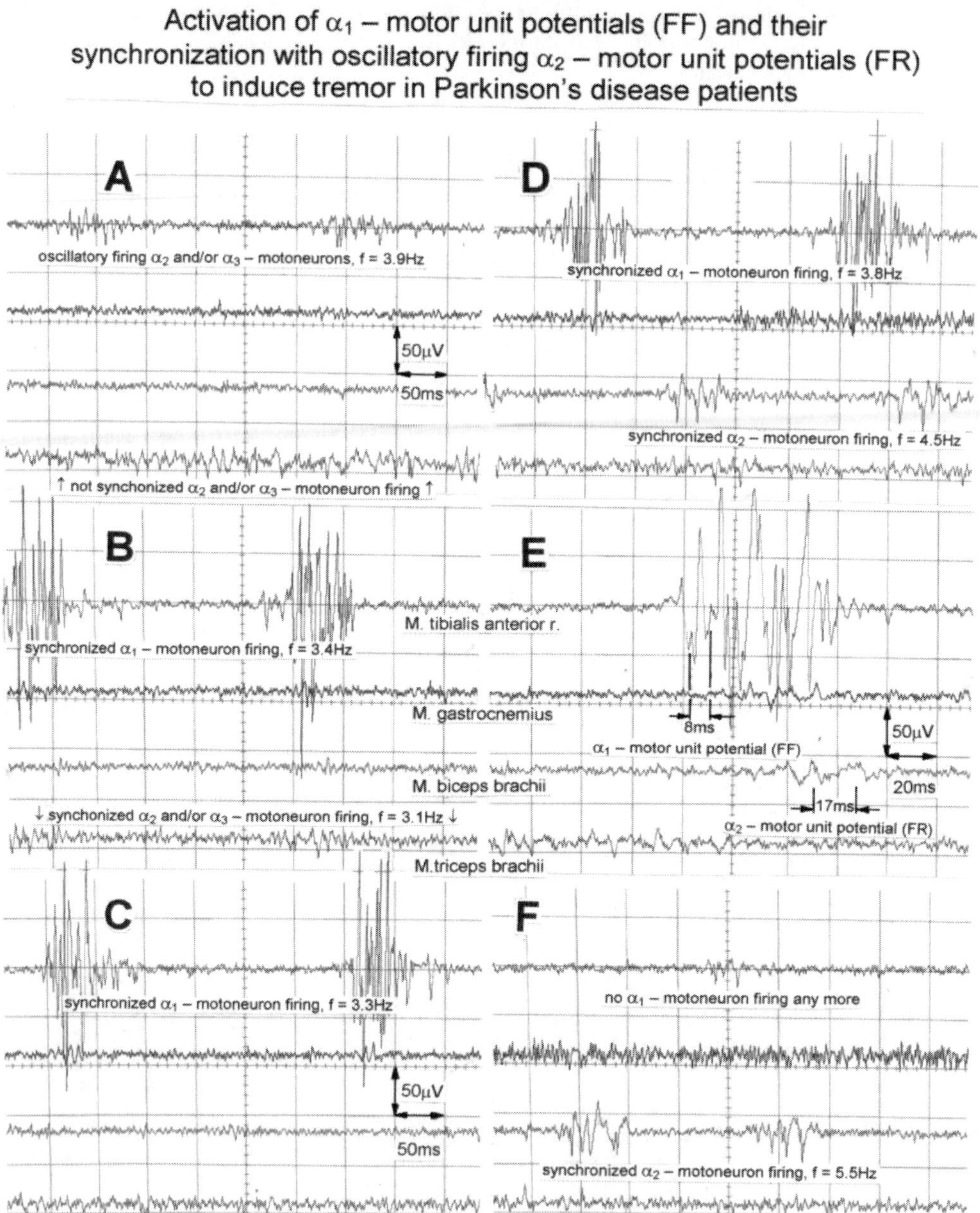

Figure 66. Time course of transient tremor induced by the activation of α_1 – motor unit potentials (FF-type) and their synchronization with oscillatory firing α_2 and α_3 – motoneuron motor unit potentials (FR and S-type). A. Synchronized oscillatory firing of α_2 and α_3 – motor unit potentials in tibialis anterior muscle (f = 3.9Hz). No synchronized firing of α_2 and α_3 – motor unit potentials in triceps brachii muscle. No movement of tibialis anterior muscle. B. α_1 – motor unit potentials are activated and synchronized with the α_2 and α_3 – motor unit potentials (f = 3.4Hz). Tibialis anterior muscle moves visibly. α_2 and α_3 – motor unit potentials in biceps brachii muscle fire partly synchronizedly with the motor unit potentials in tibialis anterior muscle. α_2 and α_3 – motor unit potentials can be seen before and after the α_1 – motor unit potentials in the activity burst. C. Similar firing as in 'B', but nearly no motor unit firing in triceps brachii muscle. D. Rhythmic firing of motor units giving rise to up and down movement (f = 3.8Hz) of tibialis anterior muscle.

Synchronized α_2 and α_3 – motor unit firing in biceps brachii muscle (f = 4.5Hz), not coordinated with the rhythmic firing in tibialis anterior muscle. α_2 and α_3 – motor unit potentials are activated in gastrocnemius muscle. E. Time-stretched motor unit action potentials: duration of α_1 – motor unit potential = 8ms (amplitude ≈ 150μV), α_2 – motor unit potential = 17ms (amplitude ≈ 15μV). F. Cessation of rhythmic firing of α_1 – motor unit potentials in tibialis anterior muscle; no visible contraction of tibialis anterior muscle any more; some α_2 – motor unit firing still occurring. Synchronized α_2 - oscillatory firing, still in biceps brachii muscle. Recordings from a 70-year-old patient with Parkinson's disease for 6 years (U.H.); tremor on both sides; measurements were performed on the more affected right side. During the measurements, the hands and feet of the patient were in the position for exercising on the special coordination dynamics therapy device.

13.5. Synchronization and De-synchronization of FF-type Motor unit Firing with Oscillatory Firing FR-type Motor Units

In Figure 66, the synchronization and de-synchronization of FF-type motor units (innervated by α_1-motoneurons) with FR-type motor units (innervated by α_2-motoneurons) is shown. In Figure 66A FR and S-type motor units (small motor unit amplitude) fired rhythmically at a common frequency of 3.9Hz in the right tibialis anterior muscle. In Figure 66B,C FF-type motor units (large unit amplitude) fired synchronized with the FR-type motor units; large FR-type motor unit potentials appeared on top of the small FR-type motor unit potentials. The common rhythmic frequency is now 3.4 and 3.3Hz. It can be seen that the large amplitude FF-type motor unit potentials slightly changed their synchronization phase with respect to the low amplitude FR-type motor unit potentials; sometimes the low amplitude FR-type motor unit potentials started earlier in the activity burst and sometimes they lasted longer than the large FF-type potentials in the burst (Figure 66B,C). In Figure 66D additionally FR type motor units started to fire rhythmically in the biceps brachii muscle at a frequency of 4.5Hz. In Figure 66F FF-type motor unit potentials no longer fired in the tibialis anterior muscle, only FR-type potentials were recorded. The rhythmically firing FR-type motor unit potentials in the biceps brachii muscle increased their firing frequency to 5.5Hz. In this patient the tibialis anterior muscle was only moving up and down when the FF-type motor units fired in addition.

In Figure 66E, the rhythmically firing motor units are displayed on a time-stretched scale. It can be seen that the FF-type motor units had motor unit action potential duration of approximately 8ms (amplitude ≈ 150μV) and the FR-type motor unit potentials had duration of 17ms (amplitude ≈ 15μV). The single nerve-fiber action potentials have a much shorter duration (approximately 0.3ms). For details of amplitudes and durations, see Figure 71.

13.6. Stability of α_1 and α_2-Motoneuron Oscillators

It was mentioned above that rhythmic muscle activity, leading to muscle movement and tremor, started with the oscillatory firing of the too little inhibited motoneurons. The tremor developed then with the synchronized rhythmic firing of the FR-type motor units to which the FF-type motor units, when activated, synchronized. It shall now be analyzed how these synchronizations can be explained by using the data obtained with the single-nerve fiber action potential recording method.

Measurements with the single-nerve fiber action potential recording method (Figures 3,4,5) from an α_1-motoneuron axon and a primary muscle spindle afferent fiber, contained in a sacral S5 nerve root from a spinal cord injury patient, are shown in Figure 67. An original recording of the α_1-motoneuron nerve-fiber action potential (α_1) and the primary afferent nerve fiber action potential (SP1) is shown in Figure 67A. The schematic pattern of firing and the definitions of oscillation period ($T\alpha_1$) and phase correlation between the extracellular action potentials of the oscillatory firing α_1-motoneuron and primary muscle spindle afferent fiber ($\alpha_1 \rightarrow$ SP1) are shown in Figure 67B. Oscillation period distributions are shown in Figure 67C,E,G and phase correlation distributions between the oscillatory firing α_1-motoneuron and its driving primary muscle spindle afferents are shown in Figure 67D,F,H. It can be seen from Figure 67C,E,G that the oscillation periods varied between 110 and 150ms (frequency variation between 9 and 6.7Hz) depend slightly on different natural stimulations (Figure 70). The phase correlation between the action potentials of the oscillatory firing α_1-motoneuron and of the primary spindle afferent fiber varied between 12.5 and 17ms. The main phase relation variation was between 2 and 3ms.

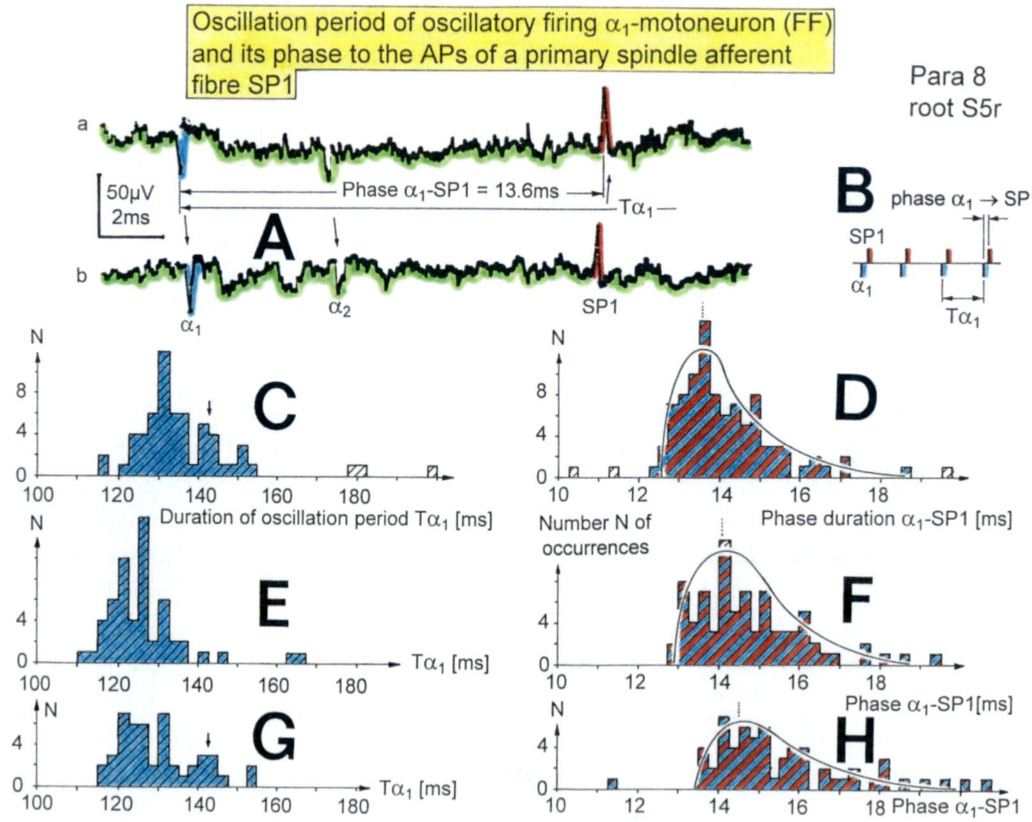

Figure 67. Distribution of the oscillation period of an oscillatory firing α_1 – motoneuron (FF) and the phase to its driving primary muscle spindle afferent fiber SP1. Recordings were taken with the single nerve-fiber action potential recording method. A. Original recording of the action potentials of the α_1 – motoneuron and SP1-fibre. B. Definition of oscillation period $T\alpha_1$ and phase $\alpha_1 \rightarrow$ SP1.

C-H. Oscillation periods and the corresponding phases following different stimulation (C;D, oscillation period and phase distributions for the time interval 65-82s. E;F, distribution for the time interval 1-30s; Figure 7B (strongest afferent drive). G;H, for 31-66s). In 'C' and 'G', the small arrows mark sub peaks in the oscillation period distributions. Para 8, right root S5.

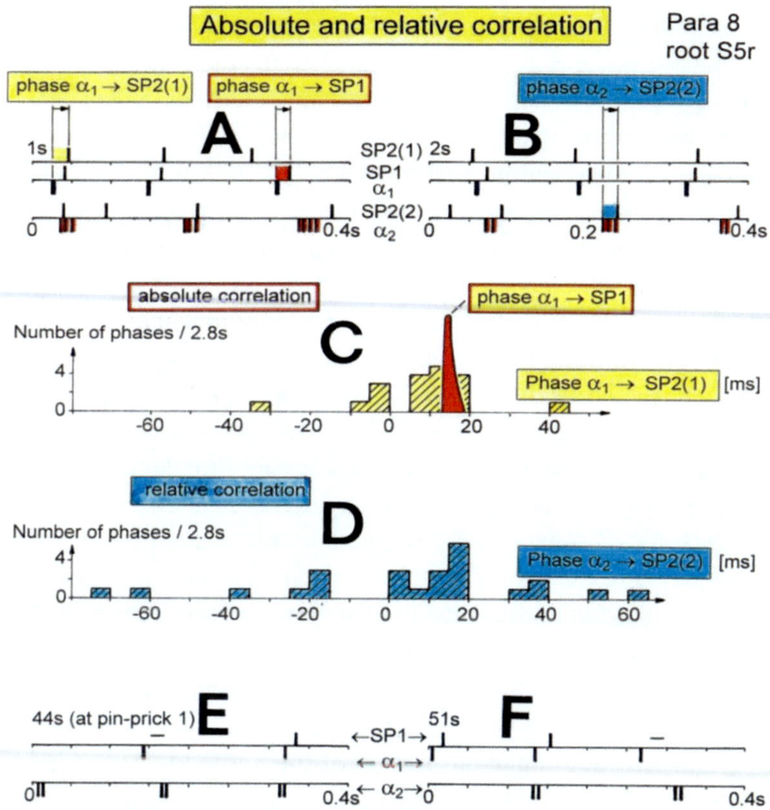

Figure 68. Absolute and relative correlation quantified by phase relations between the α_1 (FF) and α_2-motoneurons (FR) and their driving primary (SP1) and secondary (SP2) muscle spindle afferent fibers. A, B. Definition of the different phases. C. Distribution of the phases between α_1-motoneuron and the secondary muscle spindle afferent fiber SP2(1). Note that the phase distribution $\alpha_1 \rightarrow$ SP2(1) is approx. 40 times wider than that of the $\alpha_1 \rightarrow$ SP1 distribution (phase $\alpha_1 \rightarrow$ SP1 taken from Figure 65). D. Distribution of the phases between α_2-motoneuron and the secondary muscle spindle afferent fiber SP2(2). Note that the phase distribution $\alpha_2 \rightarrow$ SP2(2) is similar to that of $\alpha_1 \rightarrow$ SP2(1) (approx. 4 times wider). E,F. Note that every SP1-action potential is accompanied by a time-locked α_1-motoneuron action potential (AP). Para 8, root S5r.

In Figure 69A another original recording from the same S5 root as in Figure 67A is shown. In addition to the action potentials of the α_1-motoneuron and the primary muscle spindle afferent fiber, action potentials of an oscillatory firing α_2-motoneuron can be seen. Action potentials of secondary muscle spindle afferent fibers do not occur in this recording piece. In Figure 68A,B the different phase relations between the oscillatory firing α_1 and α_2-motoneurons and the primary and secondary muscle spindle afferent fibers are defined. It can be seen from Figure 68C,D that the phases between the α_1-motoneuron and a secondary muscle spindle afferent fiber ($\alpha_1 \rightarrow$ SP2(1)) and between the α_2-motoneuron and another secondary muscle spindle afferent fiber ($\alpha_2 \rightarrow$ SP2(2)) varied much more (phase distribution width $\approx$ 80 or 120ms) than the phase relation between the α_1-motoneuron and the primary muscle spindle afferent fiber ($\alpha_1 \rightarrow$ SP1; $\approx$ 4ms). This indicates that the primary spindle afferent fiber drove much more directly the α_1-motoneuron than did the secondary muscle spindle afferent fibers the α_1 and α_2-motoneurons. The nearly time-locked activation of the α_1-motoneuron by the primary muscle spindle afferent indicates monosynaptic activation, whereas the phase distribution width of $\approx$ 120ms between the secondary muscle spindle afferent fiber and the α_2-motoneuron indicates more polysynaptic drive and polysynaptic connections indicate that there are more neuronal networks involved. Even though the α_1-motoneuron also got a projection from a secondary muscle spindle afferent fiber, α_2-motoneurons received many projections from secondary muscle spindle afferent fibers with different synaptic strengths, quantified by the phase distribution width and shape [39,31].

The differing degrees of the phase correlation indicates that the α_1-motoneuron oscillator was more strongly connected to the peripheral input than the α_2-motoneuron oscillator in accordance with the monosynaptic stretch reflex concept. The stronger dependence on peripheral input of the α_1-motoneuron oscillator is supported by the observation that the α_1-motoneuron oscillator stopped oscillating as soon as the input from the primary spindle afferent fiber was missing for longer than 1 oscillation period (Figure 68E,F). In conclusion, the α_2-motoneuron oscillators seem to depend less on the peripheral input and are therefore more true oscillators. Rhythmic movements like tremor can therefore be better held continuously by α_2-motoneuron oscillators than by α_1-motoneuron oscillators.

The higher stability of oscillatory firing of α_2-motoneuron oscillators in comparison to α_1-motoneuron oscillators can also be seen when plotting oscillation periods consecutively against time. In Figure 70 the oscillation periods of the above α_1 and α_2-motoneurons are plotted against time and for different stimulations. It can be seen that the α_1-motoneuron stopped oscillation more often than the α_2-motoneuron. Within the time interval, from 50s (Figure 70A) to 66s (Figure 70B) ($\approx$100s time interval), the α_2-motoneuron stopped oscillating 3 times and the α_1-motoneuron 19 times for the same stimulation at the same site. In this recording (with the single-nerve fiber action potential recording method), the α_2-motoneuron oscillator showed more flexibility than the α_1-motoneuron oscillator. When pin-pricking at the anal reflex zone the α_1 and α_2-motoneurons synchronized their oscillatory firing transiently, the α_2-motoneuron adapted the firing to the one of the α_1-motoneuron by reducing the oscillation periods (increasing the frequency) to match the one of the α_1-motoneuron oscillator (time period of 54s to 56s in Figure 70A; Figure 69C,D).

The stability of both oscillators may have been reduced as the recording was taken from a patient with spinal cord injury, where neuronal network functioning is impaired.

13.7. Transient Synchronization of Premotor Spinal Oscillator in a Patient with a Spinal Cord Injury

From the schematic impulse patterns shown in Figure 69-E it can be clearly seen that the α_1 and α_2-motoneuron oscillators transiently synchronized their firing upon pin-pricking the anal reflex zone repeatedly every second. The oscillation period changes can be seen in Figure 70A. Therefore, in non-Parkinson patients, premotor spinal oscillators can also synchronize. But as soon as the repeated stimulus ceased, the oscillators quickly de-synchronized again. Also, pin-pricking the anal reflex zone (pin-prick 7, 8; Figure 55B) induced the continence automatism very strongly. Pin-pricking the peri-anal skin outside the reflex zone (for example pin-pricks 9, 10) did not induce synchronization of the oscillatory firings (Figures 70A, 69E). Thus in healthy subjects premotor spinal oscillators seem to synchronize only transiently and only if the repeated stimulus is sufficient to activate the oscillators (to self-organize).

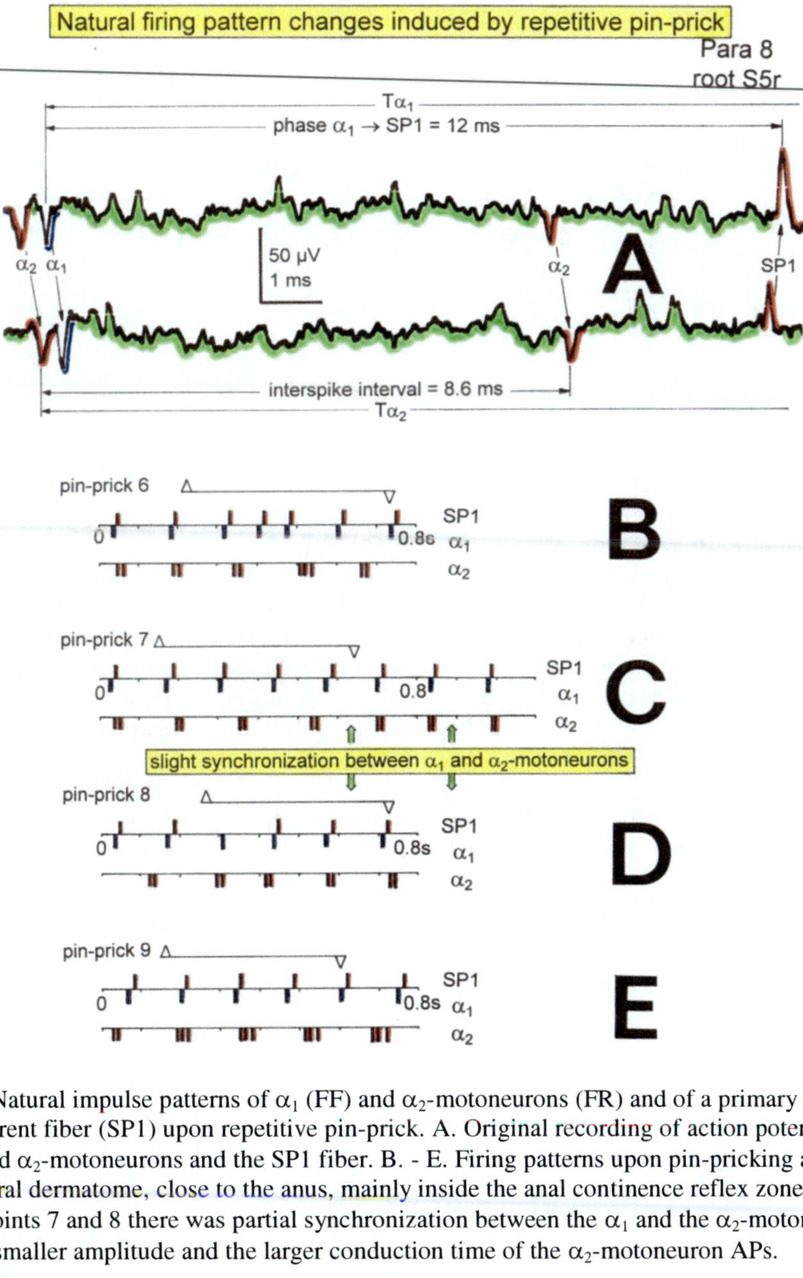

Figure 69. Natural impulse patterns of α_1 (FF) and α_2-motoneurons (FR) and of a primary muscle spindle afferent fiber (SP1) upon repetitive pin-prick. A. Original recording of action potentials (APs) of the α_1 and α_2-motoneurons and the SP1 fiber. B. - E. Firing patterns upon pin-pricking at sites 6,7,8,9 (sacral dermatome, close to the anus, mainly inside the anal continence reflex zone). Note that at pin-prick points 7 and 8 there was partial synchronization between the α_1 and the α_2-motoneurons. Note further the smaller amplitude and the larger conduction time of the α_2-motoneuron APs.

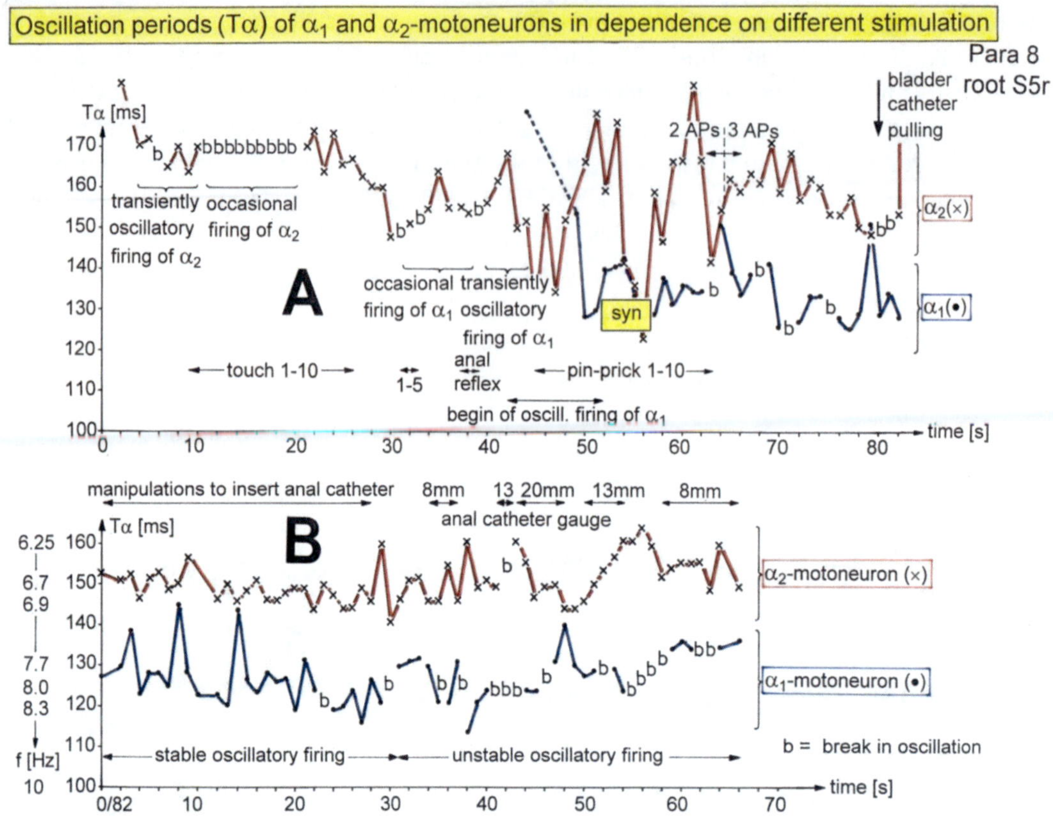

Figure 70. Oscillation periods (Tα) of an α_1 (FF) and an α_2-motoneuron (FR) in dependence on different stimulations; 'B' is a continuation of 'A'. Touch 1 - 10 = touching with a needle at different points in and outside the anal reflex zone. 8,13,20,13,8mm = changing the diameter of the anal catheter from 8 to 13 to 20 to 13 and back to 8 mm (3 gauges of anal catheter were used). syn = partial synchronization of the oscillatory firing patterns of the α_1 and the α_2-motoneurons. 2 APs / 3 APs = impulse train of the α_2-motoneuron mainly consisted of 2 or 3 action potentials respectively. Note that α_1 and α_2-motoneurons first fired transiently oscillatory before firing continuously oscillatory. Note further that the oscillatory firing was most regular upon manipulation at the anus (0-30s in 'B', anal reflex stimulation to secure continence).

In the patients with Parkinson's disease the rhythmic muscle movement or tremor started in a relaxed position and there was certainly no strong repeated stimulus. It is therefore concluded that *two kinds of inhibition were missing in Parkinson's disease patients*. First, premotor spinal *oscillators did self-organize following very mild stimulus* instead of strong, adequate stimulus (missing inhibition to stop high activation without adequate input) and second, the premotor *spinal oscillators did synchronize with other premotor spinal oscillators* without strong repeated stimulus (missing inhibition not to synchronize).

13.8. External Loop of Premotor Spinal Oscillators as a Cause for α_1-Motoneuron Oscillators Synchronizing their Firing with α_2-motoneuron Oscillators

It was shown above that several FR-type motor units synchronized their rhythmic firing to trigger muscle movement in a Parkinson's disease patient (Figure 65). But how did the FF-type motor units start to fire and how did they synchronize their firing with the FR-type motor units? It has been shown in patients with spinal cord injury that oscillatory firing motoneurons can build up an external loop to the periphery via the γ-motoneurons and spindle afferents [26]. Figure 63 shows schematically the communication of premotor spinal oscillators with the periphery. In Figure 63a sphincteric α-motoneurons oscillate following stimulation from bladder and anal catheters (no rhythmic muscle movement). The γ-motoneurons may fire partly in an oscillatory manner. When touching with a needle repeatedly every second the anal reflex area α and γ-motoneurons partly synchronized their firing. The oscillatory firing α-motoneurons built up an external loop to the periphery via the γ-motoneurons and the secondary muscle spindle afferent fibers (Figure 63b). When stopping the touching, the partial synchronization seized (Figure 63a). With the strong stimulation of pin-pricking at the anal reflex area, α and γ-motoneurons and spindle afferents synchronized their firing in addition to building up an external loop to the periphery (Figure 63e). This means that the periphery was included in the synchronized oscillatory firing including the input from the muscle spindle, joint, and skin afferents (Figure 63f).

If we transpose this external loop measurement to the Parkinson's disease patients, then the reason for muscle movement and tremor can be understood. *Because of the lack of sufficient inhibition, a number of α_2-motoneuron oscillators get activated spontaneously and they synchronize their firing because of the lack of mutual inhibition.* Slight contraction of muscles or stretching of muscle spindles, induced by the FR-type motor units, may start to induce the firing also of the primary muscle afferents. They stimulate the α_1-motoneurons to fire. Via the external loop of the premotor spinal oscillators, where muscle, joint, and skin afferents are included in the rhythmic firing, the α_1-motoneurons fire in synchrony with the oscillatory firing α_2-motoneurons. It is likely that the α_1-motoneurons also oscillated. But it was observed with the single-nerve fiber action potential recording method that also occasionally fires α-motoneurons tend to synchronize their firing with oscillatory firing motoneurons. A macroscopic recording was shown in Figure 1 of [40]. Based on the above measurements of the human nervous system, one mechanism is how tremor is generated in Parkinson's disease patients can be understood.

14. Pathologic CNS Organization Caused by Impaired Phase and Frequency Coordination due to Injury or Degeneration

Before tackling learning and relearning in detail, some pathologic patterns are shown which are caused mainly out of the impaired phase and frequency coordination in Parkinson's disease and spinal cord injury patients. Parkinson is a degenerative disease and spinal cord injury (SCI) a traumatic injury. In Parkinson's disease, mainly the inhibition in general is impaired and in SCI mainly the supraspinal control of the spinal cord is lost or strongly impaired.

14.1. FR-type Motor Units Fired Rhythmically before the FF-Type Motor Units during the Generation of Tremor

Surface electromyography (sEMG) showed that rhythmic muscle contraction and tremor was generated by rhythmic activation of FF-type and FR-type motor units, innervated by α_1 and α_2-motoneurons respectively (Figure 13). Because of the small amplitude of S type motor unit action potentials, innervated by α_3-motoneurons (for classification scheme, see Figure 8), their role in the induction of tremor is unclear. In Figures 65 and 66, it seems as if the S-type motor units fired in synchrony or coordination with the FR-type motor units.

Figures 65 and 66 show that when tremor started, the FR-type motor units fired before the FF-type motor units. In another patient, it seemed as if the tremor started with the FF-type motor unit activity. But since even the motor unit action potential amplitudes of FR-type muscle fiber motor units are five to ten times smaller than the amplitudes of FF-type motor units, it could be that FR-type motor units were not recognized in all cases. Since in Parkinson disease nerve cells die in the substantia nigra and perhaps additionally in other parts of the CNS, one can expect a large number of different pathologic CNS organizations, depending on the actual degeneration status in the given patient.

The induction of tremor in Parkinson's disease patients cannot be understood unless we differentiate between α_1 and α_2-motoneurons (and α_3-motoneurons), innervating FF-type and FR-type (and S-type) motor units, and without the specific knowledge about oscillatory firing of motoneurons in the premotor network.

14.2. Synchronization of FF-Type with Rhythmic FR-Type Motor Unit Firing

It was shown in Figures 65 and 66 that during tremor FF-type motor units were only activated in synchrony with the rhythmically firing FR-type motor units. The induction of FF-type motor unit firing was explained by the external loop of premotor spinal oscillators, which means the inclusion of the γ-loop into the rhythmic firing of the premotor spinal oscillators (Figure 63). But since the FF-type motor units fired only in synchrony with the rhythmically firing FR-type motor units, there was also network synchronization in the premotor network between the α_1 and α_2-motoneuron oscillators, otherwise there would have been also some firing of FF-type units in between the rhythmic FR-type motor unit firing at the beginning of the FF-type motor unit firing, with the building up of the external loop. It seems therefore that the *synchronization of the FF-type with the FR-type motor units has a central and a peripheral component.* This view is supported by the synchronization measurement between the intumescentia lumbosacralis and cervicalis (Figure 65B). With the synchronization of the FR-type motor units in the gastrocnemius (triceps surae) muscle, a movement artifact occurred in the triceps brachii muscle, indicating that with the synchronization of α_2-oscillators in the intumescentia lumbosacralis, also α_1-motoneurons in the intumescentia cervicalis were activated to fire in synchrony to give rise to movement and movement artifact at recording electrodes on the triceps brachii muscle.

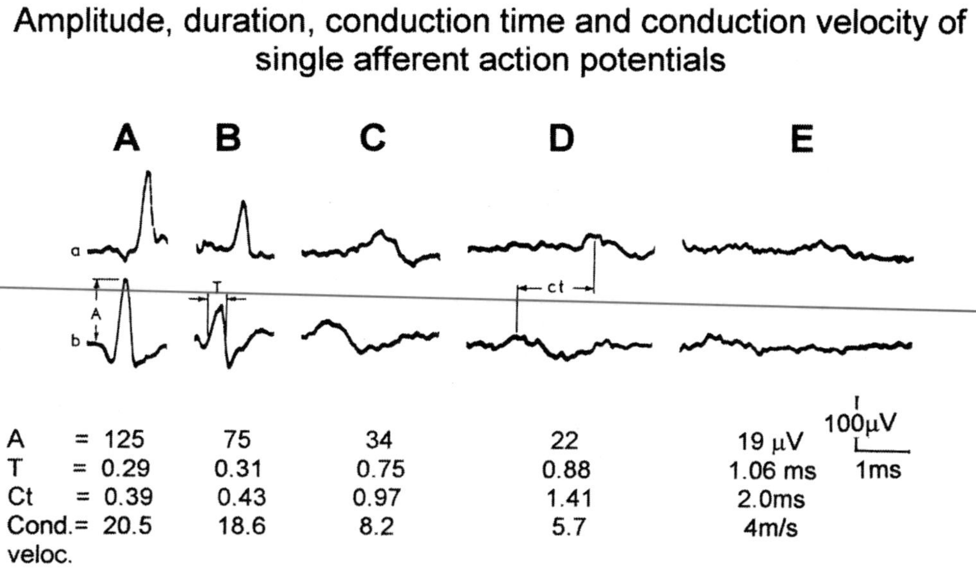

Figure 71. Extracellular single-nerve fiber action potential waveforms of different amplitudes (A), durations (T), conduction times (ct) and conduction velocities (cond. veloc.). The afferent fibers were stimulated by touch, pin-prick, and bladder and anal-catheter pulling and releasing. HT4 (brain-dead human cadaver), 56-year-old at death, 32°C (approximate nerve root temperature). Dorsal S5 root, approximate root diameter = 0.17mm.

14.3. Amplitude and Duration of FF, FR and S-Type Motor Unit Potentials in Comparison to Extracellular Single-Nerve Fiber Action Potentials

The recordings from FF and FR-types motor units in Figure 66E seem to indicate that FR-type motor unit potentials have much smaller amplitudes and are of longer duration than FF-type motor unit potentials. Extracellular single-nerve fiber action potentials have been shown to have different action potential amplitude and duration (Figure 71), depending on the conduction velocity and the nerve fiber diameter [41]. Human nerve fibers could actually be characterized and classified by group conduction velocity and group nerve fiber diameter [42]. It is of importance to know whether in general FF, FR and S-type motor units have approximately similar typical action potential amplitudes and durations, even though motor unit size may increase and decrease during re-innervation and the muscle fiber diameter is dependent on the usage. If one could distinguish better between FF, FR and S-type motor units, innervated by α_1, α_2, and α_3-motoneurons respectively, then one could better analyze natural impulse patterns of the different motor units and their innervating motoneurons. One could better understand the organization of the premotor neuronal networks and further rostral CNS structures like the basal ganglia. Human neurophysiologic research on the CNS functioning has therefore not only consequences for the development of treatments but also for the understanding of the functioning of the human CNS in general and its modulation by learning.

14.4. Tremor and Clonus in Patients with Parkinson's Disease and Spinal Cord Injury

14.4.1. Patients with Parkinson's Disease and SCI to Analyze the Human Premotor Network

When relative coordination between oscillatory firing motoneurons (and the motor unit muscle fibers they innervate) become impaired, the oscillatory firing motoneurons do synchronize their firing (instead of coordinating their firing to avoid synchronization) and muscles start to move rhythmically and fingers, hands, arms, legs, and face may start to shake. Further, the motor programs and consequently the movements are impaired, as can be measured by the coordination dynamics recording method.

Two types of CNS injuries, namely those in patients with Parkinson's disease and those suffering spinal cord injury, will be used to analyze the human premotor network. The patients with Parkinson's disease mainly show an injury to the substantia nigra due to degeneration, whereas their spinal cord is not affected. The patients with a spinal cord injury, on the other hand, have a healthy CNS apart from the injured spinal cord. Their basal ganglia, for sure, are not damaged by the injury. The differences in motor unit firing between both types of injury will provide us with information on the organization of the neuronal networks of the human spinal cord.

The importance of this research does not only follow from the possible partial repair of the malfunctioning CNS but also because of a better understanding of the functioning of the human brain. In recent publications, the conclusion was reached from the frequencies of tremor in Parkinson's disease patients that neurons in the basal ganglia fire rhythmically [43,44]. But the high activation mode of motoneuron firing has clearly been shown above to be firing rhythmically (oscillatory firing). This oscillatory firing of motoneurons originates in the spinal cord, as could be demonstrated by measurements taken from the isolated human spinal cord (in spinal cord injury patients taken from sites caudal to the complete injury) and from measurements of the isolated spinal cord in connection with the lower part of the reticular formation (in brain-dead humans). Therefore spontaneous rhythmic firing of motor units measured by surface electromyography (sEMG) in patients with spinal cord injury or Parkinson's disease corroborates previous measurements with the single-nerve fiber action potential recording method that the rhythmic firing of the motoneurons, and the motor unit muscle fibers they innervate, originates in the spinal cord. This does not mean that neurons in the basal ganglia are not firing rhythmically with similar or identical frequencies. Rhythms and rhythm coupling may occur throughout the CNS [45,46]. But it is of importance to know where certain functions and rhythms mainly originate, in spinal cord neuronal networks or supraspinally.

14.4.2. Description of Tremor and Clonus

Physiologic Tremor

The symptom of physiologic tremor can be experienced when the CNS is not functioning, optimally as for example, when it is overexcited. When one drinks too much coffee and then attempts to dissect a frog under the microscope at a magnification of 20 times, one may see one's hands shaking at a frequency in the range of 5Hz, due to the random synchronization of oscillatory firing motoneurons.

Pathophysiologic Tremor

The symptom of pathophysiologic tremor (strong tremor) can be seen in patients with Parkinson's disease. Muscles may move rhythmically. Arms, hands, fingers, and legs may shake at frequencies in the range of 5Hz. Also, face muscles may shake, depending on the degree of the disease.

Physiologic Clonus

The symptom of physiologic clonus can be experienced when leg muscles are exhausted from carrying heavy loads. Then, when you sit down with only the forefoot put on the ground the leg may start to move continuously rhythmically up and down due to consecutive dorsal and plantar flexion's of the feet. The right and left leg, mostly do not synchronize their rhythmic movements (no co-movement).

Pathophysiologic Clonus

The symptom of pathophysiologic clonus can very often be observed in patients who suffered a spinal cord injury. Continuous rhythmic plantar and dorsal flexions of the feet can be strong and continuous.

14.5. Pathologic Motor Programs: Motor Bursts Are Structured with Tremor, Clonus, and Rhythmic Motor Unit Activity

Now sEMG recordings are arranged in order to show increasing pathology and complexity of CNS functioning. In slight impairment of CNS functioning, motor programs seemed to be quite normal in accordance with repair in mild CNS injury. But in patients with Parkinson's disease or spinal cord injury, some motoneurons fired spontaneously in an oscillatory manner. The more the CNS functioning was impaired, the more pathologic the motor programs. In patients with Parkinson's disease, oscillatory firing motoneurons synchronized their firing to give rise to tremor. Tremor and rhythmic firing of motor units occurred in a motor burst of motor programs in some muscles. Some muscles were not activated at all or were continuously activated during the motor program. In patients who have suffered a spinal cord injury, muscles were not activated or activated continuously (spasticity) during the motor programs. Motor bursts were partly structured with clonus activity and the oscillatory firing of motoneurons was partly synchronized.

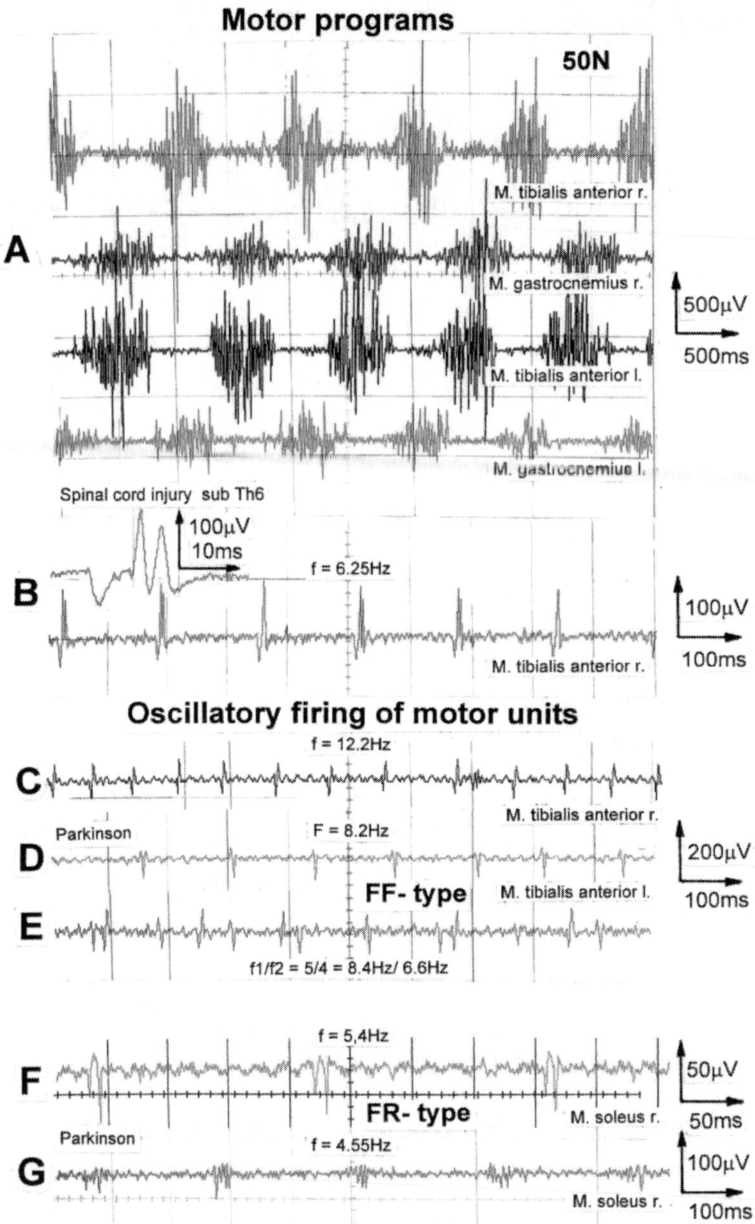

Figure 72. A. Electromyography recordings (EMG) of the motor program of a patient with an incomplete spinal cord injury sub Th6 during exercise on the special coordination dynamics therapy device at a load of 50Newton. The motor program looks normal. B. Oscillatory firing of an FF-type motor unit in the same patient in the right tibialis anterior muscle upon no exercising. Insert; time-stretched motor unit potential (re-innervation potential because of complicated structure). C-G. Oscillatory firing of FF-type (C,D,E) and most likely FR-type motor units (F,G) in 3 patients with Parkinson's disease. In E, two motor units fired simultaneously oscillatory. In F and G, the FR-type motor units fired with 2 and 3 action potentials per impulse train, respectively.

14.6. Spontaneous (Uncontrolled) Oscillatory Firing

Figure 72A shows the electromyographically recorded motor program of a patient who suffered an incomplete spinal cord injury while exercising coordinated arm and leg movements on a special coordination dynamics therapy device. The motor program looks quite normal. Still, one FF-type motoneuron of the spinal cord fired in an oscillatory manner at 6.25Hz despite receiving no stimulation (Figure 72B). In this case, the motor unit potential shows a complicated structure (Figure 72B, insert), possibly indicating the sprouting of the axon to other motor units. Thus, oscillatory firing of motor units can be observed in spinal cord injury patients with almost no stimulation.

Figure 72C,D shows single FF-type motor units firing spontaneously in an oscillatory manner at 8.2 and 12.2Hz in Parkinson's disease patients. From the motor unit waveform potential, it is clear that it is always the same motor unit that fires. In Figure 72E, we see the same motor unit firing as in Figure 72D, but additionally one further FF-type motor unit oscillating following very little or no volitional activation. The frequency ratio was f_1/f_2 = 8.4Hz/6.6Hz = 5/4. Approximately three minutes earlier the frequency ratio was f_1/f_2 = 7.9Hz/4.8Hz = 5/3. The coordination between these two oscillatory firing motor units is not clear. Figure 72F,G shows two FR-type motor units firing rhythmically in patients with Parkinson's disease with no volitional activation, mostly with 2 and 3 motor unit action potentials per impulse train at a frequency of 5.4Hz and 4.55Hz, respectively.

Since normally, motoneurons increase their firing rate with the increasing afferent input from occasional firing until they begin oscillatory firing for higher loads (Figure 12), there seems to be a lack of inhibition (as found above), especially in Parkinson's disease patients, to control the onset of oscillatory firing. But uncontrolled oscillatory firing can also be observed in spinal cord injury (Figure 72B).

14.7. Uncontrolled Synchronized Firing of Motor Units in Parkinson's Disease Patients

It was shown above that oscillatory firing motoneurons (Figure 24B) and oscillatory firing motor units coordinated their firing so as not to fire in a synchronized manner (Figures 22-24); this may be achieved by some kind of lateral field inhibition. It will be shown below in a Parkinson's disease patient that if this kind of coordinated firing is impaired, some motor units start to fire in a synchronized manner. In Figure 73B, we see mainly FR-type motor units (identified by low motor unit amplitude and FR-type frequency range (≈ 5Hz)) firing in a synchronized manner oscillating in the right soleus muscle at a frequency of 4.55Hz. In the right tibialis anterior muscle, FR-type motor units also oscillated in a synchronized manner antagonistically to the motor units in the soleus muscle. On top of the low-amplitude FR-type motor unit activity, large amplitude FF-type motor unit activity can be seen, which was synchronized. The duration of FR and FF-type activity is marked in one activity burst in the right tibialis anterior muscle. Synchronized oscillatory firing of mainly FR-type motor units can be seen further in the soleus, the biceps brachii, and the triceps brachii muscles in Figure 73D. In Figure 73C,D further synchronized activity of FR and FF-type motor units can be seen in the right tibialis anterior muscle.

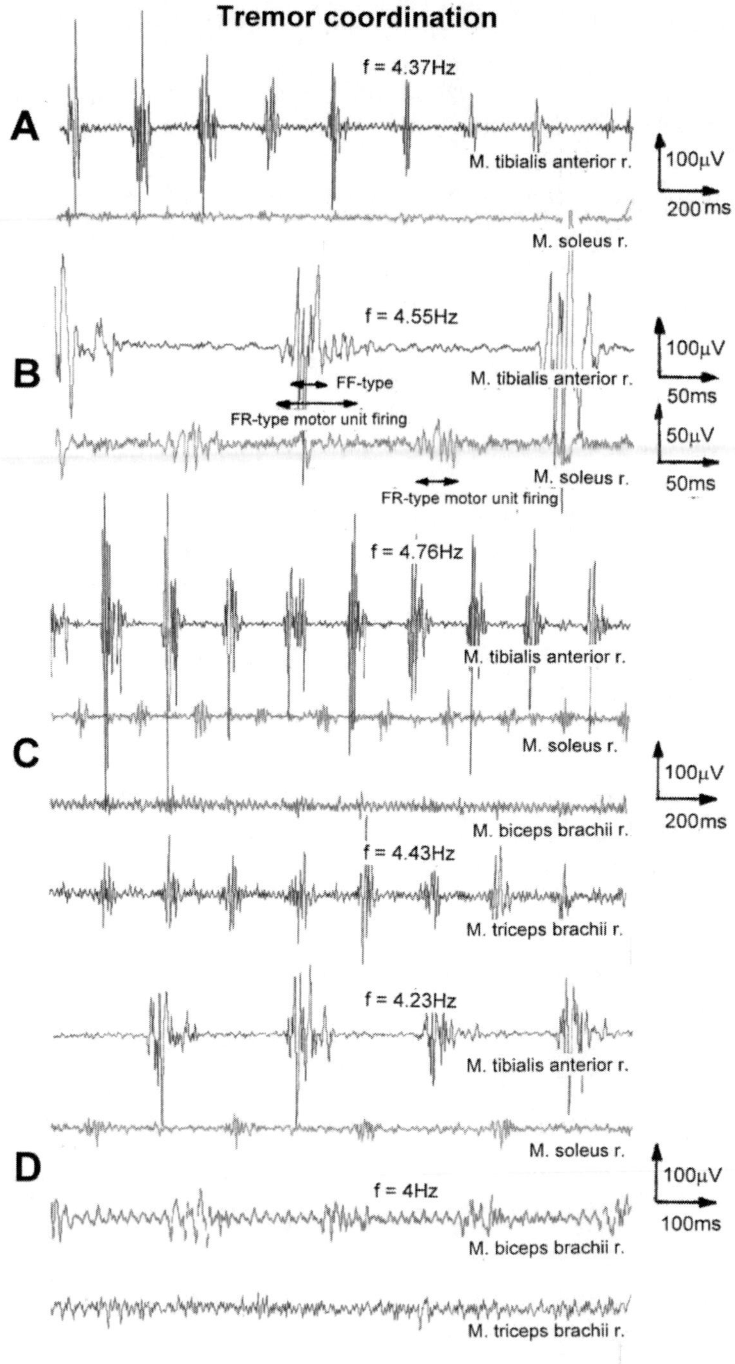

Figure 73. Electromyography (EMG) recordings of tremor coordination and synchronization of FF and FR-type motor unit firing, giving rise to tremor in a patient with Parkinson's disease. A. Ceasing of tremor activity in the right tibialis anterior muscle; no tremor activity in the right soleus, right biceps brachii, and triceps brachii muscles. B. Antagonistic tremor activity in the right tibialis anterior and soleus muscles. The EMG activity in the soleus muscle stems mainly from FR-type motor units and those in the tibialis anterior muscle from FR (small-amplitude) and FF-type (large-amplitude) motor units. No clonus activity in the other two muscles. C. EMG tremor activity in the tibialis anterior, the

soleus, and the biceps brachii muscles. The clonus activity in the right soleus and the right tibialis anterior muscles shows antagonistic coordination. The activity in the biceps brachii muscle (f = 4.43Hz) was not in coordination with those of the tibialis anterior and soleus muscle (f = 4.76Hz). D. EMG tremor activity in the tibialis anterior, the soleus, the biceps brachii, and the triceps brachii muscles. There was antagonistic tremor coordination between the antagonistic tibialis anterior and soleus muscles and between the antagonistic biceps and the triceps brachii muscles.

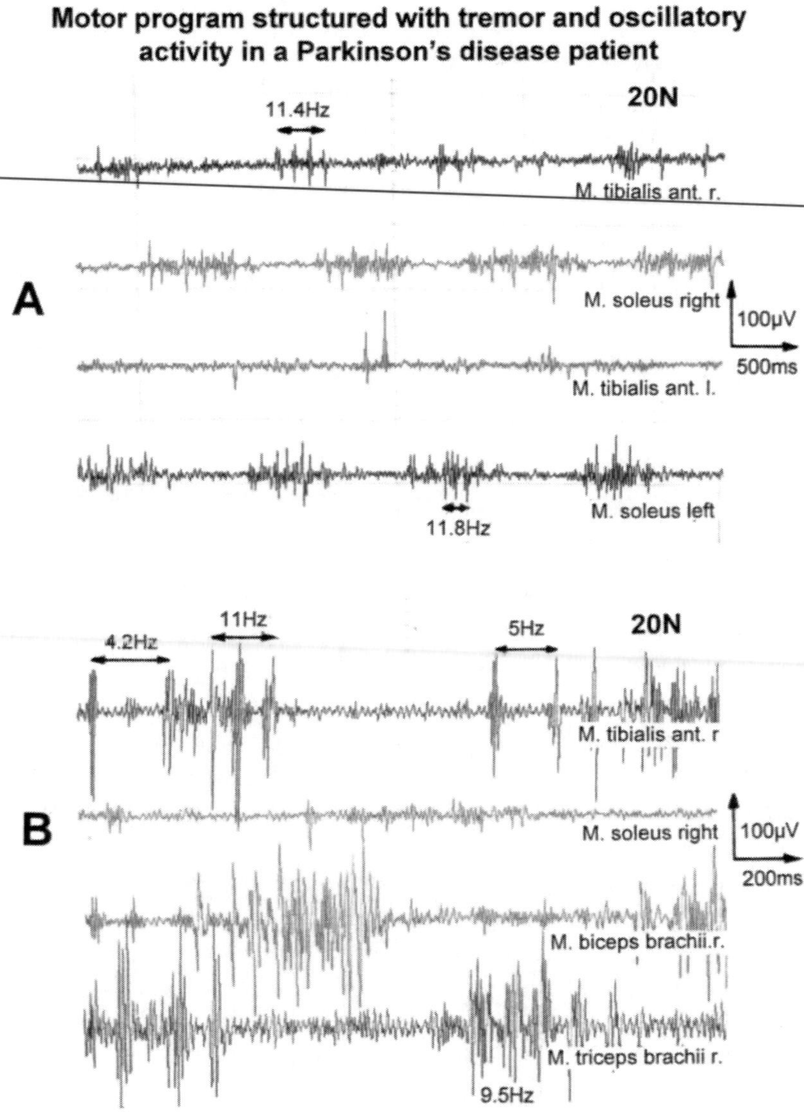

Figure 74. EMG motor programs of a patient with Parkinson's disease during the exercise on the special coordination dynamics therapy device at a low load of 20N. The motor program muscle bursts are more or less structured by rhythmicity (no rhythmic structure in normal motor bursts (Figure 1A)). In A, rhythmic activity at a frequency of 11.4 and 11.8Hz is suggested. In B, (faster sweep and partly different muscles) low-frequency rhythmicity of 4.2 and 5Hz (frequency range of α_2-motoneuron oscillatory firing) and higher frequency rhythmicity of 11 and 9.5Hz (frequency range of α_1-motoneuron oscillatory firing) might be seen.

It can partly be seen from Figure 73A that this synchronized oscillatory firing of FR and FF-type motor unit's build up and then ceases. In Figure 73A, the tremor ceases. At first, more and more FF-type units stopped firing and then the FR-type motor units stopped firing. However, when the tremor starts first the FR-type units fire synchronously in an oscillatory manner to which the FF-type units then synchronize. The process first affects the smaller units with lower amplitudes and then the larger units with larger amplitudes. There is some variation in this rule because of the different distances of the motor units to the recording electrodes.

As can be seen from Figure 73B,C,D, the tremor in antagonistic muscles is mostly antagonistic, probably due to the antagonistic proprioceptive input. Some agonistic tremor coordination seems visible. The seemingly low-amplitude agonistic tremor coordination may also be due to cross-talk of sEMG activity; this means that, for example, some tibialis muscle activity was recorded in the soleus muscle. The coordination of the sEMG-displayed tremor between different muscles may depend on the agonistic and antagonistic afferent input.

14.8. Motor Program Bursts in Patients with Parkinson's Disease Structured with Tremor Activity and Motor Unit Oscillatory Activity

The impaired coordination leading to spontaneous oscillatory firing and synchronized oscillatory firing of FR and FF-type motor units and tremor should also be seen during movements, because then coordinated organization of the CNS networks are also needed.

In the motor program bursts shown in Figure 73A, rhythmic activity can be identified, which cannot be seen in the rather physiologic motor pattern in Figure 71A. The highlighted rhythmic firing at 11.4 and 11.8Hz may indicate oscillatory firing of FF-type motor units innervated by α_1-motoneurons.

In another, more time-stretched motor program recording, such rhythmic firing can be seen more clearly (Figure 73B). Rhythms of 5 and 4.2Hz were most likely induced by oscillatory firings of an α_2-motoneuron, and those of 9.5 and 11Hz were most likely induced by oscillatory firings of an α_1-motoneuron.

Three important conclusions can be drawn from Figure 73. First, the term referring to the symptom, 'resting tremor', is not correct. Tremor not only occurs during rest, but also during movement. The reduced intensity of tremor during movements may indicate entraining of neuronal networks resulting in a reduction of tremor in the short-term memory. Second, FF and FR-type motor units, innervated by α_1 and α_2-motoneurons respectively oscillate with their 'Eigen-frequencies' for high activation during motor program burst. This was shown earlier in patients with spinal cord injury for muscles only innervated by a few motoneurons. Third, since synchronized oscillatory firing in tremor can also be seen in motor bursts during movements, two organizations, namely tremor and movement, were organized in the neuronal networks at the same time.

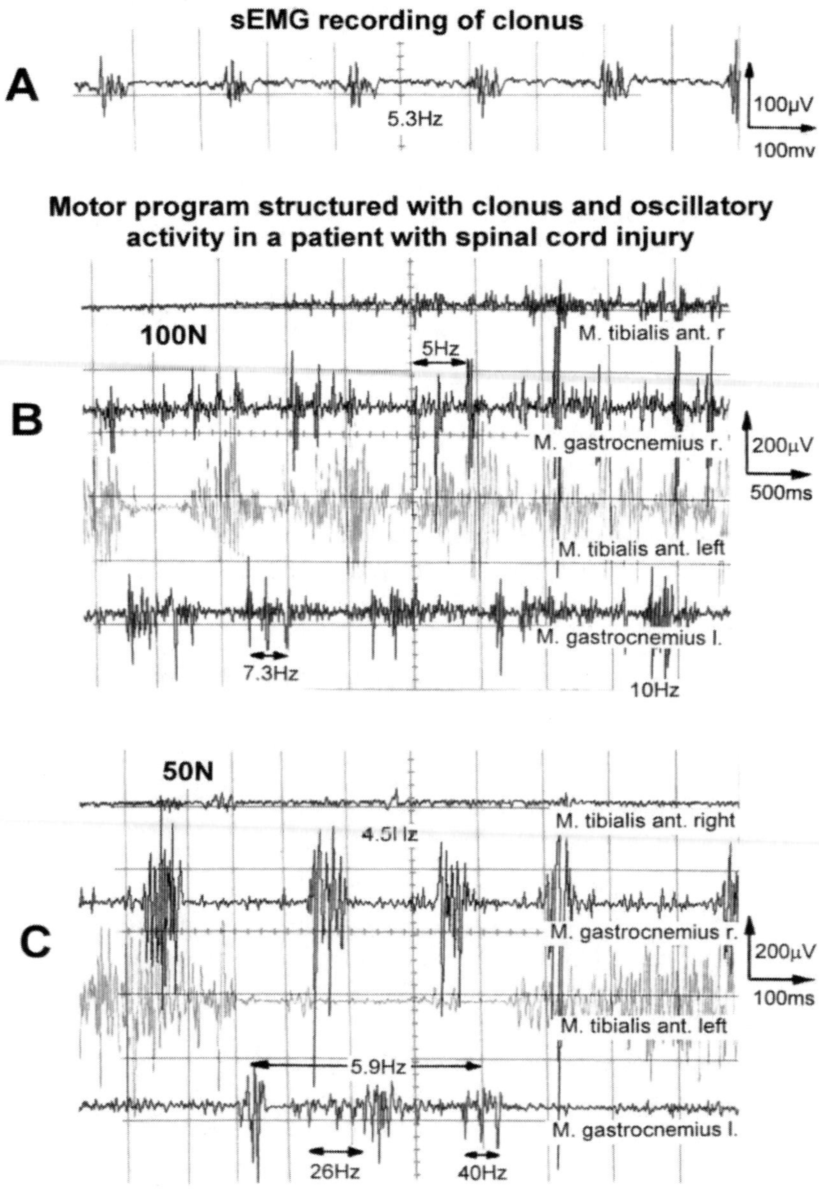

Figure 75. A. EMG recording of a clonus (f = 5.3Hz) in the right tibialis anterior muscle of a patient who suffered a complete spinal cord injury sub Th5/7; the patient was not exercising. B, C. Motor programs of a patient who suffered an incomplete spinal cord injury sub Th4 upon exercising on the special coordination dynamics therapy device at 50 and 100N (medium to high load). In B, motor program bursts are structured by rhythmicity; frequencies of 5 and 7.3Hz are suggested. No motor program in the right tibialis anterior muscle; some motor program structure in the right gastrocnemius muscle. In C (faster sweep), there is no motor program in the right tibialis anterior muscle. Mainly clonus activity at a frequency of 4.5Hz can be seen in the right gastrocnemius muscle. Two physiologic motor program bursts can be seen in the left tibialis anterior muscle (not structured by rhythmicity). In the left gastrocnemius muscle, a motor program burst can be seen, which is structured by 5Hz rhythmicity (clonus frequency, see clonus in the right gastrocnemius muscle) and higher frequency rhythmicity (26 and 40Hz).

14.9. Motor Program Bursts in Patients Who Suffered a Spinal Cord Injury, Structured with Clonus Activation and Rhythmic Firing of FF-Type Motor Units

In patients with Parkinson's disease (probably due to the lack of supraspinal inhibition), and in patients with an incomplete spinal cord injury (probably due to the damage of inhibitory tract fibers) it could be demonstrated that few (in spinal cord injury) and many (in Parkinson's disease) motoneurons start to fire spontaneously or nearly spontaneously oscillatory (Figure 72). In addition, in Parkinson's disease patients the oscillatory firing motoneurons partly synchronized their firing (Figure 73), probably because they are missing some kind of lateral field inhibition. In patients with a spinal cord injury, there was very little or no synchronization in the oscillatory firing motoneurons. Tremors therefore cannot be observed in spinal cord injury patients.

However, marked clonus, which is not functionally dissimilar from tremor (Figure 75A), can often be observed in patients with spinal cord injury. Instead of tremor of the hands and feet marked at rest, the feet perform consecutive dorsal and plantar flexions. The reason for that rhythmic dorsal and plantar flexion is probably rhythmic activation of a very sensitive regulation loop between the feet and the spinal cord rather than synchronization of oscillatory firing motor units.

In Figure 75B a pathologic motor program of a patient with an incomplete thoracic spinal cord injury can be seen. The motor program is poor, because there is no motor program in the right tibialis anterior muscle and the motor bursts in the right and left gastrocnemius muscle are structured with rhythmicity, i.e. with clonus activity. A typical clonus recording not taken during exercise can be seen in Figure 75A.

The uncoordination of motor unit firing can also be seen in the movement patterns in Figure 75. On a stretched time scale, the clonus can be seen clearly in the right and left gastrocnemius muscles (Figure 75C). In the left gastrocnemius muscle, the motor program burst is structured with three clonus bursts, which are in turn structured by rhythmic activity at 26 and 40Hz, which may originate from the oscillatory firing of FF-type motor units innervated by α_1-motoneurons.

Since tremor, clonus, and rhythmic firing of FF-type and FR-type motor units can be seen in motor programs, they all go through the same final premotor network.

14.10. Oscillatory Firing of Motoneurons Originates in the Spinal Cord

Rhythmic firing of FF and FR-type motor units was recorded with no or nearly no volitional activation (Figure 72B-G) in patients with Parkinson's disease and in patients who have suffered a spinal cord injury. These findings support earlier measurements with the single-nerve fiber action potential recording method of α-motoneurons firing oscillatory for high activation, disconnected from supraspinal drive. Recordings from isolated sections of the spinal cord, in patients with complete spinal cord injury taken from below the injury level, showed less regular rhythmic firing (Figure 37). Recordings from the whole spinal cord connected with the lower part of the reticular formation in brain-dead humans showed more regular oscillatory firing (Figures 37,45). In the presence of damage to the pre-existent descending pathways, activation of redundant pathways (and maybe redundant neurons), reorganization in spared descending tracts and building of new pathways, is required for functional repair in spinal cord injury.

14.11. Motor Bursts Structured with Rhythmic Activity

When the patients exercised coordinated arm and leg movements on a special coordination dynamics therapy device, rhythmic activity could be observed in the motor bursts for low load in Parkinson's disease patients (Figure 74) and for higher loads in patients who suffered a spinal cord injury (Figure 75). First, this rhythmic activity during motor bursts supports the conclusion that during the motor burst, highly activated motor units fire in an oscillatory manner. Second, this partly synchronized and therefore uncoordinated uncontrolled rhythmic activity was still partly controlled by coordinated afferent input, as rhythmic firing was only present during the motor burst and was not continuous. In very severe CNS malfunctioning, such firing can also occur between the motor bursts. The partial control of tremor activity was further indicated by the fact that in different muscles, the tremor was mostly coordinated with respect to agonist and antagonistic muscles, depending on the rhythmic movements of arms, legs and fingers (Figure 73B-D).

14.12. Contribution of FF and FR-Type Motor Unit Firing to the Generation of Tremor

There is evidence to suggest that the tremor in Parkinson's disease patients starts with synchronized oscillatory firing of FR-type motor units, innervated by α_2-motoneurons (Figure 65,66), to which the FF-type motor units (innervated by α_1-motoneurons) synchronize, since the tremor frequencies are in the range of the Eigen-frequencies of oscillatory firing α_2-motoneurons ($\approx$ 5Hz), and the α_2-premoter spinal oscillators have a higher stability (Figures 67,68,70). Premotor spinal oscillators probably consist of a neuronal network, with the motoneuron being a part of it [1,30]. α_2-motoneuron oscillators are less dependent on afferent input, and are therefore more stable (Figures 67,68,70,72,73,75). Moreover, the Eigen-frequencies of the muscle-limb mechanics, for example of the arm and hand, are also within the range of 5Hz. It then depends on quantitative effects, i.e. which oscillator type contributes to what extent the rhythmic movement, at which frequencies the arms, hands and fingers, shake. Since the contributions of rhythmically firing FF and FR-type motor units changed, as did the strength of activation, the tremor frequency varied during the measurements. Since FF-type motor units develop more power, substantial tremor movement was only observed if substantial numbers of rhythmically firing FF-type motor units contributed. After the start of tremor with the synchronization of FR-type motor units, first the small FF-type (and further FR-type) motor units started to fire synchronously, followed by larger FF-type motor units with, on average, larger potential amplitudes. This may indicate recruitment according to the size principle among FF-type motor units. When the tremor ceased, the large FF-type motor units stopped firing first, followed by the smaller FF-type ones, and then also the small-amplitude FR-type motor units (Figure 73A).

Only a little information was obtained concerning the contribution of S-type motor units, innervated by α_3-motoneurons, since their motor unit action potentials seem to be too small to be detected safely by sEMG, so far. They even seemed to synchronize their firing with the FR-type motor units. α_3-Motoneuron oscillators fire in the range of 1Hz (Figure 13). Higher-quality recordings are needed to identify S-type motor units, firing safely also by sEMG.

So far, we only have been considering coordination of oscillatory firing motor units. It was observed earlier with the single-nerve fiber action potential recording method that occasionally firing motoneurons had the tendency to coordinate their firing with oscillatory firing motoneurons. The problem arises: how do we measure coordination among occasionally firing neurons or motor units, since they are difficult to identify in the natural impulse traffic of many neurons or motor units? The only feature to identify occasionally firing neurons or motor units is the waveform. The advantage of oscillatory firing neurons is that they can be additionally identified by the rhythmic firing pattern.

14.13. Lack of Inhibition As One Reason for Tremor

It was shown using the single nerve-fiber action potential recording method that α_1 and α_2-motoneurons fire oscillatory, that they can synchronize their firing following repetitive stimulation, and that these oscillatory firing motoneurons can build an external loop to the periphery in the way that γ-motoneurons and muscle spindle afferents are included in the rhythmic coordinated firing (Figure 63) [26]. But the synchronization of oscillatory firing is only transient, and in non-Parkinson patients, the build-up of an external loop to the periphery could only be observed upon strong repetitive reflex stimulation. It is therefore concluded that patients with Parkinson's disease lack inhibition, so that motoneurons can start to fire in an oscillatory manner upon virtually no stimulation (spontaneously); and, secondly, they lack mutual inhibition between oscillatory firing motoneurons, so that oscillatory firing motoneurons can synchronize their firing to give rise to rhythmic muscle contractions and movements, resulting in tremor.

It is likely that the rhythmic firing during tremor starts with α_2-motoneurons because their firing is more stable, and their firing frequency range is similar to the Eigen-frequencies of the muscle-limb mechanics. By building an external loop to the periphery by the synchronously firing α_2-motoneurons primary muscle, spindles are probably also activated, which in turn stimulate α_1-motoneurons to fire. Inhibiting input to the premotor network will probably first inhibit less stable oscillatory firing α_1-motoneurons, followed by the more stable oscillatory firing α_2-motoneurons.

The results of the measurements performed in this work indicate that in patients with Parkinson's disease, and in patients who suffered a spinal cord injury, the inhibition necessary to prevent spontaneous oscillatory firing of motoneurons is missing or impaired. A second type of inhibition is missing in Parkinson's disease patients to avoid synchronization among motoneuron firing.

15. Neuronal Network Learning for Repair in Parkinson's Disease Patients

15.1. Clinical Features of Parkinson's Disease

Parkinson's disease is a complex disorder reflecting malfunctioning of the basal ganglia. Patients with Parkinson's disease typically demonstrate a poverty of movements. That is sometimes addressed as akinesia. Akinesia may involve a masklike expression of the face, stooped posture, shuffling gate, a lack of associated arm movements during walking, and 'frozen' postures. Movements of these patients are slow and frequently ineffective. During many everyday tasks, they have problems switching from an apparently ineffective motor strategy to an alternative one. Hand trembling is another common disorder, making such activities as eating with a fork or a spoon and drinking from a cup very difficult.

Histologically, the brains of patients with Parkinson's disease show a degeneration of neurons in the substantia nigra. There is also a decrease in the dopamine content of the striatum (more pronounced in the Putamen) due to the degeneration of the nigrostriatal connections. Degeneration may also be seen in other areas of the brain. It is believed that Parkinson's disease occurs because of a striatal dopamine deficiency, and this view is supported by the effectiveness of dopamine (L-dopa) therapy. On the basis of the connections between the basal ganglia and other brain structures, one may conclude that removal of dopamine projections from the substantia nigra to the striatum may lead to two types of effects: removal of dopaminergic excitation of the projection to the internal pallidum and removal of dopaminergic inhibition of the projection to the external pallidum. As a result, both direct and indirect pathways mediated by the basal ganglia lead to a decrease in the excitatory input to the brain cortex. This may explain some features of Parkinson's disease such as poverty of movements.

The four basic clinical features of Parkinson's disease are tremor, bradykinesia, rigidity, and deficit in postural reflexes.

Tremor is characterized by 3-6 Hz alternating activity of antagonist muscles controlling a joint, leading to alternating joint movements that can be seen both at rest and during voluntary movements in the joint. The frequency range of tremor is that of the Eigen-frequency of α_2-oscillators.

Bradykinesia usually refers to slowness of voluntary movements and difficulty in their initiation, although deficits in spontaneous and/or automated movements are also sometimes addressed as bradykinesia. It can affect any part of the body and be more or less generalized.

Rigidity is a sustained increase in the resistance to externally imposed joint movements.

Deficits in postural reflexes reveal themselves as decreased anticipatory postural adjustments, and an increase in preprogrammed corrections in the activity of postural muscles associated with voluntary movements, or in response to an external perturbation.

The major clinical features of Parkinson's disease are likely to reflect malfunctioning at different levels. Some are likely to be related to problems with preprogramming. Rigidity and tremor can be observed at rest. As is shown below by surface EMG, premotor spinal oscillators synchronize their firings to give rise to tremor because of missing lateral field inhibition. Inhibition seems to be impaired in general.

15.2. Learning for Repair in Parkinson's Disease Patients

It is difficult to see in detail what integrative repair mechanism could lead to repair of the missing inhibition, which is probably due to cell death, apart from optimizing the efficiency of the remaining inhibiting neurons. On the other hand, it is conceivable that systemic drug therapy to increase inhibition could reduce tremor and improve other pathologic CNS organizations. However, it is difficult to see how systemic drug therapy can distinguish between the lack of inhibition in some parts of the neuronal networks and the preserved inhibition in others, which is highly likely to be different in every patient.

Below it will be shown how coordination dynamics therapy can improve CNS functioning in tremor and in the CNS, in general. The precise capacity of coordination dynamics therapy offered for repair is not entirely known in Parkinson's disease. However, the following may shed more light on this question. The Author could find no electrophysiological measurements of how drugs improve the firing of neurons in the human CNS either. When the Author tried to measure improvement of tremor in a patient upon the first application of L-dopa, the side effects were so strong that CD measurement was not possible. It is conceivable that drug therapy and coordination dynamics therapy (movement therapy), in combination, could be more successful than each therapy alone. Drug therapy would enhance inhibition systemically and coordination dynamics therapy could work specifically in the detail. This may result in the reduction of the necessary drug concentrations and the drug side effects.

The non-invasive measurement of CNS organization by the coordination dynamics [47] to optimize drug application could help to reduce drug concentrations and side effects. The monitoring of CNS organization in Parkinson's disease patients, by coordination dynamics measurements, could help getting the progressive disease under better control.

A mechanism will be described, which reduces tremor in Parkinson's disease patients by coordination dynamics therapy [48].

15.3. Repair Strategy

It will be shown how highly coordinated arm and leg movements, dictated by a special coordination dynamics therapy device (Figure 32), can improve CNS functioning in Parkinson's disease and other patients. The coordination dynamics therapy (CDT) uses the same principle for the re-learning of movements, vegetative and higher mental functions, as the CNS uses phase and frequency coordination for its own organization between neuron firing. When exercising on the special coordination dynamics therapy device, the neurons activated for generating the specific motor program, communicate via phase and frequency coordination with the neurons, which simultaneously generate the pathologic network organization, including tremor, and entrains them by building up coordinated functional connections. So put simply, the physiologic network organization 'catches' the pathophysiologic network organization and entrains it [48]. Entrainment possibly reduces the efficiency and amount of excitation and increases that of inhibition by changing the weight of synapses and building new connections. The building of new inhibitory nerve cells can probably only be achieved following an intensive program of CDT of at least one year in duration.

15.4. Reduction of Tremor Muscle Activity in the Short-term Memory During and after Exercising on the Special Coordination Dynamics Therapy Device

When the patient was positioned on the special coordination dynamics therapy device (Figure 19), but not exercising on it, rhythmic muscle activity was recorded from the right tibialis anterior muscle. The rhythmic activity occurred at a frequency of 3.4Hz and the amplitude was ≈ 200μV (Figure 76A). No activity was recorded from the gastrocnemius, biceps brachii and triceps brachii muscles. When the patient exercised on the device, muscle motor programs were recorded from all of the four above-mentioned muscles (Figure 76B). Motor program activity and tremor muscle activity was recorded from the tibialis anterior muscle (Figure 76B). The frequency of the motor program activity during exercise was f = 0.9Hz, and the frequency of rhythmic tremor activity was now 1.7Hz. The tremor activity reduced from 3.4Hz (Figure 76A) to 1.7Hz (Figure 76B), and coordinated its firing with those of the motor program. The tremor amplitude reduced from 200μV to 100μV and the duration of the tremor activity bursts became shorter.

In conclusion, the tremor muscle activity coordinated its firing with the movement-induced motor program and reduced in general during exercise on the special CDT device. *The tremor activity was therefore brought under the control of the movement activity and was influenced by it to reduce.*

In another recording set from the same patient (Figure 76D), it seemed as if tremor reduction was more pronounced for higher (20N) than for lower load exercise (10N); this is to be expected, since exercising at a higher load will activate the CNS more integratively, so that the influence of the physiologic CNS organization upon the pathologic organization will be greater. With ongoing exercising on the device, tremor reduction could nicely be seen from the reduction of tremor arm movements. At the beginning of exercising, the arm tremor amplitude was large and was afterwards reduced. The patient had tremor of arms, legs, and face on both sides of the body. However, visible tremor reduction need not mean that the rhythmic tremor muscle activity is reduced, since the tremor amplitude could be influenced by movement kinetics. It was therefore important to show through sEMG that the muscle activity leading to tremor really had reduced.

In another measurement (Figure 77E), the tremor muscle activity reduced in the recorded sweep piece; tremor reduced from 5Hz to 1.8Hz.

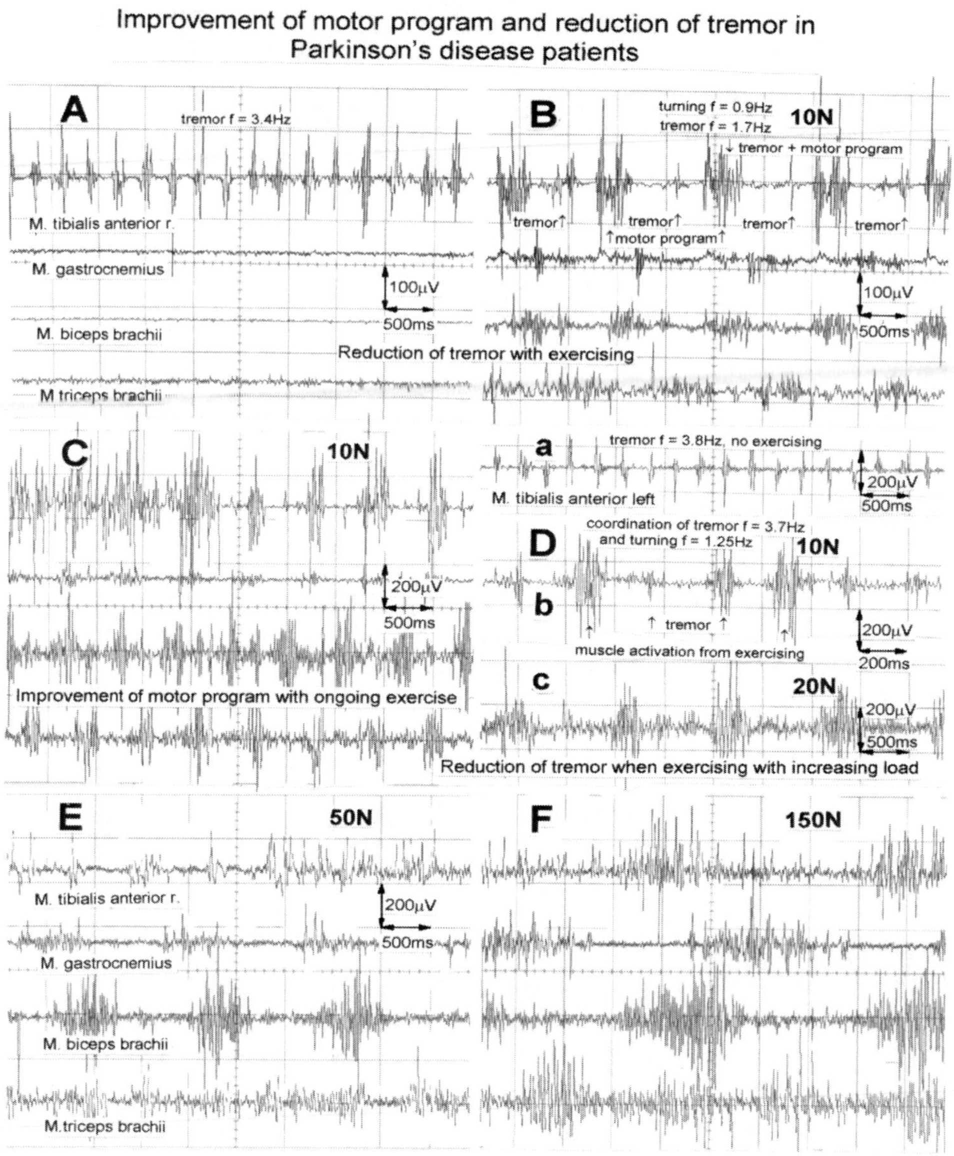

Figure 76. Improvement of the motor program and reduction of tremor in 3 Parkinson's disease patients upon exercising on the special coordination dynamics therapy device. A. Patient is in position at the special coordination dynamics therapy device, but not exercising. Non-voluntary rhythmic muscle activation with f = 3.4Hz is recorded from the tibialis anterior muscle of the right side. B. When exercising on the device at 10N, a motor program appears in all muscles recorded from. The tremor activation in the tibialis anterior muscle reduced in amount and frequency (to 1.7Hz), and the tremor muscle activation synchronized with the motor program. C. Improvement of the motor program in the tibialis anterior muscle with ongoing exercising. D. Reduction of tremor in the tibialis anterior muscle with the increasing load when exercising on the special coordination dynamics recording and therapy device: a. no exercising, tremor muscle activation full present. b. Tremor reduced in amount and frequency (from 3.8 to 1.25Hz) when exercising at 10N. c. No strong tremor muscle activation visible at 20N. E,F. Improvement of the motor program in the tibialis anterior and triceps brachii muscles upon exercising on the special coordination dynamics therapy device, increasing load from 50 to 150N. A,B,D, a 70-year-old female Parkinson's disease patient (U.H.), C, a 76-year-old female Parkinson's disease patient (I.K.), and E,F, from a 70-year-old male Parkinson's disease patient (V.H.).

15.5. Improvement of the Motor Program

In two patients with Parkinson's disease, the motor program improved after a few minutes of exercising on the device (Figure 76). Figure 76C shows that the motor program improved in the tibialis anterior muscle during the sweep piece when exercising at a load of 10N. From Figure 76E,F it becomes evident that the motor program improved substantially in the tibialis anterior muscle when increasing the load from 50N to 150N. Therefore, the motor program improved with ongoing exercise, and the load was increased if the patients could manage it.

15.6. Tremor Changes in Different Muscles

Figure 77A-D shows that the tremor muscle activity changed in different muscles depending on the position, concentration, and other influences. The patient was positioned on the special coordination dynamics therapy device but was not exercising on it. Rhythmic tremor activity can be seen in the tibialis anterior, the gastrocnemius, the biceps brachii and the triceps brachii muscles. The biceps and triceps brachii muscles fired in synchronization with a frequency of 4.6Hz (Figure 77A). The tremor muscle activity in the tibialis anterior and gastrocnemius muscles was coordinated in an antagonistic way (f = 4.5Hz), and the activity of the tibialis anterior muscle was synchronized with those of the biceps and triceps brachii muscles. Some time later there was nearly no tremor activity in the gastrocnemius muscle and the tremor activity in the biceps brachii and the triceps brachii muscles were less rhythmic (Figure 77B). Still at a later point, there was no tremor activity in the tibialis anterior muscle, but more activity in the gastrocnemius muscle (Figure 77C). The rhythmic activity in the biceps and triceps brachii muscles increased again. In Figure 77D, similar tremor activity can be seen as in Figure 77A. The tremor frequencies were slightly different. The higher muscle activity amplitudes were mainly due to a different amplitude calibration. Little tremor movements were observed in this patient in comparison to the patient shown in Figure 76.

In conclusion, it was shown that spontaneous EMG activity was changing in frequency and amplitude with time. Probably, the amplitude change of the rhythmic muscle activity was mainly due to the extent of the contribution from FF-type motor units since FF-type motor unit action potential amplitudes are much higher than those of FR and S-type motor units. The change in the frequency of the muscle tremor activity was due to contributions from all the three muscle fiber types, probably especially from the FR-type motor units since α_2-motor units have the most steady oscillatory pattern.

But changes in tremor frequency may also come from degenerative changes of the ensemble networks generating the oscillatory firing. Premotor spinal oscillators could be shown to partly lose their rhythmic properties because of CNS injury [30] (Figure 37).

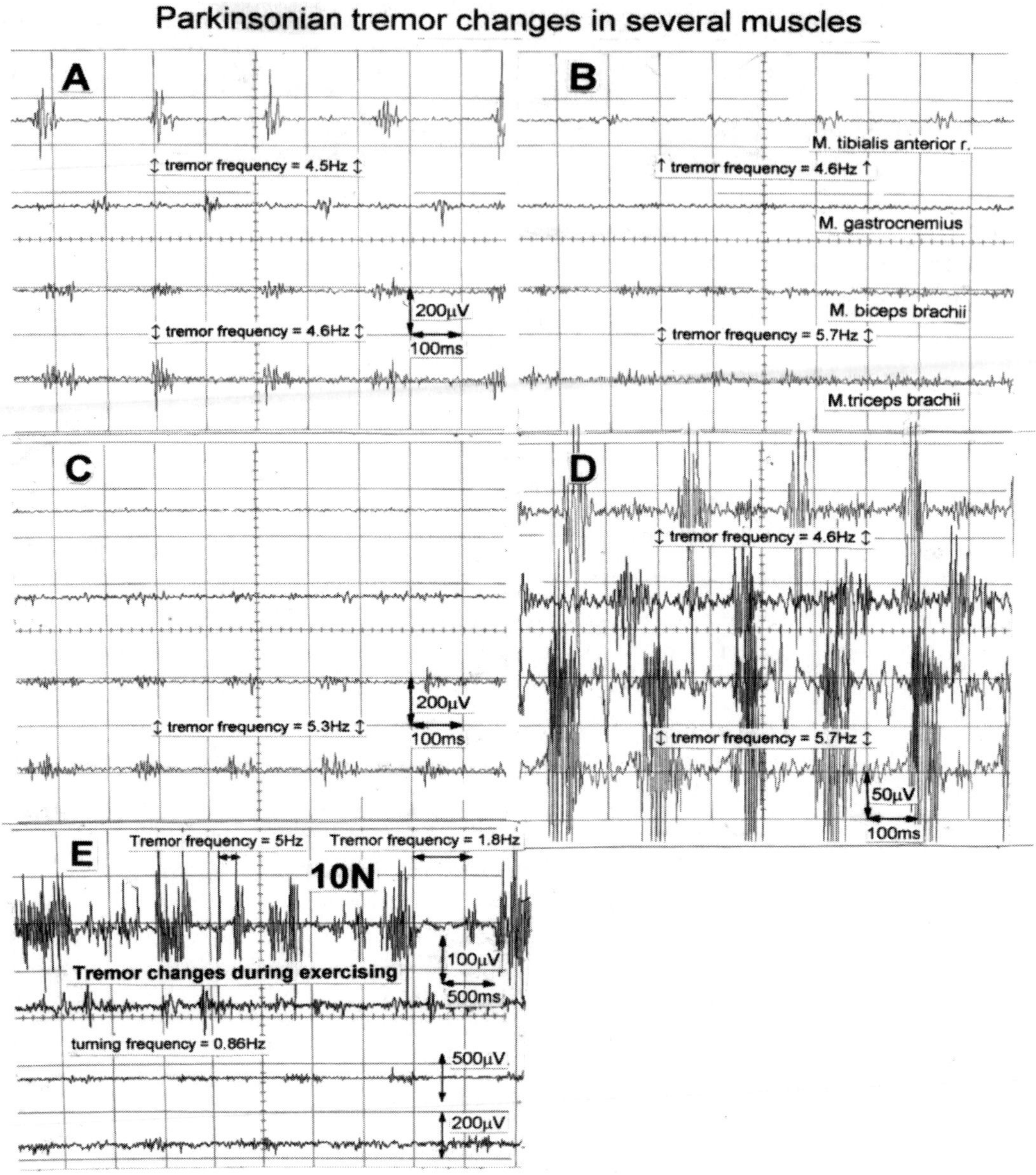

Figure 77. Tremor muscle activation changes in tibialis anterior, gastrocnemius, biceps brachii and triceps brachii muscles in a 65-year-old Parkinson's disease patient (K.V.) positioned at the special coordination dynamics therapy device but not exercising. A. Coordinated antagonistic tremor muscle activation in tibialis anterior and gastrocnemius muscle, frequency = 4.5Hz. Synchronized muscle activation in the biceps and triceps brachii muscles (f = 4.6Hz) synchronized with the tibialis anterior muscle. B. No tremor muscle activation in the gastrocnemius muscle. C. No tremor muscle activation in the tibialis anterior muscle, and activation in the gastrocnemius muscle is approximately synchronized with that in the biceps and triceps brachii muscles, unlike in 'A' and 'D'. D. Tremor muscle activation as in 'A', but different amplitude calibration. E. Tremor changes during exercising (f = 0.86Hz) at a load of 10N. In the upper trace the tremor changes from 5Hz to 1.8Hz during the exercise. In the lower traces only motor program activity can be seen at that amplification. Note that the time calibration in A through D is 100ms, and that in E is 500ms; the tremor activity has a frequency of approximately 5Hz and the motor program, due to exercising, of 1Hz; with a calibration difference of a factor of 5, tremor activity and motor program show some similarity.

15.7. Integrative Organization Mechanism to Reduce Parkinson Tremor (Learning Transfer)

It is evident from Figure 76AB that during exercise on the special CDT device the tremor muscle activity (Figure 76A) became coordinated with the motor program (Figure 76B). The tremor frequency in the tibialis anterior muscle reduced from 3.4Hz to 1.7Hz at a turning frequency of 0.9Hz, and tremor activity bursts reduced in size and duration (Figure 76B), indicating that *less motor unit firings contributed to the tremor activity*, which means that Parkinson tremor had been reduced in the short-term memory.

The question arises as to how the physiologic motor program can reduce pathologic motoneuron activation, leading to tremor. During the exercise on the special CDT device, CNS neurons organized themselves by relative phase and frequency coordination in cooperation with the movement induced re-afferent input and descending volitional input patterns. The uncontrolled synchronized oscillatory firing of motoneurons and their upstream driving interneurons, giving rise to tremor, were influenced by the physiologic network organization (learning transfer), since the activated neuronal networks for the coordinated arm and leg movements and the tremor activity overlapped. It is likely that some interneurons and motoneurons will have even contributed to both network organizations. The neurons, which are directly responsible for the coordinated arm and leg movements and the tremor activity, will be directly entrained because they serve both functions at the same time. Other neurons are only entrained by the functional projections they receive.

It has been shown that different activations can cross in neuronal networks via synfire chains [49,94]. Neurons can therefore serve several functions (organizational states) at a certain moment, if the coordinated firing is sufficiently precise with respect to time and space. In reverse logic, if coordinated firing of neurons is substantially impaired, it is difficult for the neuronal networks to organize several functional states at the same time. Injured or pathologically functioning neuronal networks may be able to generate only one network state at a time or there may be a mixture of states taking place or sudden changes from one network state to another one. A mixture of movement states, sudden changes between different movement states or patients are only being able to do one thing at a time, for example walking or speaking, can be observed in brain-injured patients.

15.8. The Mechanism to Specifically Enhance Inhibition to Reduce Tremor Activity in Parkinson's Disease Patients Is to Use the Inhibiting Neurons More Efficiently through Improving Phase and Frequency Coordination by Exercising on the Special CDT Device

A further question is, why is the network state, which we may call 'highly coordinated moving on the special device' more successful in reducing tremor than other movements? The answer is that because neurons are coincidence detectors, which respond to coordinated input more strongly than to uncoordinated input, coordinated input can reach the threshold for action potential generation with less input (fewer postsynaptic potentials) than uncoordinated input (Figure 40). Also, inhibiting neurons work as coincidence detectors. Therefore, highly coordinated afferent input, generated by highly coordinated arm and leg movements, specifically enhances the efficiency of inhibiting neurons in the short-term memory. Thus, *one mechanism to specifically enhance inhibition to reduce tremor activity is to use the inhibiting neurons more efficiently through improving phase and frequency coordination* in the CNS neuronal networks.

The integrative CNS repair mechanism to improve CNS functioning in Parkinson's disease patients lies in the improvement of phase and frequency coordination of neuron firing, especially of the inhibiting neurons, of the pathologically functioning network parts giving rise to tremor, and its functional integration in the whole CNS neuronal network. By 'linking' the pathophysiologic network organization with the physiologic one, attained during exercise on the special device, the pathologic network organization is entrained and improved.

15.9. Neurogenesis of Inhibiting Neurons by Activating the Inhibiting Mechanism at the Limit

Improvement of the CNS functioning in Parkinson's disease patients was achieved by low-intensity CDT [5] (below), in which neurogenesis contributed only a little or not at all to the improvement. It was not possible to motivate the patients to undertake intensive therapy. But if Parkinson's disease patients could embark on an intensive program of CDT, during which new nerve cells are built from stem cells, then it seems most likely that inhibiting neurons would also be built since the inhibiting mechanism in the CNS networks could be activated. The *physiologic network organization*, generated during exercising on the special device, *could inhibit the uninhibited neuron activity*, which gives rise to tremor (Figure 76B). Neurogenesis is probably induced in those network parts that are most substantially activated [50-53,31].

15.10. Integrative Repair Mechanism to Improve CNS Functioning in General

The improvement of CNS functioning through improvement of the coordinated firing of the neurons with respect to time and space does not only hold for Parkinson tremor reduction. It is also likely to hold for any malfunctioning CNS, since each injury or degeneration of the neuronal networks is followed by an impairment of phase and frequency coordination of neuron firing. Even the timed firing of neurons in the CNS of healthy humans can be substantially improved including those of athletes, since the CNS of healthy humans do not function entirely optimally as can be quantified by coordination dynamics measurements. The connectivity of the CNS neuronal networks is extremely high (each neuron gets on average in the range of 4000 projections from other neurons) that loss of neuron connections, or neurons themselves, will change the timed firing of many other neurons directly or indirectly. Large network parts will change their self-organization, and physiologic network states will become altered and pathologic network states will occur. Improvement of the coordinated firing of neurons will enhance physiologic network states, weaken pathologic network states such as spasticity, and lost physiologic network states such as walking or speaking [1] may re-appear.

The integrative mechanism of human CNS organization seems to have a bigger impact on the repair of the injured CNS than molecular mechanisms such as stem cell therapy.

15.11. The Necessity of Improving the Coordinated Firing of CNS Neurons in Any Case

It is wrongly believed that the human CNS can be functionally repaired by pure pharmacotherapy by inducing growth of new connections between neurons or building of new nerve cells. Whatever repair strategy is used, the coordinated firing of neurons with respect to time and space must be improved or re-learned. A new connection will only help in the organization of neuronal network states if the information is transmitted in coordination with others. Healthy humans also benefit from improvements in the coordinated firing of neurons with respect to time and space during the development stage of their CNS (Figure 92). Their CNS learns the timed firing of neurons by crawling, commando-crawling, standing up, walking, running, jumping, balance exercises and learning of many specific movements, and probably also by employing the higher mental functions, for example through problem-solving.

But following CNS injury, malformation and degeneration, in aging or cancer treatment (radiation and chemo-therapy), it is not enough to improve the coordinated firing of neurons, by movement-based learning or better pattern-based learning. Other parts of the brain must take over function by a functional reorganization (plasticity). The tremendous plasticity of the human nervous system, due to movement-based learning, will be demonstrated below when tackling the rate of re-learning for CNS repair. Even though the learning of the severely injured CNS is approximately 20 to 100 times slower than that of a healthy CNS [51], it seems that every CNS, including the malfunctioning one, can learn and re-learn. It should therefore also be possible to improve the functioning of the CNS in neurodegenerative diseases by learning.

15.12. The Rationale of Measuring Coordination Dynamics in Patients with Parkinson's Disease

The pathology of the coordination dynamics (CD) in the CNS of Parkinson's disease patients was analyzed above, in detail, with surface electromyography (sEMG). Then a repair strategy was introduced in order to partly repair the aberrant CNS function by learning. The physiologic network organization should 'catch' the pathophysiologic network organization and entrain it. One mechanism of repair is that the entraining reduces the efficiency and amount of excitation and increases that of inhibition (see above). But treatment needs objective assessment to quantify progress. sEMG is suitable to analyze pathologic CNS functioning, but because it is not an integrative CNS assessment, it is not suitable for quantifying treatment progress. As introduced above, CNS functioning can be assessed by the Coordination Dynamics (CD) recording trace and then the CD values give a single value for the quality of CNS functioning. The CD must first be measured and then the treatment is addressed. The coordination dynamics will be used below to measure the improvement of CNS functioning through movement-based learning (Figures 113,115).

16. Learning Enhancement by Including Vision, Hearing and Speech Concurrent with Exercising on the Special CDT Device

16.1. Efficiency of Neural Network Learning

The efficiency of CNS learning is determined by a few factors: (1) The accuracy of coordination of the exercised movements, which needs to be within approximately 5ms (Figure 16) to relearn or improve phase and frequency coordination of neuron firing and to reconnect network parts to recouple arms or legs, which may not be moving, for example a plegic arm, and to integrate it into a broader movement like walking. Co-movement (Figure 41) is an example of a reconnection of sub-networks. (2) The increase of the integrativity of coordinated CNS activation is achieved by exercising as many phase and frequency coordinations as simultaneously possible, to entrain more coordinations per instance, to reorganize the CNS more globally, and to activate and coordinate as many neurons and sub-networks simultaneously to also improve the very integrated functions of the CNS, like the higher mental functions. The inclusion of coordinated input from vision, hearing, and instructions during the performance of coordinated movements may enhance the efficiency of learning substantially, especially with respect to the improvement of the higher mental functions such as speaking (the logic of forming sentences) and hearing (see below). (3) The increase of the movement, vision, and hearing induced afferent input, and intentional and instructional impulse patterns to enhance the physiologic self-organization of the injured networks and its communication with the outer and inner world by regulatory processes.

16.2. Enhanced Learning When Combining the Exercising on the Special CDT Device with Speech Therapy

In a patient with severe cerebellum and cerebrum injury, the speech was also strongly impaired. In his coordination dynamics therapy (movement-based learning), in order to repair the traumatic brain injury [12], speech therapy was also included. The speech therapist and the intelligent patient found out that he could speak and learn better when he was exercising simultaneously on a special CDT device (Figure 78). Therefore, it is proposed that from practical learning experience, the efficiency of learning can also be enhanced if the integrativity of coordinated CNS activation is increased.

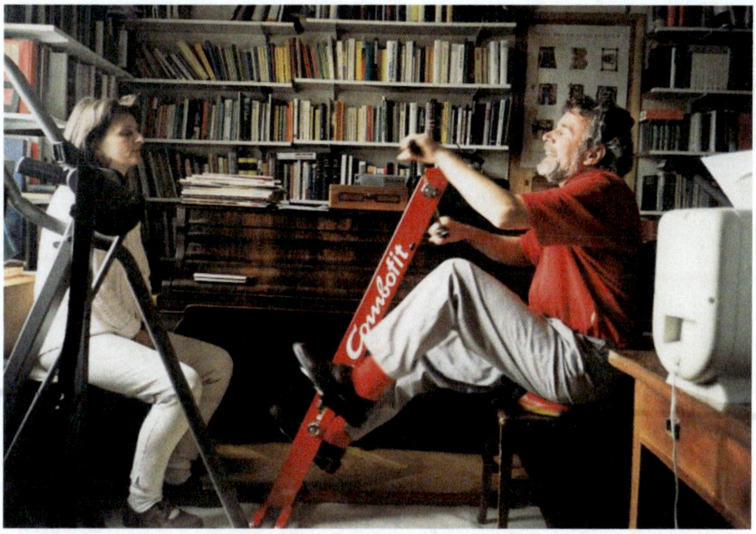

Figure 78. Speech therapy of a patient with severe cerebellum and brain injury. The patient is exercising on a special CDT device to enhance the re-learning of articulated speech.

16.3. Increase of the Integrativity of Coordinated CNS Activation by Including Vision, Speech and Hearing in Addition to Coordinated Movements

When exercising on the special CDT device coordinated visual input is included when seeing the moving leavers and pedals. The visual feedback is analog and the intellect is not involved, because the patient does not need to concentrate on the exercising. The patient turns automatically and volition is only involved for staying in that pattern, like during walking. If the patient is getting the input from flashing lights from the right and left side of the computer display in coordination with the movement, then the visual input is also patterned (Figures 79 and 80). This situation is somewhat similar to jumping on a springboard (Figure 30). Premotor spinal oscillators activating leg muscles are entrained by the Eigen-frequency of the springboard (Figure 64). In the CNS of the patient, the networks for movement and those for visual processing are activated in coordination in some similarity to walking on uneven ground, as for example in the mountains. Walking movements and vision are coordinated, otherwise one could not walk or climb in the mountains. Balance is also involved. A patient with CNS injury can safely train on the special CDT device but cannot climb safely in the mountains.

When 'right' – 'left' flashing appears on the right and left side of the display in coordination with the movement, then right-left orientation is trained in connection with speech comprehension if the patient also reads the words (Figures 79 and 80). If the patient speaks the words in addition, then the networks for speech are activated in coordination with the vision and the movements. Since the patient hears his words, the hearing networks also become activated in addition to movement, vision and speech.

If the patient cannot speak or is mute, the loudspeaker of the computer says the words in coordination with the patient's movements and the patient has to try to speak together with the loudspeaker's sound. Impairments in vision, speaking and hearing can therefore be improved by learning while using this advanced special CDT device, especially when vision, speech and hearing are only partly impaired. A physician with speech problems, following stroke, improved his speech in two weeks when using this device. This learning and re-learning of vision, hearing, and speech in coordination with movements, is especially successful when the network processing is impaired. Incomplete deaf-mutism or amblyopia can be improved.

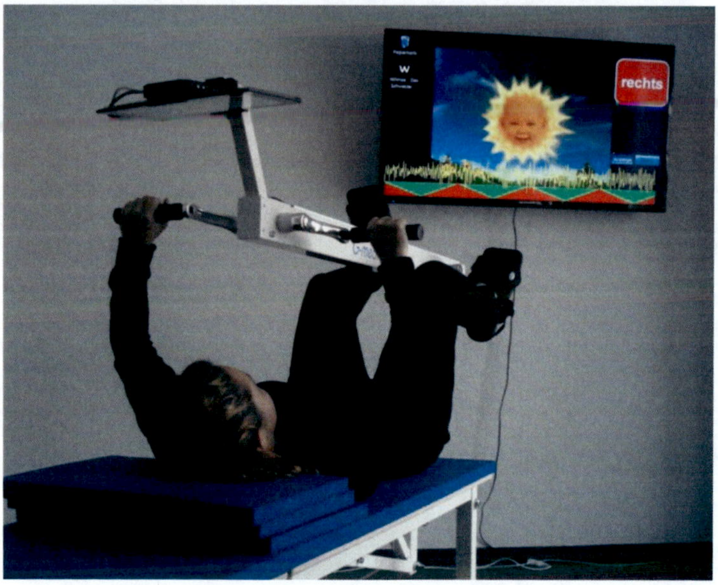

Figure 79. Special CDT device with combined coordinated input from vision and hearing. The sun with the baby face is a video film for improving motivation. The red rectangle 'jumps' from the right to the left side and backwards depending whether the patient pushes with the right or left hand. Below the arrhythmicity of exercising (yellow) and the changes from pace to trot gait and backwards (green and red) are displayed.

When exercising on the special CDT device and including vision, hearing and speech in coordination with the movements, in addition to the somatosensory and motor cortical areas, the primary visual cortex, the auditory cortex and the Broca's and the Wernicke's areas (Figure 87) are activated in coordination with their associated areas and connections. When changing the direction of exercising, the basal ganglia are especially activated because they are concerned with the initiation and facilitation of voluntary movements (see Parkinson's disease). When the patient sees different facial expressions during exercise, such as the laughing baby in the sun in Figure 79, the amygdala is additionally activated. For the importance of facial expression and recognition in connection with social life and development, see below (Figures 144 and 145). When exercising on the special CDT device while sitting on a ball (Figure 140B of [2]), the primary vestibular cortex is activated. When performing the coordinated movements the spinal cord and the brain stem are activated in coordination; especially the ARAS (ascending reticular activating system) is activated. Frankly speaking, almost the whole CNS is activated in coordination when performing CDT. Which CNS parts are preferentially activated and entrained by coordination dynamics therapy

(CDT) must be adapted to the disease of the patient. A supervising educated therapist is needed.

16.4. Synaptic Plasticity for Learning

The intelligence of children with cerebral palsy can be improved, if they are cooperative, when displaying letters, numbers, or terms like house, car or train. This progress in learning is possible, because the different inputs are processed and the corresponding neural networks are activated in exact coordination in the millisecond range (Figure 16). The sub-synaptic potentials have to overlap (Figure 40). The length of a sub-synaptic potential is in the range of 5ms. The sub-synaptic potential (endplate potential = epp) in Figure 125 C is especially long (0.5s) because of the very high membrane resistance of the slow muscle fiber. Presynaptic inhibition will also modulate the length of the sub-synaptic potential. Potentiation of synaptic efficacy by brain-derived neurotrophic factor (BDNF) is greatly facilitated by presynaptic depolarization at developing neuromuscular synapses (Figures 82 and 83, Chapter I of [1]); this potentiation depends on the relative timing of depolarization and reflects an enhancement of transmitter secretion from the presynaptic neuron [96]. As more neural networks are simultaneously activated in coordination, the rate of learning grows. For enhanced learning, the coordinated activation of the CNS must be as integrative as possible. The research on hippocampus learning comes to similar conclusions [51] (see below). Neural networks learn better if they are required to store more memories.

When the patient has to read, and say right or left when the words just appear on the display in exact coordination with the arm and leg movements given by the device, visual perception is strongly enhanced, because the attention is high and attention profoundly alters visual perception [97]. This enhanced spatial-selective attention influences the striate cortex activity during the training. It is reported that the striate and extra-striate visual cortex areas are involved in spatial attention [98].

Sensory experience is crucial in the refinement of synaptic connections in the brain during development, and probably also during repair, by learning following CNS injury. The rapid regulation of postsynaptic glutamate receptors is one mechanism for developmental plasticity and probably also for repair plasticity in the brain. Changes in NMDA receptor expression provide a mechanism by which brief sensory experience can regulate the properties of NMDA receptor-dependent plasticity in the visual cortex [99]. The finding that only one to two hours of visual experience can alter the complement of postsynaptic glutamate receptors in visual cortex in vivo [99] is of considerable interest. Such a change is fast enough to contribute to the rapid synaptic modification to improve CNS organization in the short-term memory when exercising two hours on this advanced special CDT device (Figure 79) with spatial attention induced visual input. In the visual cortex, as elsewhere, the amount of calcium passing through NMDARs can determine whether a synapse undergoes long-term potentiation (LTP) or long-term depression (LTD) [100].

16.5. Cursive versus Block Letters: Analog and Digital Learning

In Finnish schools, the learning of cursive was replaced by block letters. This was not a thoughtful decision. One should be careful not to toy with what has been instilled into the culture over hundreds of years. Writing in printed letters activates different neural network patterns than cursive writing does, and the network learning is different. Although the logic of words and sentences is the same in cursive and block letters, there are significant differences between the two. When we do cursive, the letters are continuously connected. Such continuous connection of patterns may be important for improving intelligence. When being creative one has to simultaneously connect different memory patterns. Being able to do two things at the same time is important in life. For example, to listen to somebody and come up with an answer at the same time is important for communication. Cursive writing may be beneficial to connect patterns or to activate them simultaneously, needed for learning and for doing several things at the same time. Simply walking to work and carrying a bag activates several patterns at the same time. At first, the walking automatism is activated. Secondly, the pattern for holding the bag is continuously activated. Third, we are looking for the pathway. And fourth, we may already be thinking about how to arrange things at work. When walking with a friend to work and discussing things, even more networks are activated simultaneously.

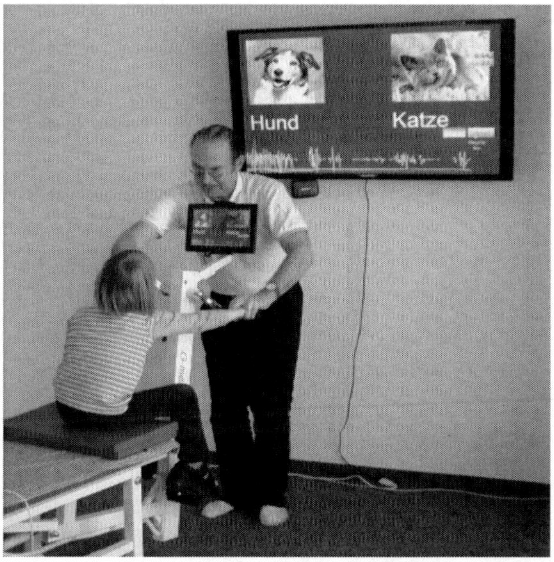

Figure 80. Device-given instructive learning when exercising on a special CDT device. Enlarging the screen gives more visual input. The larger screen can be connected via wireless connection or via an HDMI cable. Visual and auditory input is correlated to the movement. The phases of correlation between turning, visual and auditory input to the CNS can be changed in the program. In this case, the girl with cerebral palsy can see the words "dog" and "cat" in exact coordination with the turning. She could also say dog or cat or other words when appearing on the screen in coordination with the turning. In this moment, the therapist (the Author) is giving the instruction "one – two – three - …" in coordination with the hand movements. As can be seen from the Author's facial expression, he is very suggestive in giving the instructions and the patient is cooperative and is fighting for a good performance. Holding the hands of the patient is beneficial to promote the patient physically and mentally.

From Figure 125a, it can be seen that when changing the direction of turning, the coordination dynamics got better (increased frequency of exercising). Shortly before and after the exercising change, two movement patterns were activated at the same time, the one for forward, and the one for backward exercising, even though only one pattern was executed. Because two main patterns were activated, the quality of CNS organization was more integrative and therefore better. When exercising on the advanced special device (Figures 79 and 80) we have two kinds of visual inputs to the brain, the patterned visual input from the flashing light and the continuous input from seeing the moving arms and legs. So, analog and digital visual inputs enter the brain. Believing in the complexity of learning, one should not give up analog learning.

16.6. Motivation for Learning and Instructive Learning

Especially with children (cerebral palsy, Figure 15), stroke patients or patients in a coma (Figure 118 and 122), vision, speech and hearing are impaired because of the processing in the brain. Additional coordinated input from vision and hearing can improve vision, speech, hearing, and locomotion.

Not only can the integrativity of learning be improved when exercising on the advanced special CDT device (Figures 79 and 80) but also the motivation of exercising can be enhanced. On the tablet computer or computer display (Figure 80), a film can be displayed (CD or video). On the display in Figure 80, the rising sun can be seen. When the child stops turning, the video film also stops. It is astonishing that children with cerebral palsy are more cooperative in turning for a longer period of time, continuously, when simultaneously watching a film. The mood of the patient can also be influenced by the film.

In babies, the face of the mother or the sound of her heart beating can be put into the computer program. The computer offers new possibilities of integrative learning. Instructions can be given by the computer in coordination with exercise. The phase can be arranged in a way to match CNS processing and organization to enhance the physiologic movement patterns. In Figures 25 and 26, a therapist is crawling in interpersonal coordination with a patient. The instructions for learning are given by the therapist at a moment when the brain is organizing that pattern in the patient's brain to influence it. Such instructive teaching, during interpersonal movement coordination, needs a lot of concentration by the therapist. Instructive learning can be given with the help of the computer when exercising on the special CDT device, and is given by the therapist in cases of crawling, jumping, or walking.

Movement-based learning, accomplished by CDT, is successful in repairing the CNS through functional reorganization. However, in severe brain and spinal cord injuries (SCI), structural repair is also required, including the building of new neurons from stem cells. It was shown that new motoneurons could be built in SCI, but only to a very limited extent [3]. Cell physiology and genetics, especially epigenetics, have to be correlated to movement-based learning to further enhance the functional and structural repair of the human CNS.

17. Use of Animal Data on Hippocampus Learning for Repair and Learning in Humans

17.1. Limited Neurogenesis in Humans from Endogenous Stem Cells

Through CDT, a few new nerve cells (motoneurons) can be built from endogenous stem/progenitor cells in a patient with a cervical spinal cord injury. This resulted in the recovery of some finger functions [3]. Exogenous stem cell therapy, by contrast, is unlikely to be effective. This is because the administered cells have to proliferate and have to be integrated into the injured adult neural networks. But this proliferation and integration does not seem to be possible in humans, because physiological neural activity is required to stimulate the membranes of the administered stem/progenitor cells for proliferation and integration to occur (Chapter VI of [1]). Secondly, the communication distance between nerve cells during development and repair seems to be in the range of 0.001mm, which is difficult to achieve during exogenous stem cell therapy (Chapter 1 of [1]). Administered stem/progenitor cells, which are not adequately stimulated for proliferation and integration, will die.

17.2. Excitation-Neurogenesis Coupling in Learning: Comparison between Animals and Humans

17.2.1. Excitation-neurogenesis Coupling in Animal and Human Spinal Cords

Adult neural stem/progenitor cells (NPCs) can intrinsically sense excitatory neural activity and thereby implement a direct coupling between excitation and neurogenesis [51]. This finding in animals is supported by human research; new motoneurons could be built in the human spinal cord, activating finger function [3]. The dorsal flexion of the finger (extension) was generated by dorsal flexion neural networks at or in the SCI's (C5/6) spared neural networks for dorsal flexion of the hand (Figure 81). The excitatory neural activity of the remaining changed network seems to have sensed the repair of the network in which neurogenesis was also involved.

17.2.2. Activation of Excitation-Neurogenesis Coupling Is Sensed in Animals via Ca^{2+} Channels and NMDA Receptors of NPCs

The activation of excitation-neurogenesis coupling is sensed in animals via Ca_v 1.2/1.3 (L-type) Ca^{2+} channels and NMDA receptors on the proliferating precursors. The Ca^{2+} channels are opened via excitatory stimuli. Therefore, NPCs themselves can act as the signal detection and processing elements, mediating adult excitation-neurogenesis coupling [51]. The impact of excitation could also be indirectly felt on the surrounding mature hippocampal cells. There is nothing in human research, treatment (Chapter II of [1]), or human neurophysiology (Chapter III of [1]), which sheds light on how neuron production is achieved. If there are no NPCs in the human spinal cord, the NPCs would have to be stimulated in other locations such as the sub-ventricular zone of the lateral ventricle and transported via the fluid to the spinal cord. To induce migration of NPCs and direct them to the site where they are needed, integrative and specific movements would have to be trained.

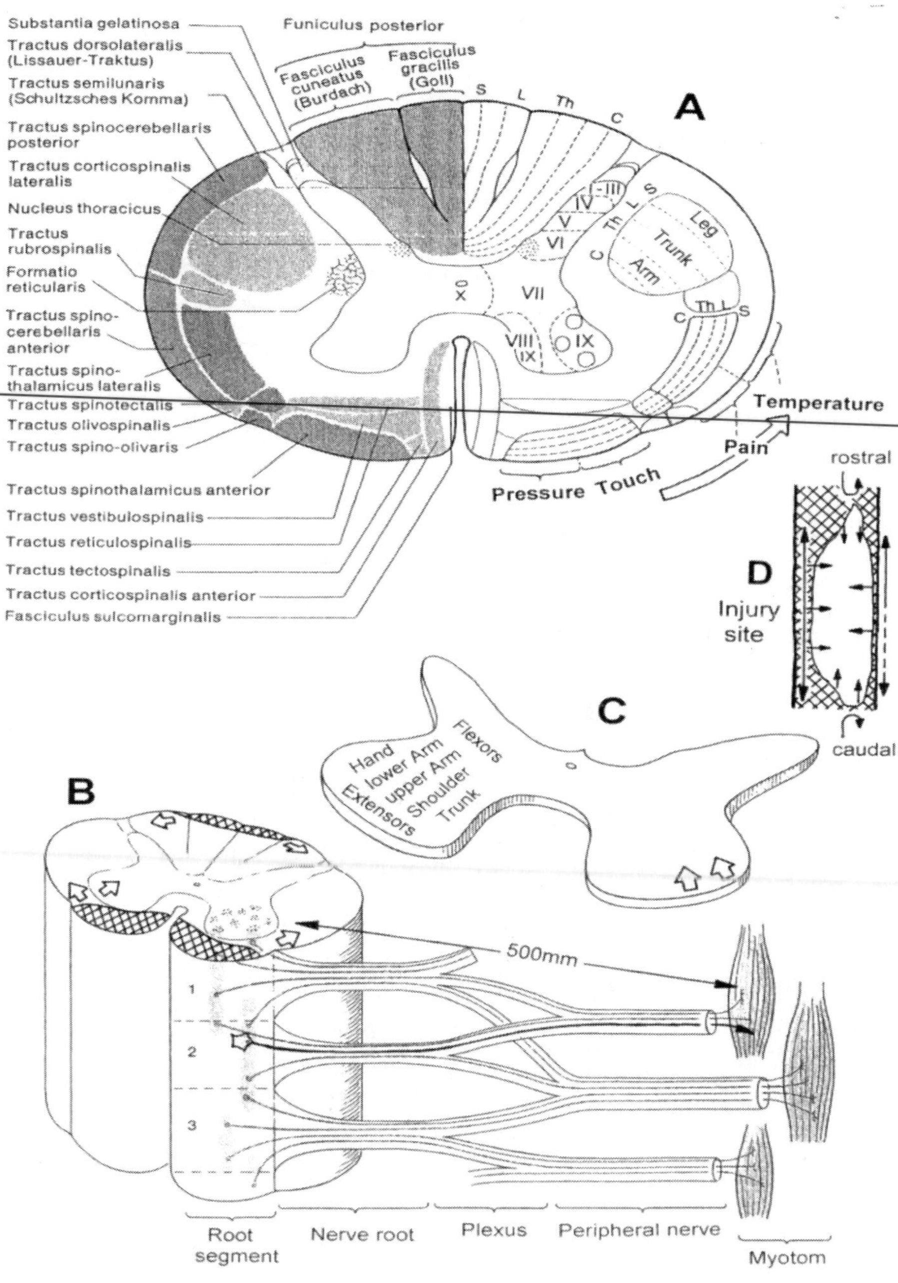

Figure 81. Schematic drawings of the human spinal cord with a cervical injury. A. Spinal cord cross-section with fiber tracts and grey matter. Note the regaining of sensitivity upon CDT from pressure to touch, pain, and temperature is indicated by an open arrow. B. Spinal cord section with suggested spared matter (cross-hatched). Open arrows indicate direction of structural repair. The axonal outgrow of a newly build motoneuron is indicated by a bended and long arrow. C. Motoneuron sites for serving different functions are indicated. Note the extensor motoneuron cell bodies are sited more adjacent to the spared spinal cord matter than the flexors and should be generated first in this case, indicated by open arrows. D. Injury site of the patient. Small short arrows indicate direction of structural activity-dependent repair. The structural repair starts from the activated spared matter into the cavity. The ascending and descending tract activity is indicated by long arrows, and the tract and network activity, rostral and caudal to the injury site, is indicated by bended arrows.

17.2.3. Excitation through Ca^{2+} Channels and NMDA Receptors Modulate Gene Expression

The excitation through Ca^{2+} channels and NMDA receptors act to inhibit expression of the glial fate genes Hes1 and Id2 and increase expression of NeuroD, a positive regulator of neuronal differentiation. The response in phenotype-determining genes occurs soon after the initiation of the excitatory stimulation [51]. There is no information from treatment in humans and human neurophysiology on this matter (Figure 81). The index finger started to work after three years of treatment. The growth of the motoneuron axon from the spinal cord to the muscle activating the index finger requires at least two years. The building of new motoneurons (and interneurons) may need six months to a year in humans.

17.2.4. Sort Survival of Newborn Neurons if Not Integrated in Adult Neural Networks

Many newborn neurons die within the first 2 weeks in vivo [51]. If, in humans, CDT is stopped for more than 3 days, progress will halt and may even go into reverse. Recovered functions start to decline. An incomplete SCI patient (C5/6) trained for six months and got a bit better. Then he stopped treatment for six months and started therapy again. It seemed that he had lost all recovered functions in the interim.

17.2.5. Activity-dependent Neurogenesis Supports the Re-learning of Lost Pattern Functions and Supports Clearance of Post-injury Developed Pathologic Patterns

How does neurogenesis allow the network to adapt to different levels of memory storage demands, which in turn are correlated with the activity level in the network? In this view, activity-dependent neurogenesis may affect future memories or the degradation and clearance of previously stored memories [51]. If memory is seen in the form of patterns of neural activity, activity-dependent neurogenesis supports the re-learning of lost pattern functions and supports clearance of post-injury developed pathologic patterns, like spasticity. Therefore, neurogenesis helps the degradation of pathologic activity patterns. The degradation of pathologic activity patterns includes those of higher mental functions. Following severe brain injury a 25-year-old male patient became aggressive. After 3 months of CDT, the aggressive behavior disappeared completely. He even began to display signs of happiness. Because of the short therapy time, the improvement in mood was probably mainly due to a functional repair by CNS reorganization, not substantial neurogenesis. In a 25-year-old patient with cerebral palsy, aggressiveness declined after 6 months of CDT. In this case, neurogenesis may have contributed to the reduction of the aggressiveness. When CDT is applied, movement, vegetative and higher mental functions always move in the direction of physiologic functioning and back to the patient's pre-existing character. Epigenetics seem to have a supervisory role over the repair in such a way as to ensure that no pathologic functioning occurs.

17.2.6. Neurogenesis Elicits More Rapid Loss or Clearance of Previously Stored Old Memories and the Newest Memories Are Recalled at a Higher Fidelity

(a) Neurogenesis elicited a more rapid loss, or clearance of previously stored old memories, and (b) the newest memories were recalled at a higher fidelity (conclusions from a model [51]). In patients, new neurons with new dendrite and axonal trees can contribute to the reduction of spasticity, which is likely to be mainly due to the network impairment and the loss of neural network complexity due to the injury. This gives an explanation as to why, with CDT, spasticity reduces the long-term memory. However, it is possible that physiologic patterns may also lose stability. Physiologic, stereotyped patterns like the trot gait coordination and walking also have to be trained in addition to complicated coordinations, in order to enhance the complexity and accuracy of neural networks. After continuous repeated performance of the high load test on a special CDT device, the pace and trot gait coordination of the Author weakened as measured by the increase of arrhythmicity of exercising (decrease of pattern stability) at the pace and trot gait coordinations (see below, Figure 125b). But a pain escape walking for years due to a damaged hip will change the walking pattern more. With new neurons, and in turn with higher network complexity, re-learned patterns can be performed more accurately and the variability of the pattern of pattern formation is higher. In poor network complexity, the rate of movement-based learning is smaller (see below) because the pattern variability is poor. New neurons will improve the accuracy and variability of patterns and will thus improve re-learning of patterns.

17.2.7. New Neurons Enhance the Accuracy of Stored Patterns Especially when Networks Had Been much More Active and Many Different Patterns Were Trained

The newest memories were recalled at a higher fidelity, despite the fact that network size was not allowed to increase in the network model. This can be intuitively understood from the lack of a favorable environment for new neurons in old memories. The synapses of the new neurons can be devoted more fully to the newer memories (patterns), which are then more accurately stored. This advantage, that the new neurons with their new synapses enhance the accuracy of the stored memories (patterns), was dramatically greater in networks that had been more active and had been required to store many memories (patterns) [51]. This finding, from a neural network model, offers one understanding of why the patient has to train at his limits (more network activity) and has to train different movements, as the networks are required to store more memories in the form of patterns. Therefore, *training different movements at the limit will increase the quality of the re-learned movement patterns* and will in turn give rise to learning transfer, so that other functions (patterns), which cannot be trained, such as continence and speech, are repaired as well. New neurons open more possibilities for learning and re-learning and most likely serve to increase the rate of repair by learning.

17.2.8. Excitation-Neurogenesis Coupling Is Influenced by Local Activity, Access to Local Activity and Ability of the Local Environment

It is of interest to determine the factors that create the conditions for excitation-neurogenesis coupling to occur. Cells with neurogenic potential exist elsewhere in the brain, including the neocortex, but contribute to neurogenesis only in the sub-ventricular zone and the dentate gyrus sub-granular zone. However, when cells from non-neurogenic areas are removed and transplanted into neurogenic areas, neuronal progeny results [51]. There are a number of factors that may contribute to the neurogenic potential to induce competence for excitation-neurogenesis coupling in the resident population of NPCs:

a. Type of (natural) local neuronal activity.
b. Access of the NPCs to the local activity.
c. Ability of the local environment to induce activity-sensing competence in the NPCs.
d. Factors intrinsic to the resident NPCs themselves.

Clinical research offers no direct answer. The naturally activated local neural networks surrounding the injury site and the cells with neurogenic potential will be important for the growing of axonal and dendrite trees and neurogenesis. It is conceivable that progenitor cells are able to identify natural activity related to repair, which would trigger neurogenesis and proliferation. The frog experiments of Chapter I of *Human Neurophysiology* [1], on development and repair, show how complicated and differentiated the establishment of seemingly simple innervation patterns can be.

17.2.9. Distances of Communication between Motoneuron Axons and Target Muscle Fiber in Frog

With respect to axon guidance and stem cell therapy for repair, it is of interest to measure the distance of communication between cells. Based on communication distances one can estimate then the distances of action of target-derived factors influencing the growth and retraction of axons during the building of innervation patterns. Attractive and repulsive factors between competing neurons for target innervation will contribute to the generation of specific innervation patterns.

Geographical landscapes of diffusing growth and inhibiting factors should be simulated in realistic culture experiments and in administered exogenous stem cell therapies in which stem/progenitor cells are applied and a homing of nerves and other cells (integration of exogenously applied cells into existing neuronal networks) is intended. For generating a specific innervation pattern the internal state of the neurons is also involved. Substances have to be transported in microtubules retrograde to the nucleus in the soma for gene expression change. In the soma, produced substances must be transported actively anterograde, in microtubules down to the nerve ending for axon growing, synapse formation, and functioning.

The communication distance between the nerve endings of the fast (Figure 82b) and slowly conducting axons (Figure 82c) with the slow muscle fiber was approximately 0.1μm; the motor endplates formed a threefold membrane. The distance of action between the target (slow muscle fiber) competing slowly, and fast conducting axons of the two motoneuron populations, was less than 0.1μm (no external membrane between the two axons) (Figures 82a and 83f). The communication distances between nerve and muscle cells are extremely short for this contact attraction and repulsion. For further details, see Chapter I of [1].

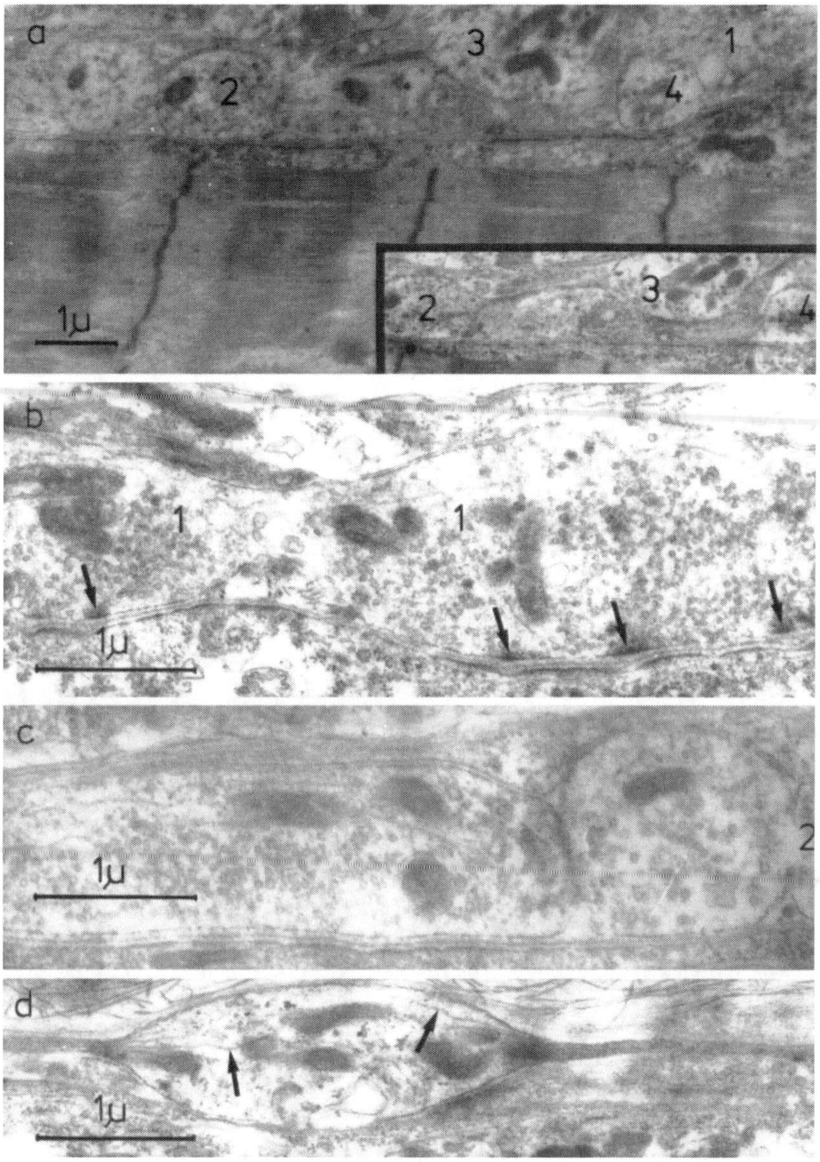

Figure 82. Electron microscopy (EM) of a slow muscle fiber with synapses of a fast (marked with '1') and a slowly conducting axon (marked with '2-4'); pyriformis muscle of a frog at the end of the metamorphosis from the tadpole to the small frog (tail length 0.5mm). By serial sections, including the sections 'a' through 'd', a picture of a part of the slow muscle fiber with the two synapses could be drawn ('e', 'f'). a. Filaments of the slow muscle fiber show no M-line. Endplates of the fast and slowly conducting axons are strongly intermingled; both have contacts with the slow muscle fiber membrane. b. Nerve ending only from the fast axon; active zones are marked with arrows; no synaptic folding opposite to the active zones. c. Synapse profile of the slowly conducting axon; no active zones and no synaptic folding. d. Axon enlargement with marked microtubules.

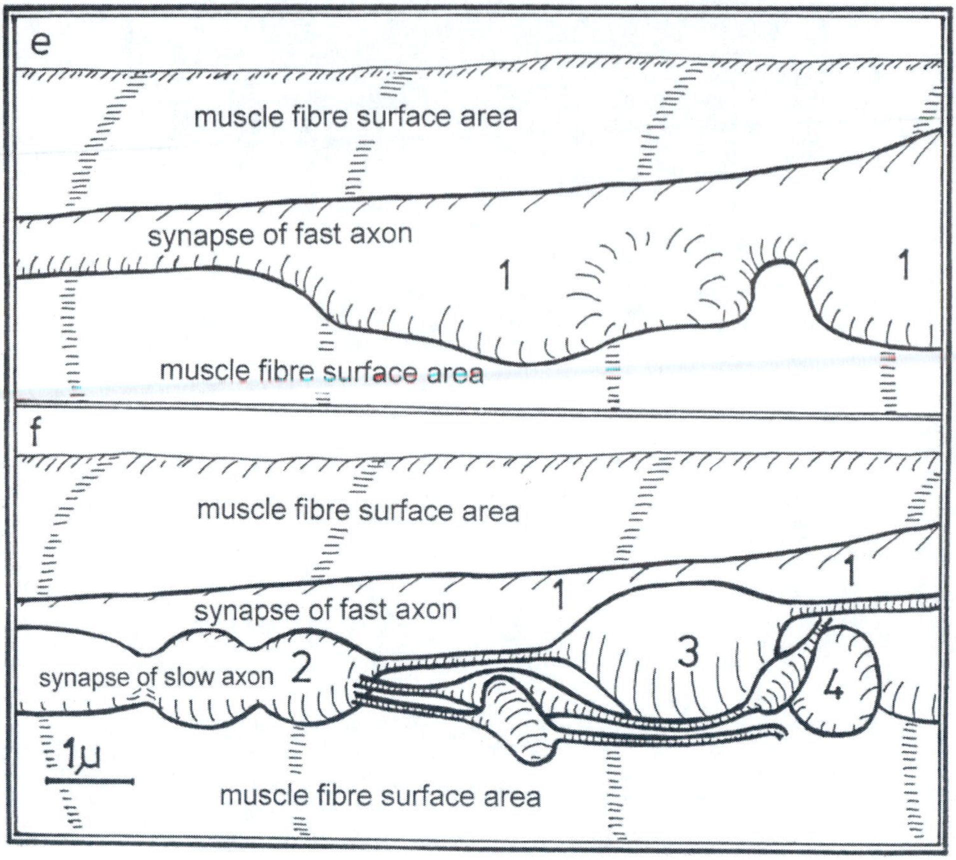

Figure 83e-f. Three-dimensional reconstruction of the synapses of the fast and slow conducting axons attached to the slow muscle fiber; obtained from EM serial sections, including the ones from Figure 82a-d; in 'e' only the synapse of the fast axon is shown; in 'f' the synapses of both the slow and the fast conducting axons are shown.

Exogenous stem cell therapy is unlikely to work because the NPCs have no access to local activity in the range of 0.1μm (Figure 82) and the new local environment may not have the ability to induce activity-sensing competence in the transplanted NPCs.

The by *electro stimulation* induced activity is far away from the natural activity patterns (see for example Figures 6 and 22), and is therefore most likely not inducing competence for excitation-neurogenesis coupling.

Olfactory transplantation, to repair the human spinal cord injury, is unlikely to work for several reasons. The distance of communication between the transplant and the remaining spinal cord in spinal injury repair would be much larger than 0.1μm. The NPCs would have no access to the local activity of the cord. The type of (natural) local neuronal activity in the spinal cord will be different to that in the olfactory nerve even if stimulated by natural movements. It is doubtful whether the local environment has the ability to induce activity-sensing competence in the NPCs of the olfactory transplant for excitation-neurogenesis coupling and growth of axons and dendrites. When tackling reconstructive surgery (below), we come back to excitation-neurogenesis coupling and the growth of axons and dendrites in relation to local activity, access to local activity, and ability of the local environment.

17.2.10. Connection of NPCs to the Activated Networks

The activation of Ca_v 1.2/1.3 channels on adult NPCs potently influenced neurogenesis. The channels open primarily in response to the depolarization provided by excitatory synaptic inputs. But proliferating NPCs in vivo are presumably not synaptically connected [51]. The mechanisms required for the activation of the NPCs are not clear. There is no answer from movement-based learning therapy regarding how the NPCs are stimulated. It can be deduced from improvements in patients that the repair, including structural repair, is constantly adapting to the needs of the specifically activated circuitry. During treatment, the CNS networks are activated integratively, but specifically those sub-networks are included in the activation, which have to be repaired.

17.3. Critical Period Plasticity

Adult-born neurons exhibit the same critical classic period plasticity as neurons in the developing nervous system [54]. When patients have a break in their treatment when training at the limit, their functions deteriorate. This can be quantified by measuring the coordination dynamics (Figures 110 and 113). This experience in patients is in accordance with the transient nature of plasticity. In order not to miss the enhanced synaptic plasticity, which occurs during the critical period in newly generated neurons of the adult brain, the treatment should not be stopped for longer than 3 days, especially when we do not know precisely when the critical periods will occur.

17.4. Neural Network and Pattern Stability

An aspect overlooked in animal research is that neural network patterns need to be stable. Such network stability becomes impaired if the structure of networks is changed. Kelso stated that the stability property of network functions is 'foreign' in research [20]. Recently, it was stated that 'the transient nature of such enhanced plasticity may provide a fundamental mechanism allowing adult-born neurons within the critical period to serve as major mediators of experience-induced plasticity while maintaining stability of the mature circuitry' [54]. In humans, with every structural change, the stability of CNS self-organization is impaired. This must then be re-established by movement-based learning (Figure 84). Chapter III of [1] showed that with CNS damage, neuronal network deterioration could be best observed in the firing patterns of the premotor spinal oscillators. During repair, instabilities can also occur. With every regained function, instabilities occurred, which were quantified by coordination dynamics values and repaired with the ongoing movement-based learning therapy (Figure 84). The incidence of instabilities with network changes is entirely logical. If one changes an electronic circuit by adding a new connection, the circuitry function will change and instabilities may occur, by starting to oscillate, for example.

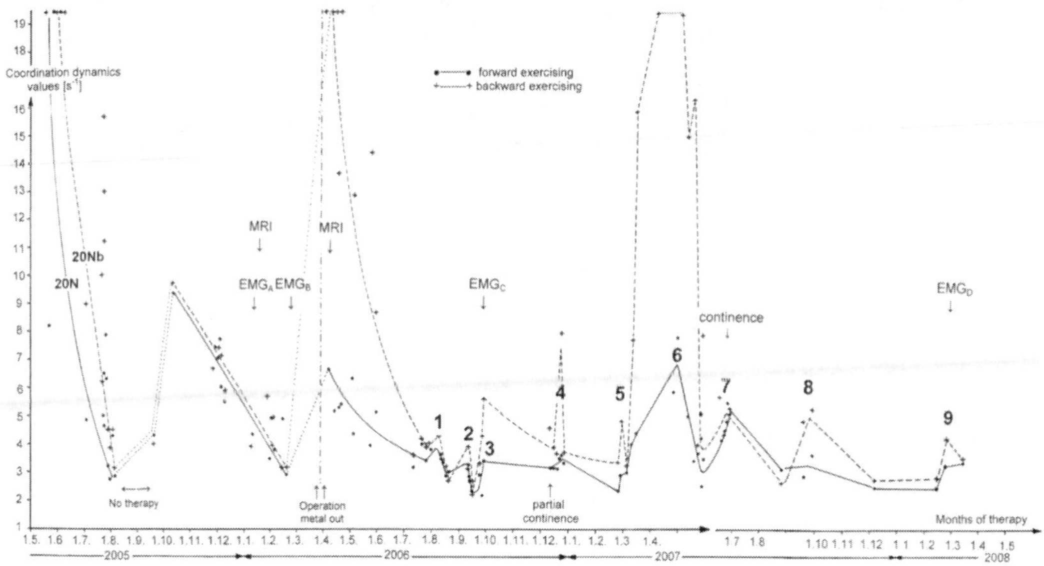

Figure 84. Relation of coordination dynamics values to therapy duration for a load of 20N and for exercising in the forward (lines and dots) and backward directions (20Nb; dashed line and crosses) in a patient with a SCI sub C5/6. Note that with no therapy the coordination dynamics values got worse (increased) and upon therapy they improved again. Upon metal removal the coordination dynamics values increased strongly. The transient coordination dynamics value increases (peaks) '1' through '9' fall together with the re-appearance of certain muscle functions or specific improvements of motor and autonomic functions and indicate therefore, most likely small bits of regeneration. After the large peak '6' of transient coordination dynamics value, increased urinary bladder functioning was re-learned.

17.5. Repair Connected to Blood Vessels

Proliferating cells and putative progenitors in both sub-granular (SGZ) and sub-ventricular (SVG) are closely associated with the vasculature, indicating that factors released from the blood vessels may have a direct impact on adult progenitors [55,57]. "Infusion of vascular endothelial growth factor (VEGF) promotes cell proliferation in the SVG and SGZ, which can be blocked by a dominant-negative VEGF receptor 2 [56]. In addition, VEGF is required for increased neurogenesis in adult mice exposed to an enriched environment or given the opportunity of voluntary exercise, which are both known to enhance adult neurogenesis. Following human SCI, autonomic innervation, especially of the skin blood vessels, is often impaired. Patients often suffer from fragile skin and easily get pressure ulcers, which heal slowly because of poor blood supply. Intensive CDT repairs this impaired blood supply within 6 to 9 months so that the skin is no longer vulnerable and the occurrence of pressure ulcers is reduced. That means movement-based learning is repairing the blood supply and the microcirculation and may have, therefore, an impact on adult progenitors in humans.

When transplanting an olfactory nerve (containing NPCs) into the spinal cord or brain, is this nervous tissue (with many neurons and NPCs) getting enough rapid blood supply not to die? And how quickly would the transplanted nervous tissue get blood supply to secrete endothelial growth factor? When the Author got a "bone, muscle and blood vessel" transplant into the head (see below), the transplant would have died if the blood supply had not been reconstructed within seven hours. Nervous tissue is much more vulnerable to anoxia than muscles and blood vessel tissue.

17.6. Repair Influences from Distant Excited Networks

Neural progenitors can be indirectly influenced by neurons outside the microenvironment that are connected to neurons within the neurogenic niche through neural circuits. Both local and distal neurons can exert direct influences on neural progenitors through the ambient levels of neurotransmitters in the neurogenic niche, or even through synaptic contact with neural progenitors. Therefore, adult neurogenesis is subject to complex extrinsic regulation [58]. CDT tries to activate the different kinds of neural sub-networks in close proximity to the injury site to start the repair from the microenvironment adjacent to the injury site (Figure 81B,D). But by the training of integrated movements, in which the injured circuits are also included in the activation, neurons (networks) outside the injury microenvironment are stimulated to influence the injured networks for repair through neural or neural network connections.

17.7. It Is Learning that Achieves Repair

It is learning and not simply training that elicits the survival effects of new neurons in the hippocampus. Learning appears to promote the survival of newborn neurons in cognitively unimpaired aged rats [59]. Learning elicits different influences on neural precursors at different developmental stages. The regulation of SGZ neurogenesis by hippocampus-dependent learning is complicated and can be affected by factors such as the age of the newborn neurons, the stage of learning and specific learning protocols [58]. When the human patient is exercising on the special CDT device, he should not just turn, but should try to turn more smoothly. He should try to reduce the arrhythmicity of exercising. The patient should try to improve the performance of the movement by learning. During learning, it is essential to concentrate and be aware of what needs to be corrected. It is also important to evaluate the performance by looking at the coordination dynamics curves on the computer screen and correcting errors in successive attempts. When the patient is well-practiced at exercising smoothly, the skill can be accomplished without conscious effort, much like in walking, swimming, cycling or skiing. Simply exercising will also improve CNS functioning. However, the rate of learning is significantly slower. This is the learning of automatic movements, in which the process is subconscious. A problem in some patients, for example in severe cerebral palsy, is that the cognitive functions are so impaired that they cannot understand that they have to learn to improve their nervous system functioning. The hope in such cases is that the simple training improves their cognitive functions to a point that they can understand that they have to improve the performance of the trained movements to improve their CNS functioning.

17.8. Microenvironment (Neurogenic Niche) Permissive for the Differentiation and Integration of New Neurons

In Chapter I of [1] it was shown that the development and the repair were very similar when the two kinds of motoneurons innervated two kinds of target cells (muscle fibers) (Figures 82,83). In adult frogs, a slow muscle fiber can be innervated by two motoneurons. By cutting one axon, the membrane changes its properties in the denervated part. It can now generate action potentials, which cannot be generated in the still innervated part of the muscle fiber. By partial denervation [60], therefore, we obtained innervation and denervation membrane properties in the same cell; but with more than one cell nucleus. In the developing frog, the muscle fiber has an approximate length of 1mm. The distance of action of the neurotrophin is shorter than 0.5mm. In order not to influence neighboring muscle fibers (Ø ≈ 15μm), the distance of action of the neurotrophin, secreted from the motoneuron, must be shorter than 15μm for building up the specific physiologic innervation pattern. It is likely that the distance of action of the neurotrophin is in the range of 1μm, which is governed by the distance between the neuron and the muscle fiber (Chapter I of [1]).

One may assume that a similar detailed neurogenic landscape is present in the neural networks of the human CNS. It is difficult to see how such microenvironment can be generated during the administration of neural stem/progenitor cells for treating patients with SCI or Parkinson's disease. Before treating human patients, the microenvironment should be clarified in humans and grounded on a scientific basis. To induce competence for excitation-neurogenesis and excitation-repair coupling, the following should be explored in humans:

a. What kind of local neural activity is needed?
b. How can the access of the NPCs to the local activity be achieved?
c. What is the ability of the local environment to induce activity-sensing competence in the NPCs?
d. What is the intrinsic state of the transplanted NPCs after injection?
e. Proliferating cells and putative neural progenitors in both subgranular zone of the dentate gyrus and subventricular zone of the lateral ventricles are closely associated with the vasculature, indicating that factors released from the blood vessels may have a direct impact on adult neural progenitors [55]. How can blood vessel supply be achieved for the administrated stem/progenitor cells?
f. Astroglia is supposed to induce neurogenesis from adult neural stem cells [61]. During stem cell therapy in spinal cord injury, the astrocytes will not contribute to the stimulation of neurogenesis, because outside the spinal cord matter there is no astroglia. How can the potential contribution from the astrocytes be simulated in a physiological context?

17.9. The Necessity of Adequate Activation of Networks to the Repair of the Human CNS

New neurons can be built in the animal and human CNS [3] (Chapter II of [1]). Electrical activity has been shown to regulate development in a variety of species and in various structures [62], including the spinal cord and cortex. Within the mammalian cortex specifically, the development of dendrites and commissural axons in pyramidal cells is activity-dependent [63]. Excitatory stimuli act directly on adult hippocampal neural stem/progenitor cells to favor neuron production [51].

In Chapter I of [1], it was shown that there was a close similarity between development and repair in the frog peripheral nervous system. The case reports and group studies of Chapter II of [1] indicate, on the other hand, that there is only some similarity between development and repair in the human CNS.

Assuming that data concerning the animal CNS development, learning and memory formation in the animal hippocampus partly holds in human development and repair, it may be understood how the natural impulse patterns can change the neural networks for repair through movement-based learning.

If it is only the overloaded networks, which induce CNS repair, then those networks, which must be repaired, and those which will take over function must be sufficiently stimulated. The load in neural networks is increased by going to the limits of exercising and by improving the coordinated firing of neurons through improving the coordination of arm and leg movements, since neurons work as coincidence and more generally as coordination detectors (Figure 40). Further, by training complicated coordinated movements, neural network patterns can be reached, which lie deeply in the complexity of CNS organization. For functional repair, especially in spinal cord injury, the function of tracts has to be changed. By training different movements, different kinds of tracts will become overloaded when going to the limits of these coordinated movements. By training impaired functions (with support), the corresponding damaged circuits and the functionally connected networks will become stressed and receive the stimulus for repair. As the building of new motoneurons in a human patient indicates, stressed networks in particular seem to also be the stimulus for neurogenesis. The building of new neurons is a powerful repair strategy, especially when certain kinds of neurons are missing. But as the human research indicates, the neurogenesis is very limited in the human CNS and the benefit from such repair strategy needs more than a year of intensive training to contribute to functional recovery.

17.10. Selective Requirement for Natural Activity in Specific Neurogenesis and in Shaping the Integration of Specific Neurons into Damaged Adult Neural Networks for Repair

It was discovered that 30% of all cortical interneurons arise from a relatively novel source within the ventral telencephalon, the caudal ganglionic eminence [65]. Owing to their late birth, these interneurons populate the cortex only after the majority of other interneurons, and pyramidal cells are already in place and have started to functionally integrate. In mice, it was shown that activity is essential within three days of birth for correct migration, and that after this period, glutamate-mediated activity controls the development of their axons and dendrites [64].

The treatment of patients (Chapter II of [1,2]) and mice data [64] indicate that during development and repair, selective activity is necessary for activity-dependent neuron migration and neuron axonal arborization, and neuron dendrite trees besides weight changes of synapses. Further, functional and structural repairs can lead to the repair of physiologic CNS functioning, but they can also lead to pathologic functioning like epilepsy and cancer. It seems therefore, that for a repair or build-up of physiologic CNS patterns, natural physiologic activity patterns are needed, which are generated by movements, vegetative or physiologic cognitive function patterns. Some natural impulse patterns in afferents, efferents, and sub-neural networks (network oscillators) in humans were identified in Chapters III and V of [1]. To reach and train natural organizational patterns in the depth of phase and frequency coordination complexity, extremely coordinated and complex arm, leg and trunk movements have to be trained, like the movement patterns performed in Figures 14 and 15. For selective activation and repair of the human CNS, it is necessary to first understand how the human CNS functions and how its patterns are changed by the natural impulse patterns of receptors, informing the CNS about changes in the outside world (Figures 6,12,49-51).

18. Epigenetic Modification for Repair by Movement-based Learning

18.1. Epigenetic: Adaptation and Repair by Learning

If genomes were the sole determinant of evolutionary development, humans would not have developed any further than mice, and plants would have as much chance of ruling the world as man. Mice and Arabidopsis genomes share the same number of genes, approximately 25,000. In fact, it is the epigenome that has evolved tremendously since the appearance of the first multicellular organisms [50]. The term epigenetic describes the interplay between genes and environment resulting in phenotypes, epigenetic landscape, and adaptation.

Simple organisms, such as bacteria, increase their rate of spontaneous mutations to enable the survival of species in a changing environment. Multicellular organisms use complex mechanisms coordinated by the central nervous system (CNS) to behaviorally adapt to changing environments without having to undergo the difficult process of mutating their genome [66]. Their behavioral adaptation depends on learning, sub-served by epigenetic mechanisms. Uniquely, humans have the capacity to partially repair their severely injured nervous system by learning. It was shown that in a patient with cerebral palsy, speech could be achieved by learning transfer (Chapter II of [1]). When the patient learned to exercise by herself on the special coordination dynamics therapy (CDT) device, the patient's CNS partly learned from the device's complex coordinations between pace and trot gait. Her CNS repaired to an extent sufficient to generate the speech pattern, even though it was predicted by different speech therapists that she would never learn to speak. Cognitive function also improved by learning transfer. Continence, which is impaired in very severe brain injury, was also fully repaired in the patient.

Animals have a high capacity for spontaneous repair of their nervous system by the growing nerve fibers (Chapters I, IV of [1]). Humans, on the other hand, have a high capacity for repair by learning, including the re-learning of the fundamental principles of CNS organization mechanisms such as 'phase and frequency coordination' among nerve cell firings. A limiting factor for movement-based learning in humans is structural repair.

Although through measuring we know that there is very limited neurogenesis in the human CNS (spinal cord), it is possible to build other neurons (interneurons) through exercising to the *limit of endurance*. This was indicated by the regeneration of the spinal cord from rostral to caudal following injury; surface electromyography showed with the appearance of the first small motor unit potentials that the partial spinal cord repair was probably achieved by neural network repair and not by the regeneration of long tracts such as the pyramidal tract.

Figure 85. Alpenveilchen (cyclamen) painted by the Author's father the painter Edmund Schalow in 1948. Note the lack of material to paint or draw on. It is astonishing that plants have approximately the same number of genes as humans.

To enhance the rate of repair of the human CNS by learning, ways must be found to enhance structural repair, especially neurogenesis. There seem to be two immediate possibilities – the adding of stem/progenitor cells (NPCs) or influencing the epigenetic by learning. In the hippocampus, it was shown that neuronal network activity itself increased neurogenesis [51]. If we assume that cells with neurogenic potential exist elsewhere in the brain, albeit to a lesser extent, then the activity itself guides neurogenesis for repair in accordance with the principle that local cellular environments are important in controlling neurogenesis.

18.2. Gene Expression Pattern Triggered by Excitation in Proliferating Adult NPCs

For weight changes of synapses, growing of axons and dendrites, neurogenesis and building of glia cells during the process of activity induced structural repair, proteins are needed, which must be generated by appropriate transcription from genes. It seems that natural patterned activity-dependent depolarization of the plasma membrane triggers activity-dependent gene expression programs [51]. Ca^{2+} signaling through $Ca_v1.2/1.3$ channels and NMDA receptors can activate a broad array of rapidly responsive transcription factors. Moreover, excitation could in principle induce release of autocrine factors from the NPCs themselves, leading to recruitment of a host of additional signaling pathways to nuclear transcription factors [51].

18.3. Regulation of Epigenetic Modification for Repair by Movement-based Learning

To generate repair in the nervous system, it is likely that permanent changes in gene expression patterns are achieved through permanent changes in chromatin remodeling without changes in DNA sequence. The concept of chromatin remodeling addresses a key challenge of how stable changes in gene expression are induced [66] in neural networks to produce long-lasting changes in repair. DNA methylation is one of the many epigenetic modifications that can alter gene expression. Dynamic and reversible DNA methylation may be essential for learning and memory formation and could transmit repair influences onto adult neurogenesis.

Understanding the complex epigenetic regulation of neural activity and adult neurogenesis is integral to designing therapeutic approaches to restore neurogenesis and cognitive functions. It will also give a tremendous insight into understanding how certain environmental or pathological influences, such as stress, physical activity, depression and epilepsy regulate adult neurogenesis [50].

Research in movement-based learning, on the other hand, has to identify how epigenetic mechanisms can be efficiently modified by the performance of specific, corresponding movements or learning processes to improve development (correction en route, which is particularly salient to the treatment of cerebral palsy) and the repair of the human CNS and to avoid pathologic CNS changes like epilepsy and cancer.

18.4. Movement-based Learning and the Critical Postnatal Period

The establishment of adult patterns of DNA methylation during the postnatal period suggests that critical periods exist after which levels of gene expression are relatively stable [66]. However, movement-based learning in adulthood is capable of targeting the epigenome and altering gene expression and hence repair; but the efficiency of repair may be much lower. CDT should therefore be started as early as possible in the event of CNS injury or malformation. Movement-based learning was administered to a 19-month-old child with a severe CNS injury. The child was nearly blind because of impairments in visual processing, he was afraid of moving and was clearly distressed by doing so. However, exercising on the special CDT device with the mother and physiotherapist was possible. With this administered movement the frequency of crying was reduced, especially when being moved for longer than ten minutes continuously. It was as if his CNS was benefiting from this passive exercise and he was being made to feel more comfortable through the improvement of CNS organization in the short-term memory. Even though it is difficult to treat such patients, it is very important to treat them efficiently as early as possible, because the critical period for improving or repairing vision may be missed and the patient may never have a reasonable prospect of repairing his vision.

Young infants can laugh and cry at the same time CDT. They laugh because they like the improvement of CNS organization and they cry because they are getting exhausted. In order to alleviate such exhaustion, interval training is needed.

18.5. Movement-based Learning in the Prenatal Period

The early environment, consisting of both the prenatal and postnatal periods, has a profound effect on gene expression and adult patterns of behavior [66]. Movement-based learning upon exercising on the special CDT device is normally not possible in the prenatal period. But in premature born babies, the CNS is probably still in a state of prenatal development. A movement-based learning therapy should therefore in principle be beneficial for a repair of the premature CNS, injured for example by hypoxia. Figure 86 shows the special CDT device to administer movement-based learning to a premature born infant. Such a small device may even fit into an incubator.

Figure 86. Special Coordination Dynamics Therapy device for young infants and baby's including premature born ones. Age range of the device 0 to 3 years. Movements can be performed in the sitting (A) and lying position (B). The 18-month-old healthy boy is already quite big for the device.

18.6. Early Treatment in CNS Injury

If there is some similarity between development and repair, and the early environment following injury has a profound effect on gene expression and patterns of behavior, then early treatment following CNS injury should be very beneficial for repair. In spinal cord injury, the treatment should be started in the spinal shock phase; the special device in the lying position is as safe as lying in bed if the trunk is not made to rotate. In stroke patients, the therapy should also start as early as possible, especially to improve blood supply to the brain. Putting a patient with severe brain injury into artificial sleep to reduce oxygen consumption is of doubtful therapeutic value. The improvement of blood supply by early movement based learning therapy may be more successful than trying to reduce oxygen consumption by putting the patient into artificial sleep. Further, through this conservative approach, the therapist partly loses control over the patient. He cannot speak to the patient anymore and cannot monitor the patient's consciousness and brain function. I had a patient with a very severe brain injury. After a bicycle accident, the patient could speak and was able to tell the paramedics his name and address. He was then put into an artificial sleep. When the artificial sleep was terminated, the patient could not speak any more, indicating that the injury had worsened. Children could be taken out of comas sooner following severe brain injury when CDT was administered to them in the coma stage [2].

Below it is reported about a 21-year-old male patient who suffered a very severe brain injury in a car accident. After several operations in a university hospital, he was placed in a rehabilitation center and provided with twenty minutes physiotherapy per day. There was the fear that if the treatment is not sufficient efficient the patient may die because CNS functioning got continuously worse. After three-months of conventional rehabilitation the patient was taken home and CDT was commenced 11 months after the accident. Following 10 months of CDT the patient came out of the vigilant coma 21 months after the accident. By learning it seems possible to even re-learn consciousness.

18.7. Can CDT Influence the Epigenome to Functionally Repair Genetic Defects?

There are two further matters of interest, which concern movement-based learning. CDT has never caused long-term pathologic CNS changes and even in patients with Down's syndrome (trisomy 21), physiologic CNS functioning could be improved. It seems as if CDT is even partly able to repair genetic defects by influencing the epigenome of the patient. A possible mechanism of such a repair could be that the epigenome is modified by movement-based learning in the way that the gene with the defect is not recruited by the epigenome. It seems worthwhile to correlate movement-based learning with epigenome changes. Can the modification of the epigenome by movement-based learning block defect genes and recruit different, compensatory genes?

18.8. Dynamic and Reversible Modification of the Epigenetic Landscape in Comparison to Modification of the Landscape of Pattern Formation

The way of dynamic and reversible modification of the epigenetic landscape is probably essential for learning, memory formation, and repair of the functional organization of the injured CNS. Movement-based learning is one of the ways of changing epigenetic landscapes. This is because during development, movement-based learning is used for developmental "correction en route" and because movements like running, jumping and training balance will have influenced the human genome and epigenome in the last 5 million years, since these movements were essential for getting food, and to escape from predators.

It seems likely that CDT has the capacity to change the epigenetic landscape for repair partly even in defects of the DNA sequence. However, not all defects in DNA can be repaired by changing the epigenetic landscape.

It was shown that speech could be induced in a patient with severe cerebral palsy by improving the phase and frequency coordination of neuron firing, through exercising on the special CDT device. But the Wernicke's and Broca's areas (speech centers) were not destroyed (Figure 87). These special neural networks for speech were probably only damaged functionally (Chapter II of [1]). If these special neural networks for speech were completely destroyed by injury or malformation, the repair of phase and frequency coordination would not have been sufficient for the induction of speech. An enormous number of neurons and their functional integration into the existing networks would have been necessary for repair. Such a repair is for the time being out-of-scope for this movement-based learning therapy.

Eight months after birth it was recognized that this patient, Popi, was disabled because of the occurrence of strong convulsions. The convulsions were treated. Computer tomography and magnetic resonance imaging (MRI) were performed. The given prognosis was that Popi probably would never be able to walk nor speak. One year after birth, the right hand and right leg were not functioning as well as the left. The patient was developmentally delayed and still could not speak. She could only generate vowel sounds.

She did however develop further. At an age of two, she could say a few words. Her level of function then worsened. She stopped speaking. The mother asked for a new MRI. The neuro-pediatrician refused because he thought that nothing would have changed. Following the mother's insistence, an MRI was performed at the age of four. It appeared that the cyst in the cerebrum had enlarged and there had most likely been high brain pressure for one to two years. A shunt was installed. Schizencephaly was diagnosed. The cyst was probably responsible for the high pressure. The mother was advised to have another child. However, the mother started to fight for her daughter. For two to three years, speech therapists from Greece and the USA tried to make the child speak, but without success. They concluded that it was impossible that the patient would learn to speak. At the age of six, CDT was commenced. Speech was induced and motor, higher mental, and vegetative functions improved. When she became able to perform supported jumping on springboard, full continence was achieved. For further details see [1].

If there is similarity between the modification of the landscape of pattern formation of movements (according to the system theory of pattern formation) and the modification of the epigenetic landscape, then changes in the epigenetic landscape can be repaired, but defects of DNA sequences can only be functionally repaired by epigenetic modification (recruitment of genes) in some cases, especially when the epigenome can substitute other DNA sequences for repair. There seems to be some similarity between the organization of the genes, accomplished by the epigenome and organization of the CNS according to the system theory of pattern formation. In genetics, it is still not entirely clear how the problem of how dynamic organization is accomplished. What controls the organization?

There is a similar problem in human neurophysiology. Although we know a lot about single neurons, little is known about how the tremendously complex CNS organization of the many billions of neurons is accomplished, and how it is made to function physiologically.

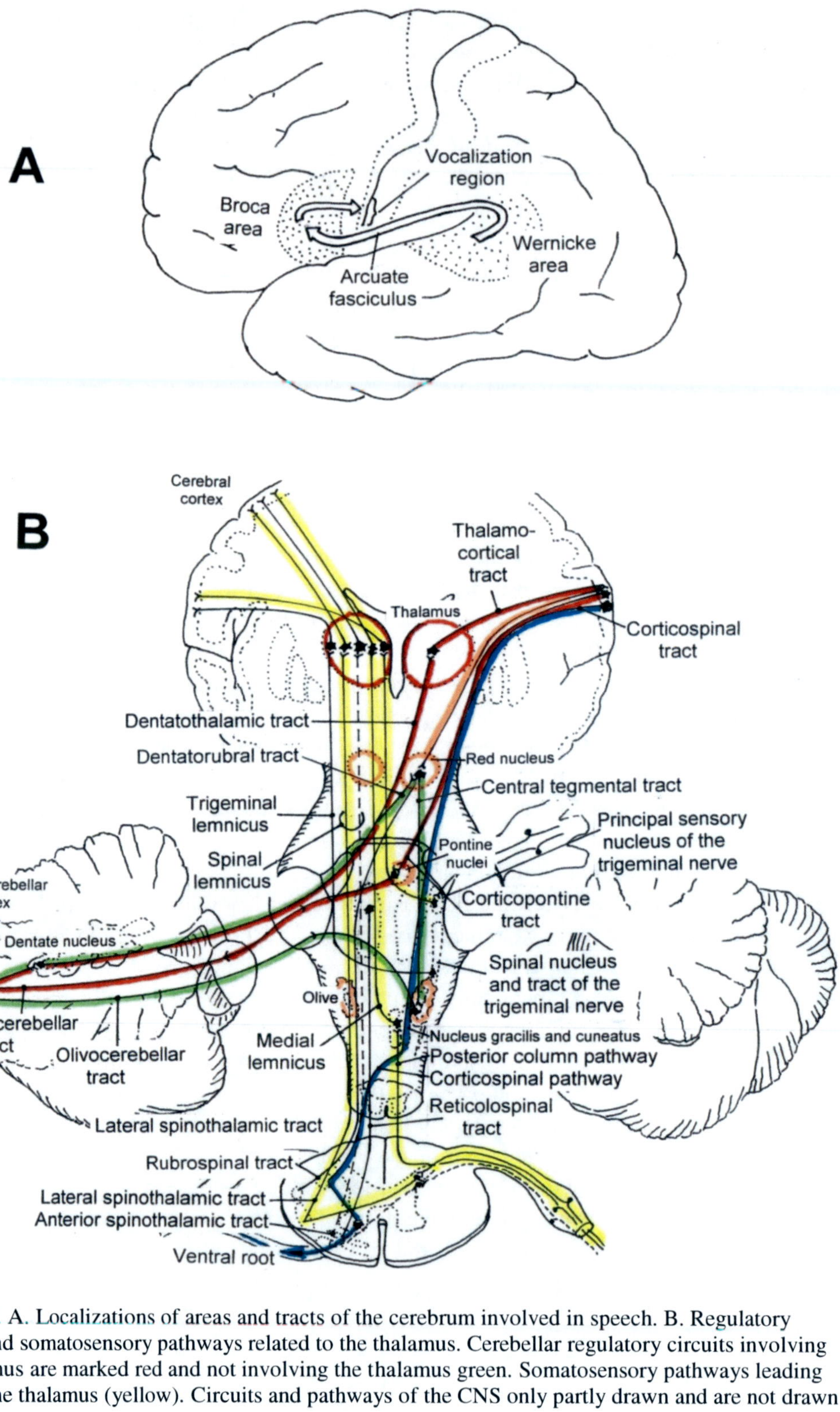

Figure 87. A. Localizations of areas and tracts of the cerebrum involved in speech. B. Regulatory circuits and somatosensory pathways related to the thalamus. Cerebellar regulatory circuits involving the thalamus are marked red and not involving the thalamus green. Somatosensory pathways leading through the thalamus (yellow). Circuits and pathways of the CNS only partly drawn and are not drawn to scale.

18.9. Interaction of Neural Activity and Genetic Programs During Development and Repair

The traditional model of brain development broadly consists of two phases during which a rudimentary wiring of the nervous system is laid out, and a later phase during which the rudimentary connections are refined. In this model, the developmental events that underlie the coarse wiring are the result of predetermined genetic programs and occur independent of neural activity, whereas the refinement is a result of interactions between the nervous system and the outside and inside world. In this traditional model, movement-based learning can be seen as a refinement of impaired development. However, the repair of urinary bladder functions and the induction of speech in a five-year-old girl with a malformation of the cerebrum and thalamus damage (Figure 88) following CDT through learning transfer cannot just be seen as a refinement of this coarse wiring of the CNS. Therefore, this traditional model has to be re-assessed to accommodate the discovery that neural activity and genetic programs interact to form the composition and organization of neural networks during all stages of development [62] and repair.

Neural networks are quite resistant to perturbation, suggesting that redundancy is built into neural circuits to ensure that the natural activity in the networks is maintained. Pharmacological or genetic disruptions of crucial network components lead to the expression of alternative circuit mechanisms that generate activity similar to the natural endogenous pattern [62]. But in severe brain injury, such redundancy is insufficient and movement-based learning by the training of coordinated movements is necessary to support the building and maintenance of natural activity in the circuits. Spasmolytic drugs alter the natural activity in the developing, and in networks undergoing processes of repair, and may not be helpful for improving development and repair.

18.10. Relative Contribution of Cell Intrinsic versus Non-intrinsic Fate Determinants

Even though many of the components of cell identity, such as general intrinsic physiological properties and cell position, are likely determined at the progenitor level, other components such as choice of synaptic partners are likely influenced by local environmental cues.

Therefore, these two mechanisms may contribute to different aspects of cell specification and integration. While earlier decisions, as it were, such as the decision to be a certain neuron are likely established at the progenitor level, later restrictions occur over time as they migrate, integrate into the neural networks, and establish the first contacts with other cells. Genetic profiling of cells at critical developmental stages will bring new insight into interneuron development and subtype specification [39]. A complex set of developmental steps are involved in the integration of interneurons into cortical network. These require the interplay of intrinsic genetic programs and their modulation by cell to cell and matrix to cell interactions.

In Chapter I of [14], it was shown that the targeted slow muscle fiber was twice chosen to be a synaptic partner. The target muscle cell was first innervated by fast growing thick axons of large motoneurons and then slow growing thin axons of small motoneurons (a second group of motoneurons) migrated to the synapses of the fast conducting axons, pushed their synapses away, and established its own synapse of different structure and changed membrane properties and Ca^{2+} secretion properties from the sarcoplasmic reticulum of the target slow muscle fiber. There may be, therefore, several critical periods of innervation and cell intrinsic and extrinsic determinants are intermingled.

Abnormal development of cortical interneuron subtypes during late embryogenesis due to environmental perturbation coupled with genetic abnormalities might represent a primary cause for many neurodevelopment disorders [67]. But the networks can be changed by training certain movements. An autistic child can walk, run and jump, but cannot perform a certain movement for a longer period of time than a few seconds. The property of the CNS to stay in a certain organizational pattern for sustained periods of time is absent. By forcing the child to stay in the movement pattern while exercising on the special CDT device with fixed feet and holding the hands in position during turning (so that the patient cannot take the hands from the handles) and giving instructions during coordinated movements, one can often see in the eyes and the face of the child that the therapy is succeeding in reaching deeply into the CNS and improving its organization in the short-term memory. One can observe a marked reduction in anger and frustration in the child. This is because the therapy renders the child more comfortable within its CNS through the improved organization. The child can learn to sustain a movement or cognitive task for longer periods. It is, however, very physically and mentally tiring for the therapist to ensure that the child exercises regularly.

In Figure 39, running stability on a treadmill was measured in a patient with severe traumatic brain injury. It was shown that the stability of this movement pattern increased with CDT (Figure 40). The stability of the higher mental functions also improved.

Subtle imbalance in the ratio of excitatory versus inhibitory levels within the cerebral cortex may underlie many neurological and neurodevelopmental disorders such as epilepsy, autism spectrum disorders, and schizophrenia [67]. CDT can improve CNS functioning including cases of neurological and neurodevelopmental disorders. It was shown that a young woman improved her mild epilepsy by exercising on the special CDT device [2]. In a young boy, the convulsions reduced, following a temporary increase, after exercising on the special CDT device [17]. Before the discovery of anti-schizophrenic drugs, a working therapy was administered to patients, which only helped a little. CDT is much more effective in repair of schizophrenia than the working therapy of the past.

18.11. Activation of Tumor-suppressor Gene by Exercise

If movement-based learning can suppress genes by modifying the epigenome, it may also be able to activate genes with anticancer effect.

In the long term, if CDT is performed intensively, the CNS can always be made to function more physiologically and efficiently. Unspecific pathophysiologic CNS functions like spasticity or tremor are reduced. In the framework of the system theory of pattern formation, the stability of physiologic functions increased (the potential well became deeper) and the stability of pathologic states like spasticity decreased (the potential well became less deep) (Figure 31). In Parkinson's disease, physiologic network organization should 'catch' the pathophysiologic network organization and entrain it (Figure 76). One mechanism of repair by learning is that the entraining reduces the efficiency and amount of excitation and increases that of inhibition. Since the nervous system is involved in nearly all the body's functions, it could well be that CDT enhances specific physiologic functions and inhibits non-specific pathologic functions. Cancer is an uncontrolled unspecific growing of cells. It may therefore be plausible that movement-based learning has an anti-carcinogenic effect.

It was reported that exercise might exert an anti-cancer effect by arousing a tumor-suppressor gene [68]. Significant changes in the DNA methylation status of 43 genes occurred in women who engaged, after having completed treatment for breast cancer, in a six-month moderate-intensity aerobic exercise program. A correlation between gene expression level and overall breast cancer survival was found in three of these women. One of these three was L3MBTL1, a putative tumor-suppressor gene. In the exercising patients, L3MBTL1 was demethylated, indicating an increase in gene expression. Elevated expression of this gene may be associated with a lower risk of recurrence and improved survival [68].

19. Learning Seen Through Measurements of the Human CNS

Neural network learning is approached from measurements in the human nervous system and not from models of motor learning. The benefit of this approach is that the human nervous system can be improved in its function, and in the case of an injury, it can be repaired. I have tried to find out how humans can live longer with a better quality of life by learning.

Motor learning needs to address several aspects of the problem:

1. What is the learning method?
2. What is being learned in an information processing sense?
3. How can learning be quantified? What cellular mechanisms underlie neural plasticity?
4. Credit assignment problem, which concerns the difficulty of directing training signals to appropriate sites in the networks and at appropriate moments in the learning process, in order for learning to be adaptive.
5. Available learning information for guiding the learning process in the organism.

19.1. Learning Method

The learning method is a movement-based learning method called coordination dynamics therapy (CDT). Different specific movements are trained like the different automatisms and the exercising on a special CDT device. The learning includes vision, hearing and speech, instructive training, and neuro-feedback.

19.2. What Is Being Learned?

The improvement of movements and autonomic functions is learned. Patients learn to walk or jump again and learn to make their bladder function again. Improved cardio-vascular performance is learned. Surface EMG measures improved motor programs. Improvement of sensitivity can be learned and measured; patients report about it. Urodynamics measure the quality of urinary bladder functioning.

Within the framework of System Theory of Pattern formation, the learning is quantified by the improvement of coordination dynamics values. The improvement of CNS functioning is measured via movement pattern change when exercising on a special CDT device. Movement attractor layouts are changed with learning.

19.3. What Cellular Mechanisms Underlie Neural Plasticity?

Little is known about the cellular mechanisms underlying CDT. Surface EMG measures improvement of motor programs during learning. Single-motor unit measurements assess the improvement of coordinated firing of motoneurons. Changes of membrane properties and weights of synapses are not measured. Also short and long-term potentiation and depression were not measured. Neurogenesis of motoneurons can be measured by sEMG and by movement improvement in the patient.

19.4. Credit Assignment Problem

It is believed that the whole CNS neural network learns. The learning is distributed. The inferior olive, premotor networks, basal ganglia, and cerebral cortex might interact with the cerebellar cortex to create an advantageous environment for overall learning. It is not possible so far to direct training learning signals to appropriate sites in the networks. The improvement of the coordination dynamics, which means the coordinated firing of neurons, indicates an improvement of the coordinated firing of neurons in general. The phase and frequency coordination of neuron firing improves and can be quantified by the coordination dynamics values.

19.5. Learning Information Available for Guiding the Learning Process

Learning information is obtained from the coordination dynamics values and sEMG, improvement of movement performance and from the case history (anamnesis) of the learning patient or subject.

Having discussed anatomical, electrophysiological, kinesiological, cell physiologic and epigenetic details of the coordination dynamics therapy (CDT) and the assessment of CNS functioning by the coordination dynamic, we shall now proceed to discuss the rate of learning for repair following the application of CDT in healthy humans and in patients with CNS injury. The rate of learning will be measured and compared in order to better understand the repair of the human CNS. A correlation between neuronal network functioning and movements and other macroscopic patterns of human CNS organization is partly achieved. With respect to the healthy CNS, through CDT, we are coming closer to the dream that the spirit may rise above the body by influencing the neuronal networks (the hardware) through movement and higher mental functions (the software).

Chapter II

Rate of Neuronal Network Learning in the Healthy and Injured Human CNS

Abstract

Based on human anatomy, knowledge of neural network organization from measurements of natural impulse patterns for the communication of the CNS with the outside world with two electrophysiological recording methods, the single-nerve fiber action potential recording method and surface EMG, and using the System Theory of Pattern Formation to quantify integrative functioning of CNS organization, a method for measuring neural network learning of the human CNS was developed. This assessment of CNS functioning is called Coordination Pattern Dynamics. It is measured how well the CNS can organize movement pattern change of a repertoire of specific complicated arm and leg movements given by a special device. This special coordination dynamics therapy (CDT) device cannot only quantify CNS functioning by a single value, but it can also be used for movement-based learning. Especially the phase and frequency coordination among neuron firings can be improved in the healthy human when the subject is exercising on that device. In patients with brain injury, the injury impaired phase and frequency coordination can be improved by exercising on this special CDT device, which means by movement-based learning.

This special CDT device is used to improve CNS functioning by learning and to measure the rate of learning in the healthy and injured human CNS. First, motor learning is measured during normal and deviant motor development by using the low-load coordination dynamics values. Improvement of CNS functioning from five months up to 18 years is measured. With the high-load coordination dynamics values, improvement of CNS functioning is measured in older healthy pupils and in patients with mild and severe CNS injuries and their rate of learning is compared. It is shown that movement-based learning in severely injured humans, the CNS can be reduced by up to a factor of 100. What a healthy subject learns in one year; a patient with CNS injury may need 100 years.

1. Principles of Neural Network Learning

1.1. Recapitulation of General Principles for Neuronal Network Learning

The purpose of coordination dynamics therapy (CDT) is to repair/improve by learning the healthy, injured, or malfunctioning human central nervous system (CNS). The concept of CDT builds upon the theory of self-organization, pattern formation of neuronal networks, coordination dynamics (CD), and human neurophysiology [1,2]. Patterns of coordination are generated by the coordinated firing of single neurons and neuron assemblies and are characterized macroscopically by low dimensional collective variables or order parameters whose dynamics are function-specific. Loss of stability leads to switching of patterns. Fluctuation and differential stability govern the switching dynamics among multiple coordinative patterns. If a certain coordinative pattern has a very high stability, it may seem as if this pattern is 'hard wired' and is generated by a pattern generator. The patterns generated in the neuronal networks are a direct consequence of cooperative and competitive interactions between the intrinsic coordination dynamics of the neuronal networks, the intentional impulse patterns and the movement induced afferent input patterns.

The strategy to improve the functioning of the injured or malfunctioning CNS is partly based on measurements on phase and frequency coordination of firing of neurons and neuron assemblies. This phase and frequency coordination on the neural level, the neuron assembly level and macroscopic level, i.e. movement, is partly lost following CNS injury. Coordination dynamic therapy tries therefore to re-establish phase and frequency coordination between the firings of neurons of the neuronal networks by (re-)learning coordinated movements. Coordination dynamics therapy is therefore a learning method, which is based on measurements in the human CNS and has an attendant neurophysiologic theory.

An indispensable part of a therapy is to have tools available to measure the improvement. The improvement of CNS functioning can be indirectly measured by the improvement of movement performances, and can be measured directly by the coordination dynamics by using the coordination dynamics recording method.

The CDT is based on five new developments/concepts in human neurophysiology: (1) Self-organization of neuronal networks. The self-organization takes place by the relative coordinated firing of neurons with respect to time and space. The CNS is viewed as a neuronal network, which can be changed in its functioning by re-learning. (2) The organization takes place by a specifically changing relative phase and frequency coordination between the firings of neurons. (3) The building of new nerve cells in the adult human CNS, and their proliferation with respect to functional aspects, which is led by learning processes including motor learning of coordinated rhythmic movements. (4) The human CNS can learn and store information relating to all manner of things integratively as a second mode of functioning. This means that brain function is not strictly localized as previously thought. This aspect of brain function can be used to reorganize the injured CNS and it is borne out in the clinical work undertaken in CDT [2]. (5) Each CNS can substantially be improved in its functioning through movement-based learning (CDT).

In severe CNS injuries, the restoration of physiologic functioning is only possible if the relearning is integrative, coordinated, and efficient. The reorganization of the pathologically functioning CNS by relearning can be achieved by: (1) The use of special instruments such as the special CDT device to relearn the lost phase and frequency coordination between the firings of neurons and the lost coordination between the movements of arms, legs and trunk. These movements may relate more to functional reorganization. (2) The training of automatisms, postures and old learned movements, which may remain largely unaffected by the injury. This may relate more to structural reorganization.

The efficiency of CNS re-organization by relearning is determined by at least four factors: (1) The accuracy of coordination of the exercised movements, which needs to be within approximately 5ms (Figure 16) to relearn or improve phase and frequency coordination of neuron firing and to reconnect network parts to recouple arms or legs which may not be moving, for example a plegic arm, and to integrate it into a broader movement like walking. Co-movement (Figure 41) is an example of a reconnection of sub-networks. (2) The increase of the integrativity of coordinated CNS activation by exercising many phase and frequency coordination's as possible simultaneously to entrain more coordination's per time, to reorganize the CNS more globally and to activate and coordinate as many neurons and sub-networks simultaneously to also improve the very integrated functions of the CNS, like the higher mental functions. The inclusion of coordinated input from vision, speech, and hearing, during the performance of coordinated movements, may enhance the efficiency of learning substantially, especially with respect to the improvement of the higher mental functions as speaking and hearing (see below). (3) The increase of the physiologic movement induced re-afferent input and intentional impulse patterns to enhance the physiologic self-organization of the injured networks and its communication with the outer and inner world by regulatory processes. (4) The intensity of therapy. At least four hours therapy per day, 5.5 times per week for six months (at least 15,000 coordinated integrative movements per day) must be performed. The 'adaptive machine' CNS must be forced to adapt.

Instructive training, interpersonal coordination, and co-movements are used to enhance physiologic movements in the patient. The movements, which are often performed during coordination dynamics therapy, are creeping, crawling (commando-crawling), standing from a seated position, jumping on springboard, walking, running and exercising on the special coordination dynamics therapy device in the lying, sitting and standing position.

1.2. Improvement of CNS Organization in the Short-term Memory

Since substantial reorganization of the human CNS needs many months to several years, it is of importance to have diagnostic means available to give the patients a prognosis concerning the time needed to reorganize their CNS. CDT is a learning method. Therefore, many of the ordinary principles of learning will hold. For example, at school it is believed that if a pupil can retain something in the short-term memory, then it will, with many repetitions at least partly go into the long-term memory. The same thinking holds here also for motor learning. If a stroke patient can relearn in the short-term memory, then with many repetitions, the things learned are expected to also pass into the long-term memory. Therefore, when a patient asks whether it is worth undergoing CDT, the CD are measured and this can show the patient that the organization of his CNS can be improved in the short-term memory.

1.3. Increase of the Integrativity of Neuronal Network Learning by Including Vision and Hearing and Speech

The increase of the integrativity of coordinated CNS activation, by exercising as many phase and frequency coordinations as simultaneously possible, to entrain more coordinations per instance, to reorganize the CNS more globally and to activate and coordinate as many neurons and sub-networks simultaneously to improve the very integrated functions of the CNS, like the higher mental functions. The inclusion of coordinated input from vision, speaking and hearing during the performance of coordinated movements may enhance the efficiency of learning substantially especially with respect to the improvement of the higher mental functions, such as speaking and hearing.

1.4. Speech Induction by Learning Transfer from Coordinated Movements in a Patient with Cerebrum, Thalamus, and Corpus Callosum Malformation or Injury

1.4.1. Learning Transfer

By learning transfer not only the trained movements can be improved, but also trained functions like vegetative and higher mental functions. CDT has been able to improve, by learning transfer, autonomic and higher mental functions in cerebral palsy [9] and autonomic functions (cardio-vascular performance, breathing, continence and micturition) in patients with spinal cord injury. Here it will be shown that speech can also be induced in cerebral palsy by learning transfer from movements.

It has long been known that practice of one part of the body in performing a skilled act increases the ability of the bilaterally symmetrical part in the same act [69]. In coordination dynamics theory it was found that patterns of coordination between symmetrical parts transfer spontaneously. The spontaneous transfer of learning points to an abstract nature of learning [70].

The present knowledge of transfer of learning so far is as follows: (a) The amount of transfer seems to be small and positive unless the tasks are practically identical and (b) the amount of transfer depends on the similarity between tasks [71]. Earlier publications and this chapter indicate that the amount of transfer of learning has by far been underestimated at least for patients with CNS injury. Also, the tasks or patterns need not be similar. Jumping on springboard and exercising on a special CDT device, for inducing learning transfer from movements to urinary bladder functions, are quite different network patterns than continence and micturition [15,16]. The speech is also a completely different pattern than exercising on the special CDT device, which mainly induced the speech in the cerebral palsy child, analyzed here.

For the repair of the urinary bladder [15], jumping on springboard contributed substantially to the repair by simultaneously activating the neuronal networks for generating the jumping movement, as well as the continence and micturition patterns; some neurons will also have been involved in both network organizations, mainly sited in the lumbosacral spinal cord. On the other hand, the neuronal networks for generating movements and speech are different, even though sounds and (self) instructions can improve rhythmic movements like walking. To a large extent, the learning transfer will have an abstract nature imminent in the self-organization of the human CNS. Since following CNS injury, the phase and frequency coordination between neuron firing becomes impaired; the improvement of this phase and frequency coordination of neuron firing will contribute substantially to the transfer of learning for CNS repair. When a patient performs very coordinated movements on special CDT devices, his CNS can learn from the device imposed movements, via coordinated arm, leg, and trunk movements, and better coordination between neuron firings for CNS self-organization. This improved phase and frequency coordination not only gives rise to better organization of the trained movement, but other patterns will also benefit from the improved timed firing of neurons, because during the integrative process of self-organization most CNS neurons are involved, not only those that directly generate, for example, the pattern of muscle activation.

Learning transfer is a powerful tool to repair the human injured, malformed, or degenerating CNS. But to further develop treatment (discipline Neurotherapy), we have to understand the organization changes of the human CNS following injury and during repair by learning. Since the thalamus relays and processes input to the primary sensory areas of the cerebral cortex, as well as information about motor behavior to the motor areas of the cortex, including speech muscles, it is likely and shown that the improvement of functional organization of the left thalamus, primarily gave rise to speech induction and not so much to the function improvements in Wernicke's and Broca's areas with their connections. For learning, we also need to have knowledge of the regulatory circuits involved in CNS organization.

1.4.2. Regulatory Circuits and Pathways Involved in the Coordination of Movement, Motor Learning, and Memory Which May Be Impaired upon Thalamus Injury

As will be analyzed, the most severe CNS damage is caused by the ischemic injury of the left thalamus and not by the malformation of the cerebrum due to schizencephaly.

The cerebellum, involved in the coordination of movement, motor learning, and memory, cannot perform its task properly because the regulatory circuits involving the pontine nuclei project through the thalamus (neocerebellar cortex → dentate nucleus → dentatothalamic tract → thalamus → thalamocortical tract → cerebral cortex → corticopontine tract → pontine nuclei → pontocerebellar tract → neocerebellar cortex) (Figure 87B) and were impaired. The exercising of volitional coordinated movements may therefore be only partly successful for repair.

But the cerebellar regulatory circuits for motor coordination involving the red nucleus and the olive (triangle of Guillain and Mollaret) (neocerebellar cortex → dentate nucleus → dentatorubral tract → red nucleus → central tegmental tract → olive → olivocerebellar tract → neocerebellar cortex), and not projecting through the thalamus, was probably fully working in this patient (Figure 87B). This motor learning for long-term memory is mainly subconscious and automatic and was stimulated by automatic movements like walking. The regulatory circuits involving the thalamus, the red nucleus, and the olive (neocerebellar cortex → dentate nucleus → dentatothalamic tract → thalamus → thalamocortical tract → cerebral cortex → corticorubral tract → red nucleus → central tegmental tract → olive →

olivocerebellar tract → neocerebellar cortex) (Figure 87B) was also impaired, because it also projects through the thalamus.

Upon performing coordinated arm, leg, and trunk movements, the neuronal networks of the intrinsic apparatus of the spinal cord (especially in the cervical (arm movements) and lumbosacral enlargement (leg movements)) and the reticulo-spinal system can be activated, controlled, and coordinated by the corticospinal, the corticorubral, and other tracts.

Damage of the thalamus not only impairs the function of the cerebellum; it also impairs the functions of the ascending reticular activation system (ARAS), which contributes to consciousness. Ascending fibers project from the reticular formation to the thalamus and further on to the cerebrum. The reticular formation includes input from the spinal cord (reticulo-spinal system) and cranial nerves of which only the input from the trigeminal nerve is pictured in Figure 87B. If the level of consciousness is lowered by reduced impulse traffic in the ARAS, then the consciousness network state may switch more easily between healthy and unhealthy consciousness attractors. It may become difficult to keep the patient in the healthy consciousness state. To keep a patient in a healthy consciousness state or to keep a patient continuously in a certain network state is difficult in many different brain injuries, because in all CNS injuries the geographical landscape of attractors of patterns are malformed, the attractor basins are shallower, and the network state fluctuations are larger so that the attractor states are less stable and the network state switches easily between the attractors unintentionally. The lack of continuous concentration on a certain task can be explained by a pathologic geographical landscape of attractors in combination with a too high pattern fluctuation.

Since the thalamus is the largest subcortical collecting point and an important center for integration and coordination for all exteroceptive and proprioceptive sensory inputs (Figure 87B); somatosensory deficits will occur due to damage of the left thalamus. The afferent impulses of different modalities, from different regions of the body, are wrongly integrated and false affective coloration may be given. Elementary phenomena such as feeling, pain, pleasure, and well-being, may be changed.

Due to the increased brain pressure for a time longer than one year, in this patient, the reticulo-spinal system may also be slightly damaged. The reticulo-spinal system is the phylogenetically oldest system descending into the spinal cord from the region of the brain stem. This system is contributing strongly to reorganization and has remarkable regulatory effects at the segmental level.

The reticulo-spinal and the proprio-spinal systems are only partly drawn in Figure 87B, even though they are essential for learning transfer from coordinated movements to speech functions.

For the repair of the injured human CNS, the functional anatomy of the human must be taken into consideration. It is not sufficient to think only of a reorganization of neuronal networks or an enhancement of neurogenesis or growing of nerve fibers.

1.4.3. Coordination Dynamics in Speech

Sound emerges as a distinctive and well-known pattern. The many different articulators (tongue, lips, jaw, etc.) whose motion changes the shape of the vocal tract, which in turn creates sounds, allows us to talk and listeners to understand. It takes the coordinated action of about thirty muscles in the tongue, the respirators, and so forth for a baby to say “ba”. It was shown that the relative phasing among the movements of the articulators is a collective variable that characterizes stability and change in syllable structure. A key collective variable turned out to be the relative phasing between the opening of the glottis and the closing of the lips [20,72]. Relevant information for perception of dynamic speech patterns lies in the collective variables that characterize the pattern's generation.

Even though the relative phasing among articulators is crucial for speech production and improvement, the substantial progress to induce speech, and make further speech therapy possible in the patient of this report, was achieved by enhancing the stability of the speech pattern, probably mainly by improving the timed firing of neurons in the left thalamus within the context of the whole CNS.

1.4.4. Speech Induction by Learning Transfer from Coordinated Movements in a Patient with Cerebrum, Thalamus, and Corpus Callosum Malformation or Injury

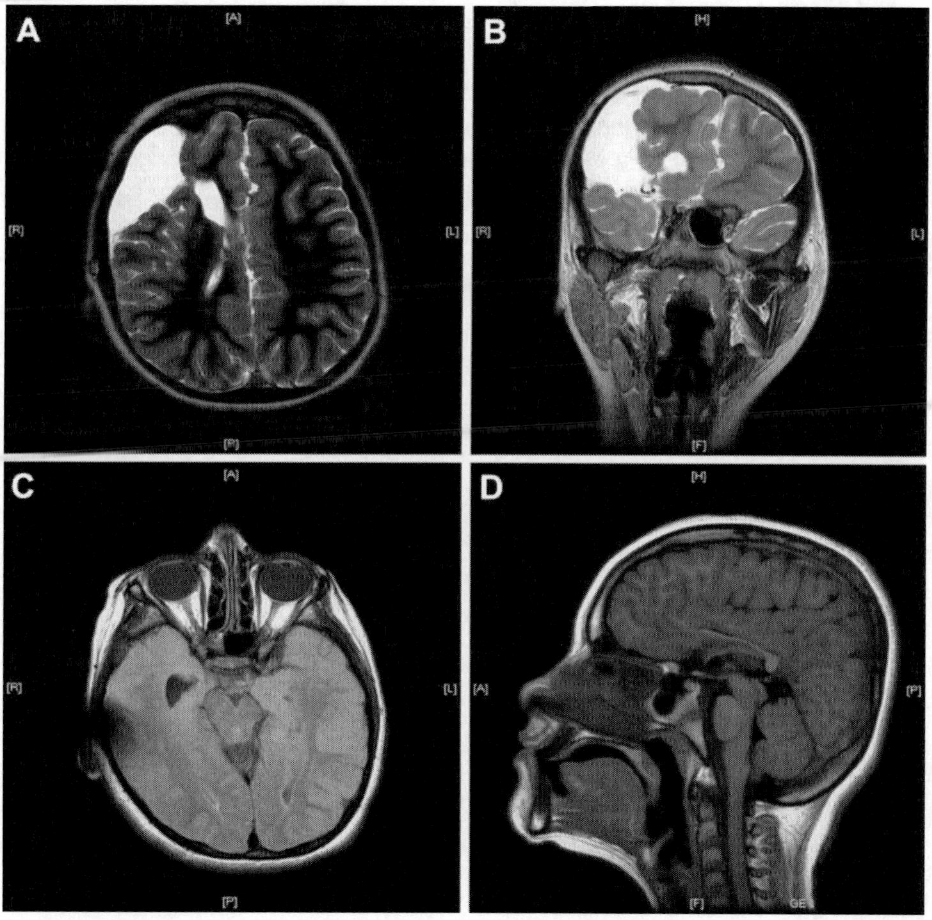

Figure 88. Magnetic resonance images (MRI) of a 7-year-old cerebral palsy child with cerebrum (A,B), thalamus (C), and corpus callosum (D) malformation and/or injury. The cerebrum malformation and/or damage seem to show a schizencephaly. But, as is indicated in the text, the real severe CNS injury with strong deficits in the higher mental functions seems to originate in the damage of the left thalamus (C). The damage of the thalamus may also impair the ascending reticular activating system (ARAS). The third ventricle cannot be seen.

Coordination dynamics therapy (CDT) was administered to a 5-year-old cerebral palsy child with a malformation of the right cerebral hemisphere (dominant side) and corpus callosum, an ischemic damage of the left thalamus (Figure 88), and high brain pressure for approximately 2 years. At the beginning of therapy, the motor functions of the right body side were strongly impaired, most likely mainly caused by the damaged left thalamus and not by the malformed cortex (see below). Her higher mental functions were very low; she could not speak, and was not fully continent. Speech therapists had previously tried unsuccessfully for 2 years to induce speech. Upon an intensive CDT period of 3 months at an age of 6, speech was induced along with improvements in motor and vegetative functions. The neuronal networks of the cerebrum, thalamus, and cerebellum (and other CNS structures) became able to generate the language pattern with ongoing therapy, most likely because the neural coordination among neurons, neural ensembles, and regional specialized sub-networks improved strongly upon instrument imposed, coordinated movements (learning transfer from coordinated movements to speech pattern). Within the framework of the System Theory of Pattern Formation for Repair, the change from aphasia to dysarthria can be understood in the way that the stability of the speech pattern increased (and the fluctuation of the pattern decreased) so much that the patient could speak a hundred words at the beginning. With further therapy the speech, the higher mental functions, and the movements improved further; the patient also became fully continent. The dysarthria, probably due to the impairment of motor coordination of speech muscles, mainly caused by the pathologic processing in the thalamus, is treated further by CDT and speech therapy.

1.5. Learning Is Hampered by the Deficiency of the Neuronal Structure of the CNS

Below it will be shown that the rate of learning depends strongly on the severity of the CNS injury.

The learning process may be hampered by a deficiency of the neuronal structure on which they are dependent. Stimulating the patient to use uncommon modes of operation may enable him to achieve results, which he cannot achieve spontaneously. If an adult patient or child cannot jump in anti-phase on springboard (Figure 30) continuously, he may learn it by including for example a swing in between the jumps. Elements of function must also be learned, which are necessary for more complex motor patterns. If one recognizes which learning processes are deficient, one may then be able to offer the infant a cue to start other learning processes or help him eliminate errors and to progress in the right direction. If treatment strategies become more personalized to the needs of individual infants or children, it may become more difficult to evaluate the effectiveness of any single treatment program. But if CDT would be 10 times more efficient than conventional treatments, differences in effectiveness would be found.

2. Normal and Deviant Motor Development and Repair Following Movement-based Learning

2.1. Neural Network Learning Is Hampered by Deficiencies of Networks and Lack of Network Variability

Neural development must be considered in any analysis of the repair of the human Central Nervous System (CNS). Firstly, there is some similarity between repair and development and the repair strategies should attempt to recapitulate development and learn from normal motor development. Secondly, because of CNS injury or adverse perinatal events, such as hypoxic ischemic encephalopathy, bronchopulmonary dysplasia, or bacterial meningitis we may find retarded, accelerated, or deviant development of motor and other functions. Some functions may not develop at all, while others may only show a decrease in variability. Combinations of these outcomes may occur within single infants and result in individual motor and other patterns. The *deviant patterns are characterized by a lack of variability and assume a thoroughly stereotyped appearance*. The *learning processes in patients* with CNS injury may be *hampered by the deficiencies of the neuronal structures on which they are dependent*. One can expect that the rate of learning of injured CNS is smaller than that of a healthy one. The on-going neuronal network development of the CNS in infancy, childhood, and adolescence is assessed by the Coordination Dynamics (CD), and the function of the healthy CNS improved following Coordination Dynamics Therapy (CDT).

2.2. Longitudinal and Cross-sectional Study

The first part of development is studied longitudinally in two sibling infants using EMG and motor performance assessment (CD). From 5 years of age on, the CD data were collected from pupils from Year 1 until Year 12. Because of inter-individual variability of the cross-sectional study, certain aspects of sequence of development may not come out as pronounced as in a longitudinal study. A general longitudinal study would be better for analyzing development, but it is difficult to organize. Further, the measurement of the CD in the longitudinal study improves CNS functioning and changes the development itself (Figure 92).

For comparing rates of neural network learning a healthy pupil, and patients with scoliosis and severe traumatic brain injury will be considered. It will be apparent that human development cannot be recapitulated by repair. Still, there seems to be some CNS organization strategies (to achieve coordinated firing of neurons, change of motor functions), which are common to both development and repair.

2.3. Neural Development

CNS development is mainly measured in two periods. In the developmental period from six months up to seven years of age, the healthy infant Jürgen has been followed up longitudinally. From 5 up to 18 years of age, CNS development will be assessed by the coordination dynamics of female and male pupils from one school (preschool plus classes 1 through 12). It turns out that the interpersonal variation of CNS development was so marked that in spite of the large sample size, the true development of function and neuronal network complexity became only partly visible. Even though a longitudinal study of motor development of infants may offer more specific data, the evaluation of a developmental period of twenty years for a study is difficult to organize. Combining data from several schools together (even though further data were collected) was not improving the results, since the CNS development seems to be affected by many environmental factors, including the time of the year, the social situation of the pupils at home and at school, the variation of the rate of development, and physical activities.

Later on, long-term repair of the injured CNS is studied in several case reports. It is then attempted to partly compare CNS development with CNS repair to see whether the repair can recapitulate development, and to better understand development and repair of the human CNS.

2.4. Postnatal Development and Repair

With respect to CNS repair through learning, the stepping automatism of the postnatal development is of great importance. The heel strike induced the step automatism of the newborn (Figure 89) partly re-appears following spinal cord injury. The achievement of the partial reappearance of the stepping automatism during supported treadmill walking is one of the first stages in relearning to walk.

Figure 89. Automatic stepping in a newborn infant. The 5-day-old infant, Juliane, performing primary automatic stepping; slight backward posture. The heel of the right foot touched the ground first. B. Infant Juliane, 8-day-old, performing automatic stepping.

3. Intrapersonal Development from Six Months to Seven Years of Age

3.1. Special CDT Devices for Babies and Children up to the Age of Six

It is only relatively recently that special CDT devices became available for babies and children up to the age of six (Figures 15,86). Coordination dynamics therapy can thus be started shortly after birth in both the healthy and pathologic case. If a baby may have suffered prenatal hypoxia during pregnancy, CDT can be started after birth with a special CDT device (Figure 86). Even premature babies can be treated efficiently by movement-based learning.

The clinical research presented below was partly undertaken when small CDT devices were not available.

3.2. EMG Motor Programs in Young Infants

At the age of five months, the infant Jürgen walked more or less in a stepping automatism fashion (Figure 18a). The EMG motor program was not fully developed. The antagonistic action between the right tibialis anterior and right gastrocnemius muscles was not developed (Figure 18b). Three months later, supported walking began to resemble walking somewhat more (Figure 18c), but there was still no antagonicity between the EMG activity of the tibialis anterior and gastrocnemius muscles (Figure 18d). At the age of five, Jürgen could walk nicely by himself and the motor programs likely showed antagonistic action between the tibialis anterior and gastrocnemius muscles. Since the EMG motor program varies among different trials, motor development is better measured by coordination dynamics (CD).

3.3. Neuronal Network Complexity Is Needed for Complicated Coordination Patterns

At six months, the infant Jürgen was too small to be able to turn on the special coordination dynamics therapy (CDT) device. At the age of one, the infant was introduced to the device and he liked it. By the age of two, Jürgen was tall enough to exercise on the baby CDT device and he really enjoyed it. However, he was not able to exercise continuously. He repeatedly got stuck after 1 to 3 turns. He solved the problem himself by taking his hands or the feet from the device and avoiding, in this way, the complicated coordination patterns of arms and legs (with respect to coordination) for which his CNS had not yet developed the necessary complexity (page 92 of [1]). Jürgen's case thus nicely demonstrates that high complexity of neuronal network organization is needed to be able to perform coordinated arm and leg movements with changing coordination between arm and leg movements (pattern change). For turning only with the legs (fitness bicycle), no high network complexity is needed. By the age of three, he was able to exercise on the special CDT baby device (Figure 90A); even though the rhythmicity of turning was poor, especially in the backward direction. By five years old, Jürgen was able to turn on the larger children's CDT device in the forward and backward direction (Figure 90B). The exercising in the backward direction was less smooth. The assessed CD will give further insight into the maturing of the CNS during development.

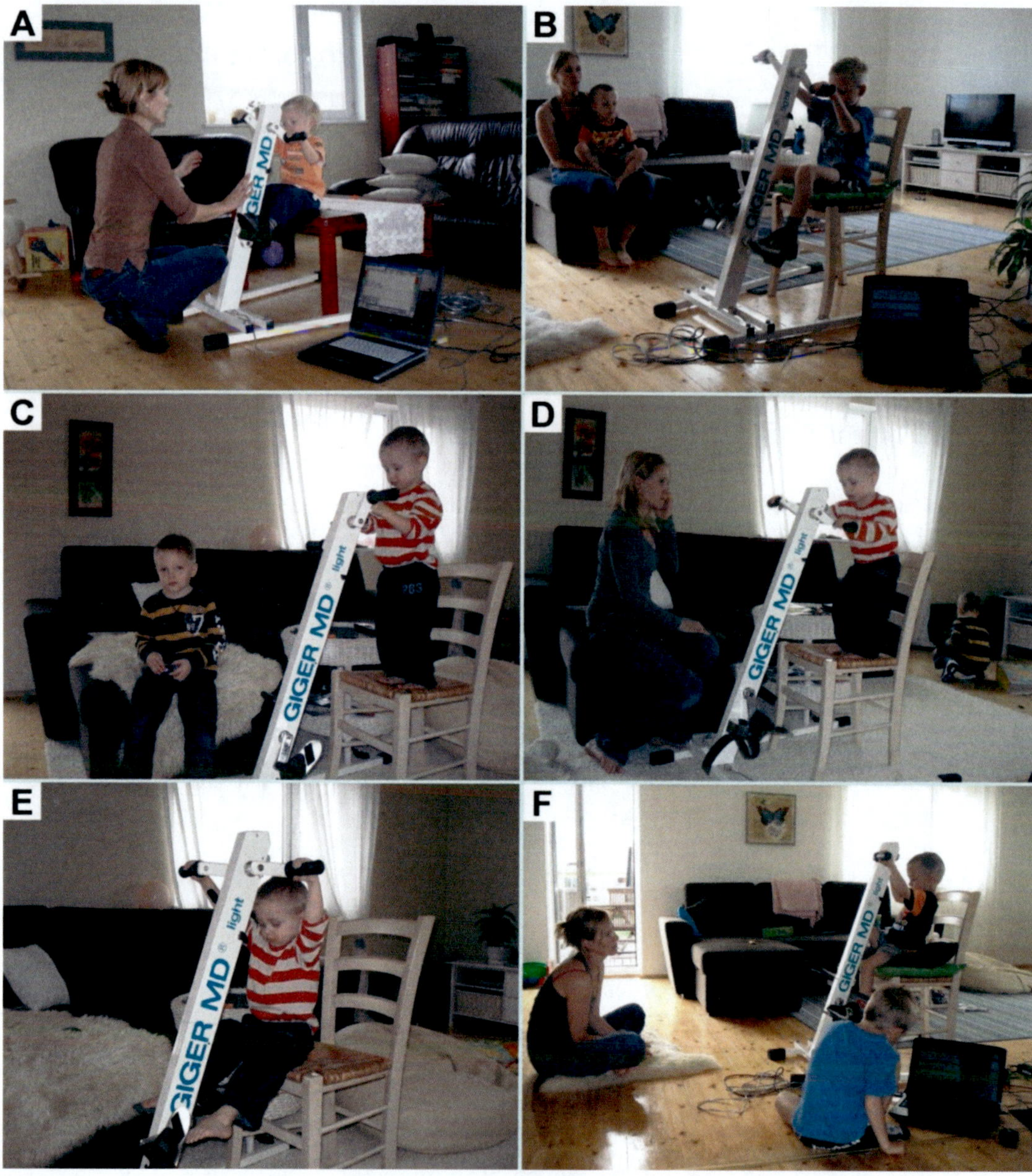

Figure 90. A. Exercising of the 3-year-old Jürgen on a special baby CDT device. B. At an age of 5 Jürgen can turn already on the larger children`s device. B-F. Jonas: Early learning to use a special CDT device. B. From a safe position, 2-year-old Jonas watches how the brother Jürgen is turning. C. First, he pushes all illuminated buttons. D. Then he turns the leavers for the arm movement. E. Now he is trying to also use the foot pedals. He is not depressed that he cannot exercise because he is still too small for this device (a smaller device was not available). F. At an age of 3, he is able to exercise.

The learning situation for the younger brother Jonas, who is four and a half years younger, was different. He observed his brother Jürgen exercising (Figure 90B). Of course, he wanted to emulate what his brother was doing. But the only available children's device was too big for him. So, he climbed onto a chair and pushed the entire device's illuminated buttons (Figure 90C). Then he turned the handles of the device (Figure 90D). And then he tried hard to really exercise with arms and legs (Figure 90E). Unfortunately, the instrument was still too large for him. He could not get his feet fully on the pedals. One can see in the face of the two-years-old Jonas how he is experiencing and learning the outside world when playing with the device. At three years old, Jonas became big enough to exercise on the children's device (Figure 90F) and his CNS could be measured (Figure 91A,B). The now 7-years-old Jürgen is looking on the display of the computer to see how the CD traces change when Jonas exercises. This shows how the brothers learn from each other. Later on, Jonas enjoyed exercise on the Special CDT device in different sitting and walking performances (Figure 14).

With respect to repair of the injured CNS, it is very helpful if the young patient can learn from a brother or a sister and from healthy pupils when going to school. If possible, therefore, cerebral palsy children should attend a normal school and not a school for disabled children. Disabled children have to be motivated and given the opportunity to copy healthy children.

There was actually a problem when assessing the five year-old infants in the pre-school or kindergarten. Some of them were afraid or were too shy and some of them refused to be measured. But in safe surroundings at home with the mother, and motivated by the mother, the measurements can more accurately demonstrate normal functioning of the CNS.

3.4. Coordination Dynamics (CD) Assessment between Three and Seven Years of Age

The generation (increase) of CNS neuronal network complexity during development can be quantified when the child is performing complicated coordinated arm and leg movements and the quality of performance is measured. The child is made to exercise on the special CDT device and the rhythmicity of turning (the quality of performance, the CD) is measured. The best values are taken over a period of 90 seconds of exercise.

Original CD recordings of the three year-old Jonas are shown in Figure 91A,B for forward and backward exercising. For comparison, recordings from Jürgen are shown when he was three years old (Figure 91C,D). It can be seen that Jürgen had slightly better (smaller) values for forward (31.4 against 41.3) and backward exercising (51.2 against 58.4) than Jonas. The probable reason is that Jürgen could exercise more smoothly as he had previously trained a few times on the smaller baby device while Jonas had not. From Figure 91A it can be seen that Jonas could turn around better in the easy coordinations pace (P) and trot gait (K). The frequency of turning was higher.

Figure 91E,F shows original recordings of the now seven year-old Jürgen. The performance of these movements improved strongly in the intervening four years. The values for forward exercising improved from 31.4 to 11.8 and for backward exercising from 51.2 to 15.4, that means they improved by approximately a factor of three. As shown in Figure 92, Jürgen could also turn better than the average pupil of the same age, assessed by the cross-sectional study.

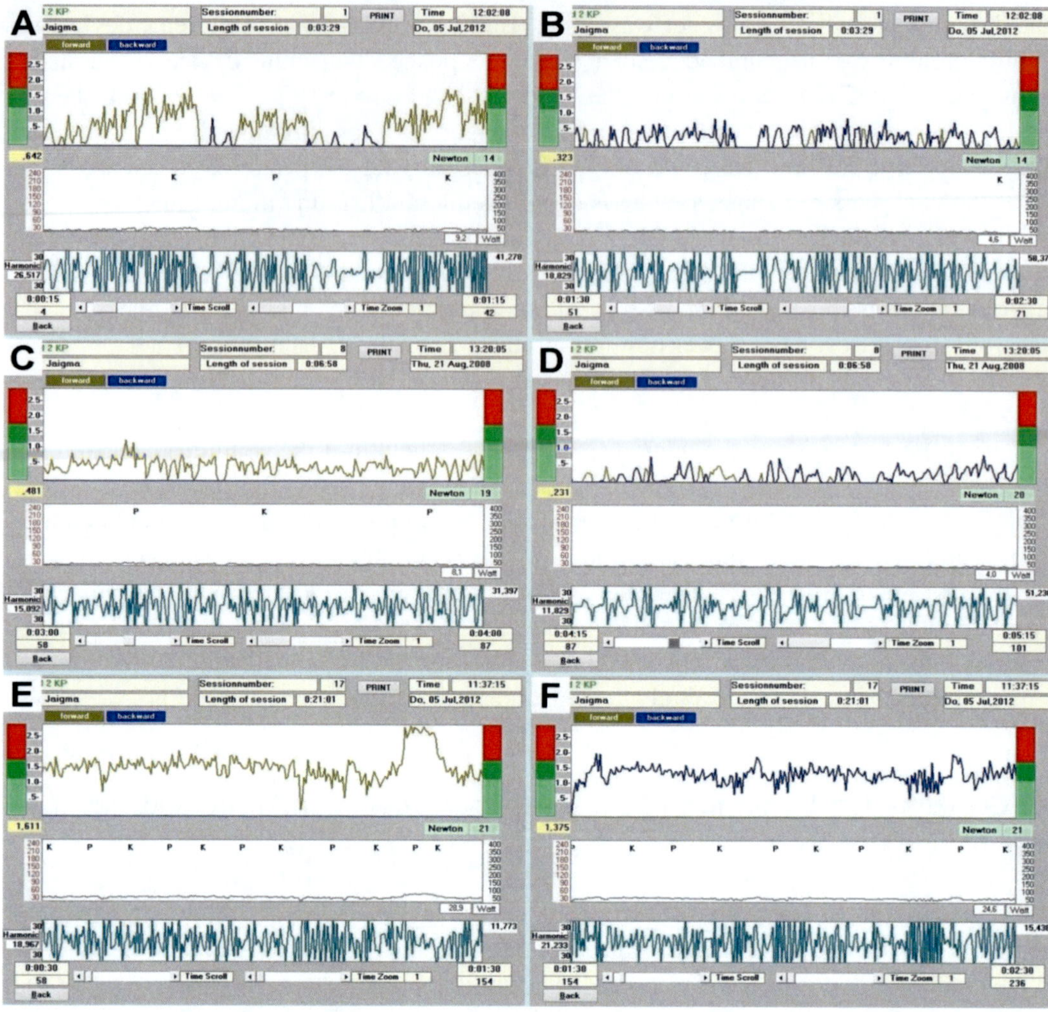

Figure 91. Coordination dynamics recordings of the 3-year-old Jonas (A,B) and the 3-year-old (C,D) and 7-year-old (E,F) Jürgen. Note that the 3-year-old Jonas and Jürgen (A,B,D) made mistakes in the direction of exercising. For example, instead of turning always in the backward direction, they turn a few times in the forward direction (B,D) as one can see from the different color. Notice further the strong improvement in Jürgen's turning (faster and more smoothly) as the age increased, from 3 years (C,D) to 7 years (E,F).

This indicates that if healthy pupils would occasionally exercise at school, for example in sport lessons, on the special CDT device, their CNS would improve its functioning more quickly, which would probably also enhance their higher mental functions by learning transfer. The motivation of the lead staff of the gymnasium to strongly support these measurements was actually to find out how physical performance can be improved in the pupils, and how this physical improvement correlates with higher mental and social functions. A correlation between physical performance and higher mental and social functions could not be found, even though it is believed that physical variability correlates with mental variability: to a healthy mind belongs a healthy body and vice versa. A possible reason for this failure is that CNS functioning depends on many influences. The young lady with a cervical SCI (Figures 19,29) stated after more than five years of CDT that her intelligence was enhanced.

3.5. Transient Rapid Exercise

Before exploring CNS development further through a cross-sectional study, we shall explore an interesting movement performance phenomenon observed in Jürgen. When he became able to exercise sufficiently fast, he also occasionally turned very fast for brief periods (Figure 91E,F). It was as if he was being directed to do so. Such transient fast exercising of children may be a strategy of the CNS to improve its coordinated firing of additional neurons. During repair, the patient can exercise on a special CDT device to improve the impaired phase and frequency coordination of neuron firing. During normal development, the CNS may use the fast performance of different movements, for example fast running or crawling, to continuously improve the phase and frequency coordination of neuron firing, which is necessary because the developing network is in a state of frequent of change and transitions. With the increase of connectivity and complexity changing all the time, the coordinated firing of neurons has to be improved continuously. Following severe cervical spinal cord injury, it was shown that with every bit of regeneration of the spinal cord (re-appearance of function), the CD values became transiently worse (Figure 84). During exercise on the special CDT device the need of the CNS for fast transient movement can be clearly seen.

4. CNS Development between 3 and 18 Years of Age, Quantified by Low-load Coordination Dynamics (Cross-Sectional Study)

4.1. Problems of Assessment

It was shown in Chapter II of [1] that girls and boys have very similar low-load (20N) coordination dynamics (CD) values during development. The values were, therefore, lumped together to get bigger group sizes to better observe the maturing of the CNS with increasing age. The development of the human CNS, quantified by the CD for 20N, is shown in Figure 92. As can be seen from the curves for forward (solid line) and backward exercising (dashed line), the values decreased continuously from 3 to 18 years of age. The decrease is initially dramatic and then only slight. There may have been a transient increase of CD values at an age of 10 to 11 years, which is marked with an arrow. The size of the age groups is indicated by the numbers at the curves; it varies mainly between 32 and 50 pupils per group. At an age of 14, the group size was only 14 because it was difficult to motivate the pupils. The group sizes for 5 and 6 years (6 and 10) are small because it was difficult to motivate the girls and boys at the pre-school or kindergarten; they were shy and a bit afraid in general. As can be seen from Figure 92, the CD values decrease strongly between 3 years and 9 years of age and then only slowly. The evaluation of CNS functioning in the first 5 years of life is most interesting, because of the fundamental CNS changes (Figure 92), but is also the most difficult to obtain. Further, it was difficult to motivate the infants to exercise on the special CDT device for 10 to 20 min because of their difficulties in performing these complicated coordinations between arms and legs. Due to their still under-developed neural network complexity, their CNS had great difficulty in generating these complicated coordinations between pace and trot gait. The control of the supraspinal CNS centers onto the spinal cord neural networks is not sufficiently mature to vary them with enough sophistication to permit them to make these complicated movements. The infants also could not understand why it was so difficult to exercise the instrument; it looked so easy. Other movements, once observed, were quickly learned. Parents are needed to motivate the child.

4.2. Enhancement of Normal CNS Maturation upon Exercising on the Special CDT Device

As can be seen from Figure 92, the CD values of the infant Jürgen are better than those of the average child, and were also better than the values of his brother because he had exercised on the special CDT device several times before. The functioning of his CNS improved faster because of the exercise on the device. Therefore, CNS maturation can be influenced by movement-based learning to improve faster. It is known that CNS functioning and improvement can be modified by cultural influences. It is shown here that Jürgen could turn better than the average child for forward and backward exercising by approximately 30%. The forward values improved from 17.8 to 12.0 and backward values from 22.8 to 15.3. This improvement was partly also possible because Jürgen could start to exercise on the smaller baby device from an age of two onwards. Exercising twelve times made his CNS function better than the average child by 30%.

4.3. More 'Correction En Route' of Abnormal CNS Maturation by Exercising on the Special CDT Device

Such improvement of CNS function is especially important in cerebral palsy children, whose development is retarded. It is actually not only the rate of development, which has to be enhanced; *retarded, accelerated or deviant development has to be corrected 'en route' by therapy*. This is possible in the very important first five years of life, because special CDT devices are available for all ages (Figures 15, 86). The first five years of life are very important for correcting deviant development. As mentioned earlier, the sooner treatment starts, the higher the rate of repair will be, and the stronger the 'correction en route' will be. The performance in Jürgen improved by 30% after a total of just four hours exercising in total (12 sessions of 20 minutes) between the age of two and seven years. It will be shown below that the rate of repair by learning is much slower in the injured CNS. If the infants cannot turn by themselves, the therapist has to help. It is also easier to exercise on the smaller devices, only the integrativity of functions is reduced. For older pupils, it will be shown below that the integrativity of CNS organization increases with increasing load. It is the integrated function that substantially induces learning transfer.

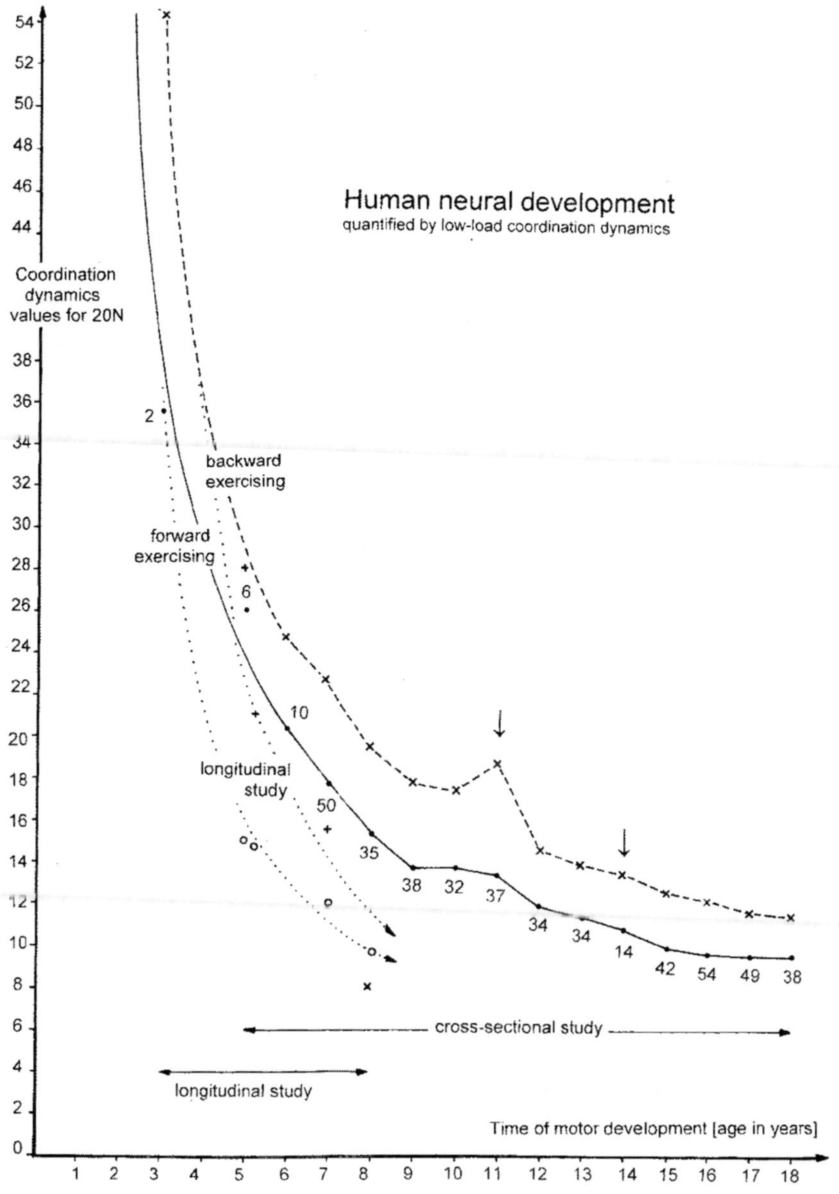

Figure 92. Coordination dynamics values of boys and girls (lumped together) to quantify human neural development. Note that the coordination dynamics values transiently increase at 11 years and that the group size is small at an age of 14 (puberty), as is indicated by arrows. Note further that the coordination dynamics (CD) values for backward exercising during the whole developmental period are worse (higher) and that in the longitudinal study (dotted lines) Jürgen's CD values become smaller (better) than the average ones of the cross-sectional study due to the repeated assessment.

4.4. Symmetry

As can be seen from Figure 92, the pupils could exercise, at all ages, better in the forward than in the backward direction. Natural maturation did not correct this missing symmetry. The symmetry can be corrected by movement-based learning, as will be shown below. Symmetry is important, especially in repair. This is particularly apparent in stroke patients.

4.5. Mistakes Made in the Direction of Exercise During Normal Development: Stability Maturation

As can be seen from Figure 91A-D, at the age of three, Jonas and Jürgen made mistakes in the direction of turning. The intention was, for example, to turn in the backward direction. But by mistake, they turned transiently in the forward direction. At the beginning, they needed instructions from their mother to stay continuously in one direction. With further development, Jürgen noticed the mistakes by himself and corrected the false direction of exercising. By the age of seven, Jürgen did not make these mistakes any more (Figure 91E,F). Also in the cross-sectional study, some young infants sometimes made similar mistakes. The ability to consistently turn in a certain direction develops only with time.

The ability to stay in a certain movement pattern has two components: the stability of the movement itself, i.e. the deepness of the potential well and the size of the variability of phase and frequency coordination (Figure 28), and the stability of the pattern intention to convert the intention into the movement pattern. Initially, the child makes mistakes. They are not noticed nor corrected because the intention is forgotten. Instruction is needed. With further neural network maturation, the normal child makes mistakes, but corrects the mistake by itself. The pattern intention is working. Later on, the child no longer makes these basic mistakes and stays stable, in the pattern of forward or backward exercising.

4.6. Movement Stability Impairment Following CNS injury

Following brain injury (not spinal cord injury) and in cerebral palsy, the patients sometimes cannot maintain the direction of exercising. With therapy, they learn to stay in a certain direction of exercise. The rate of learning depends on the severity of the injury. Above, the stability (Figure 39) of the pattern 'running on treadmill' (Figure 38) was measured in a patient with severe brain injury.

4.7. Autistic Children: Abnormal Infantile Development

Autistic children change their intention every few seconds. They cannot perform a certain movement for a longer time because their intention changes all the time. They can keep movement patterns themselves like running, jumping or crawling quite well; it is the pattern intention, which is very unstable. It seems that their intention changes are entirely unpredictable. They are not cooperative. But it seems that they can learn, at least to a certain extent, to stay in a certain pattern of movement or higher mental function. However, the therapist has to be very patient.

4.8. High-load Coordination Dynamics Assessment to Measure Rates of Learning

CNS development between the ages of three to eighteen years was measured by the low-load coordination dynamics value. Exercising against 20N for 21 minutes was easy to do. High-load testing was not popular among the pupils, especially repeated high-load exercising, because it was hard work. The argument that it is healthy did not particularly impress the pupils. But a few pupils did this high-load exercising, which lasted approximately one hour (including the low-load test for warming up). Their data will be compared with those of patients and rates of movement learning, i.e. rates of repair, measured.

5. Motor Learning in the Healthy CNS

5.1. From Development to Movement-based Learning

It was shown above that CNS function improved in Jürgen, compared to the average youngster, at the age of seven by 30% by exercising 4 hours on the special CDT device for twenty minutes on twelve occasions over five years. Jürgen could turn more easily (better (lower) CD values) for lower load than 20N or using the smaller baby device. Increasing the load made the turning with arms and legs too difficult. That means for exercising against higher load, it is more difficult for the CNS to learn the different patterns between pace and trot gait. Also, it is more difficult to coordinate finger or limb movements for higher loads.

5.2. Exercising against Higher Loads Is a Good Measure to Quantify Movement-Based Learning

For older pupils and adult volunteers higher loads can be used to measure better movement-based learning in the normal CNS. The activated functions are more integrated, if the volunteer is exercising against 150 or 200N. For the generation of such power, more muscle fibers and more upstream neuronal networks are activated. The whole cardio-vascular performance is activated including respiration. As well as the somatic nervous system, the vegetative nervous system is activated and trained. Because of the activation of the vegetative nervous system, urinary bladder continence functions can be improved and repaired.

5.3. Assessment Is also a Form of Therapy

The repeated measuring of the low-load and high-load coordination dynamics is analogous to CDT because healthy subjects move around in their everyday life and participate in some activity. It therefore becomes possible to compare this kind of motor learning between healthy subjects and patients with CNS injury. The measurement of the CD improving CNS functioning (the measurement influences the value to be measured) does not present a problem, because changes occur due to measuring in both the healthy and injured

CNS. The measuring of the CD values in the patient and healthy subject also constitutes therapy, so no time is wasted due to the diagnostic process. It is in fact good training, because the patient is intent on improving their values. The diagnostic is an efficient therapeutic hour for improving CNS functioning, fitness and health.

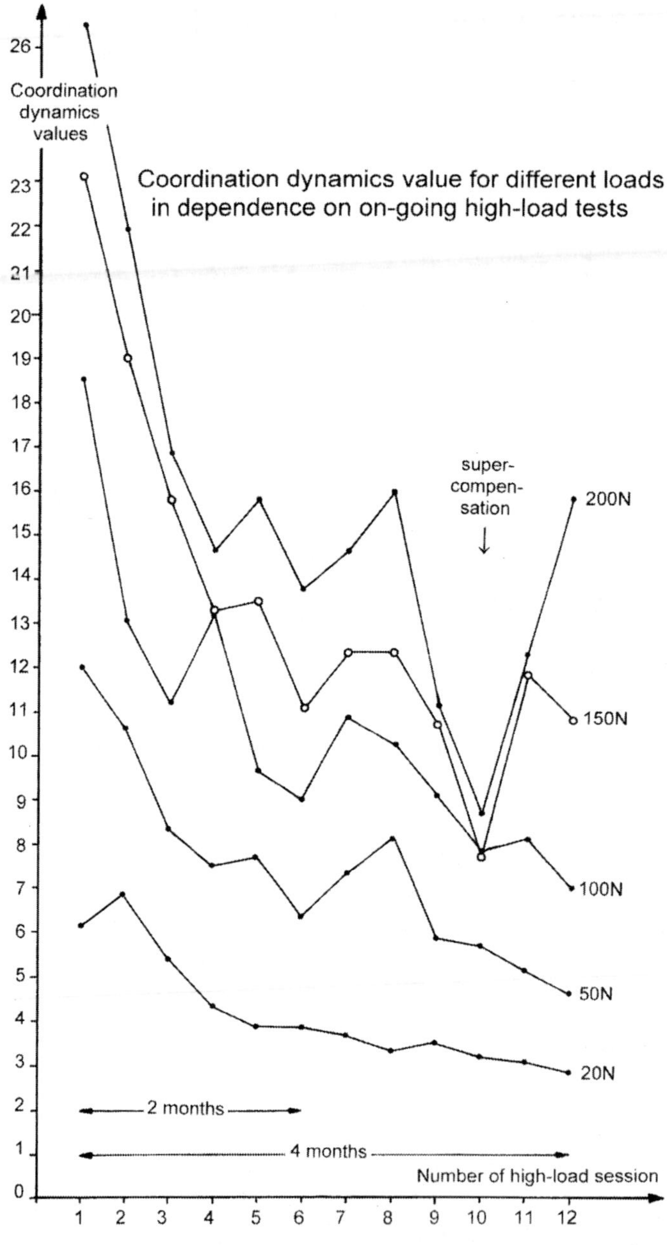

Figure 93. Improvement of coordination dynamics (CD) values against different loads due to 12 repeated (low- and) high-load coordination dynamics testing and exercise (low-intensity motor learning) in the forward direction in a 15-year-old pupil. The mean time interval between 2 sessions was 11 days and the number of turns per session was approximately 3500. Note the phase of super-compensation around the 10th session; only the CD values for 150 and 200N improved (decreased) strongly transiently.

5.4. Movement-based Learning for Different Loads in the Healthy CNS

Figure 93 shows the improvement of the CD values for a 15-year-old pupil for increasing load over the course of repeated sessions. It can be seen that the CD values of the boy reduced (improved) for all loads between 20 and 200N. After the first measuring sessions the CD values reduced (became better) strongly and from then on improved only a little. This strong improvement of the CD values will be used to quantify movement-based learning and is used later for comparisons with patients. All curves for the loads between 20 and 200N had the same characteristic. With the sixth session, after a period of two months, the strong improvement of CNS functioning for all loads ceased. As will be seen below, this similarity of improvement for all the loads will be different in patients. For the 9th to 11th sessions, the performance of exercise was transiently very good for the high loads of 200N and 150N. Such phases of transiently good performance are called in sport 'super-compensation'.

In a 17-year-old female athlete, this fast motor learning was achieved in one month or less (Figure 94). This female athlete learned faster than the normal pupil. The super-compensation was already reached during the 6th to 9th session.

But when this and other pupils did not get better any more, their motivation to improve ended. They also had good reason to stop the repeated measuring, as they had to prepare for examinations. They did not exercise often enough to reach the phase of slow improvement. In mild epilepsy the fast and slow phase of learning can be seen, as well as the occurrence of several phases of super-compensation [2]. The slow phase of learning can be seen when the Author exercised every day for more than one month (see below, Figure 116).

Between the 6th and 7th session, the female athlete pupil also had to study for examinations and did not sleep enough at night. She suffered from stress caused by the examinations, and had difficulty with pain in the bones from her training of high-jumping. Her performance, therefore, got much worse from the 7th to the 8th session onwards, even though she had no pain while exercising on the special CDT device. This case shows how dramatically performance can worsen if there is stress, not enough sleep, and some health problems. In another pupil, who had had only 5 hours of sleep the previous night, the performance also worsened. In the female athlete, this impairment was very strong. During the 11th session, she could exercise better for 200N than for 150 and 100N, which is not normal. Her nervous system was completely mixed up. Her fitness had not declined, because she was continuing with her sports training. For tennis players, transient worsening of performance can be dramatic, especially when motivation is reduced. Health conditions, motivation, and amount of therapy also influences the rate of CNS repair in patients.

5.5. Exercising Coordinated Movements at High Load to Improve Deep Network Complexity by Learning

To improve CNS functioning, the developing child or the patient has to exercise on the special CDT device in the forward and backward direction to improve symmetries of network organization. By turning both slowly and rapidly, the movement performance can be improved through the movement-induced afferent input generated by the slow, medium, fast

(fast fatigue resistant), fast contracting muscle fibers (fast fatigue), and their corresponding upstream neuronal networks including the premotor spinal oscillators (Figure 13).

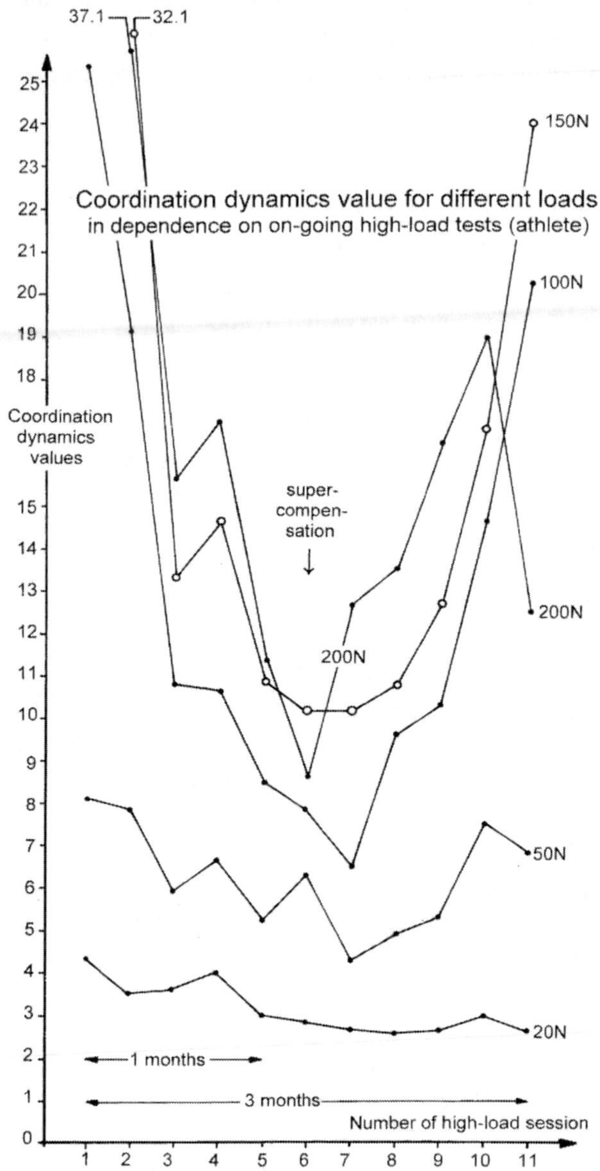

Figure 94. Improvement of coordination dynamics (CD) values against different loads due to 11 repeated (low- and) high-load coordination dynamics testing and exercising in the forward and backward direction in a 17-year-old female pupil (athlete). The mean time interval between 2 sessions was 9 days and the number of turns per session was approximately 4000. Note the strong increase again of the high-load CD values for 100, 150 and 200N following reduced sleep, examination stress, health problems and lack of motivation. In the first session the CD values for 150 and 200N were with 32.1 and 37.1s^{-1} respectively, very high and reduced strongly with the next two assessment sessions to 13.3 and 15.6s^{-1}, respectively.

However, exercising against different loads must also be performed to enhance the integrativity (many neural sub-networks are activated simultaneously and coordinated together), and complexity of neuronal network organization. By increasing the load during longer periods of exercise, the slow muscle fibers (and the upstream networks) (Figure 13) are trained which, for example, a marathon runner needs. To get sufficient oxygen, to regulate temperature and cardio-vascular performance, the vegetative nervous system needs to be trained and improved in its functioning in general. This is especially important because of the substantial learning transfer needed to repair trained functions, which cannot be deliberately trained in the injured and uninjured CNS.

The high-load assessment and training was performed every one, two, or perhaps four weeks, to see how the CD improved for high-load exercising. The headmistress of the high school also performed this high-load assessment and training and her body felt very good afterwards. A sports teacher at the school, by contrast, failed the test (he refused to continue at the load of 150N), probably because he did not have sufficient fatigue resistant slow or medium fast muscle fibers (as 100m runners often do not), and insufficient enough motivation to push himself through the high loads. For this high-load assessment and training, one needs the slow muscle fibers, the medium fast and fast (fatigable) muscle fibers (Figure 13). Women can therefore perform this high-load exercise as well as men.

5.6. Improvement of High-load CD (Movement-based Learning) and Super-compensation in Older Healthy Pupils

Note that for force changes between 20 and 200N the power in Watt changed stepwise in the author (A), but not in the pupil who showed a transient frequency reduction (D). Note further that for the load of 200N the pupil could exercise quite well on the special coordination dynamics therapy device for the easy coordinations close to pace and trot gait, but not for the difficult intermediate coordinations; the frequency of turning decreased strongly and the coordination dynamics increased strongly (worsened) for the intermediate coordinations, which means that the pupil's CNS had problems organizing the motor control for the difficult intermediate coordinations of arm and leg movements.

Figure 95 shows such high-load assessment strategy. The exercise load was increased from 20 to 200N and then decreased again (Figure 95A). A fit adult or a pupil with a very well-functioning CNS can turn at the same frequency for all loads (no load escape (Figure 95A)) with low CD values (Figure 95B,C). An unfit subject with suboptimal network organization will reduce the frequency of exercise or even stop exercise to escape the load, and the CD values will increase (get worse) for higher load exercise (Figure 95D,E,F). In adding up the CD values for the different loads, a value for the quality of high-load performance is obtained. The whole high-load test supplies further information with respect to movement-based learning. Repeated assessments show the improvement of fitness and CNS organization. The assessment of healthy pupils can show improvements in CNS organization by going deeply into the network complexity of CNS organization when exercising at high loads.

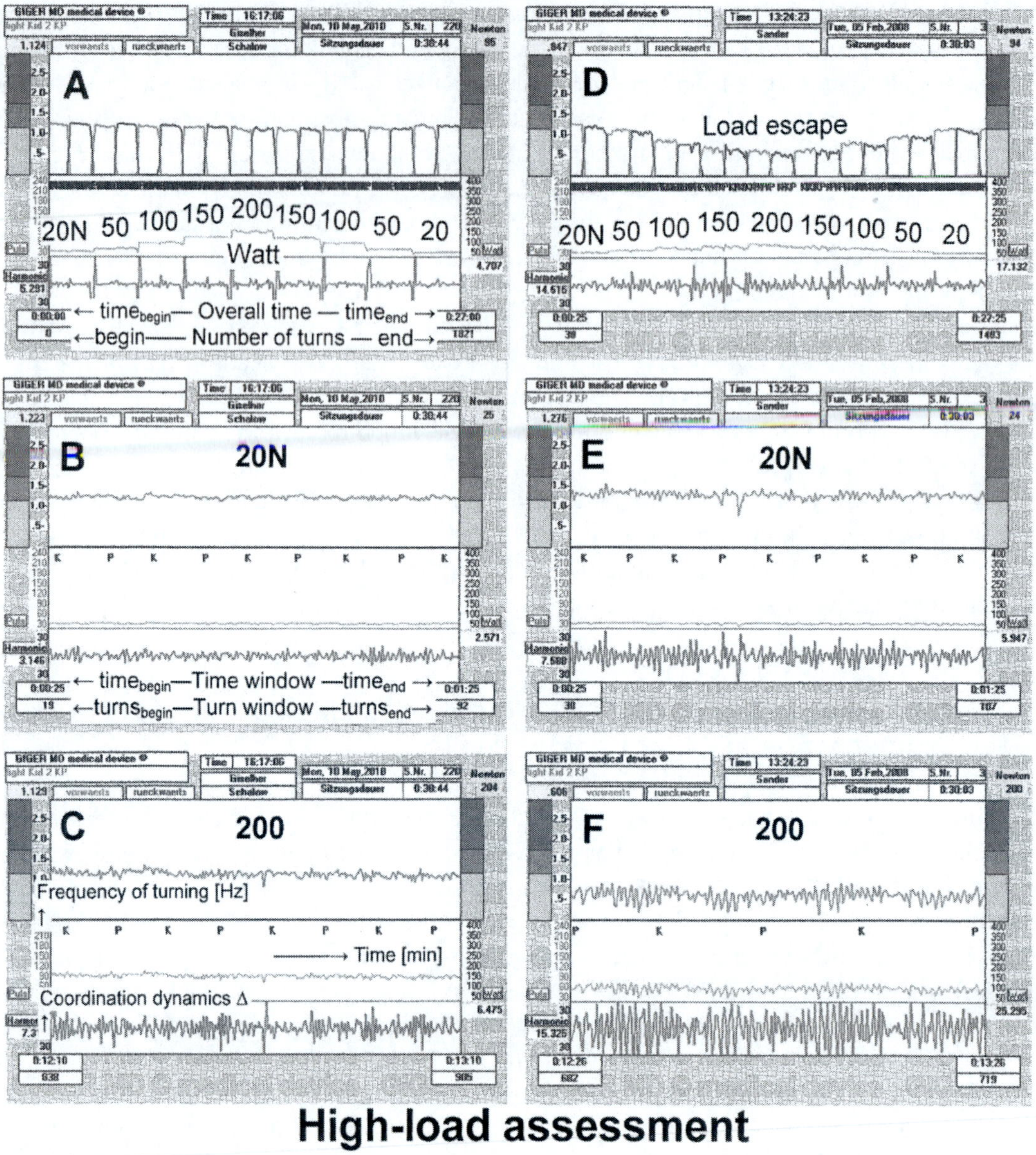

Figure 95. High-load coordination dynamics recordings for well and balanced functioning CNS of the author (A-C) and a 15-year-old pupil (D-F). In the recordings A through F, the upper traces represent frequency of turning, the middle traces show load changes in Watt, and the lower traces show the coordination dynamics. In A and D, the whole record of 27min duration is shown during force changes in 3min steps (1.5min forward and 1.5min backwards) from 20 to 200N and backwards, and in B,C,E,F, a time window of 1min is shown when the load was approximately 20 and 200N (P and K are the pace and trot gait coordination positions).

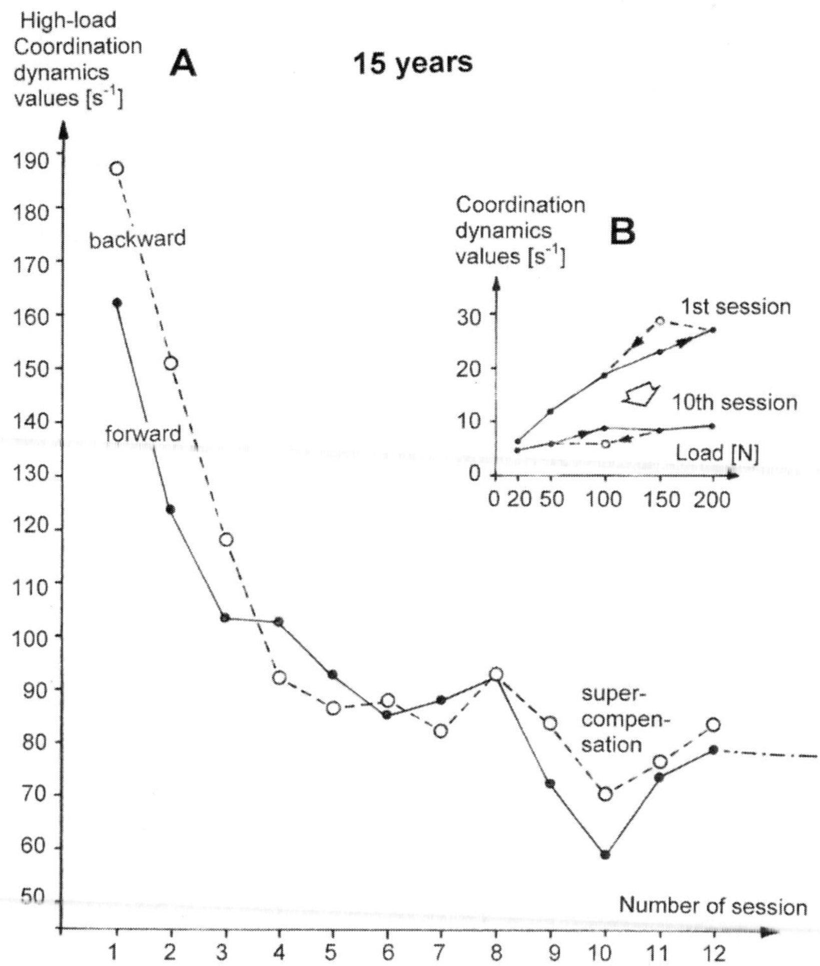

Figure 96, A. Improvement of high-load coordination dynamics (CD) values (for evaluation see B) due to 12 times repeated (low- and) high-load coordination dynamics testing and exercising (low-intensity motor learning) in the forward and backward direction in a 15-year-old pupil. The mean time interval between 2 sessions was 11 days and the number of turns per session was approximately 3500. Note the transient improved exercising (super-compensation) at the 10th session. The dashed-dotted line suggests the further continuation of the high-load CD values. B (inset). Coordination dynamics values (Δ) in dependence on load increase (solid line) and decrease (dashed line) for the 1st and 10th session for exercising in the forward direction. The hysteresis-like curve for the 10th recording shifted to smaller coordination dynamics values. The high-load CD values were obtained by summing up the single CD values, Δ (high-load CD value = Δ20N + Δ50N + 100N + Δ150N + 200N + Δ150N + Δ100N + Δ100N + Δ50N + Δ20N + Δ20N).

Figure 96 shows the high-load CD values with on-going assessment (and no additional training on the instrument) of a 15-year-old male pupil (same as in Figure 93). In Figure 96B the single values of the first and tenth sessions are shown. As can be seen, the hysteresis-like curves improved with on-going sessions. Plotting the high-load CD values (the sum of the single CD values at different loads) against the number of sessions, a curve for forward and backward exercising is obtained (Figure 96A). First, the high-load CD values improve strongly and then reach a plateau with a slight decline (Figure 96A). Before the plateau is reached, a phase is passed through in which the performance is best. This phase of super performance is called 'super-compensation' and occurred in this case during the 10th session. After this super-compensation phase, most pupils stopped. The high-load testing of a 17-year-old female athlete is shown in Figure 97 (this is the same pupil as in Figure 94). Her super-compensation phase was reached during the sixth session. Her plateau values were with approximately lower than that of the 15-year-old pupil ($80s^{-1}$). An 18-year-old pupil had a plateau value of approximately $60s^{-1}$, the lowest value (Figure 98). It seems that the high-load CD values also slightly decreased with age (CNS maturing). The low-load and high-load CD values depend strongly on the fitness of the person. The 15, 17 and 18 year old pupils (Figures 96-98) were sporty and healthy. The Author, a bit older than the pupils, reached after many years of exercising a high-load value of approximately $30s^{-1}$ (Figures 116 and 126).

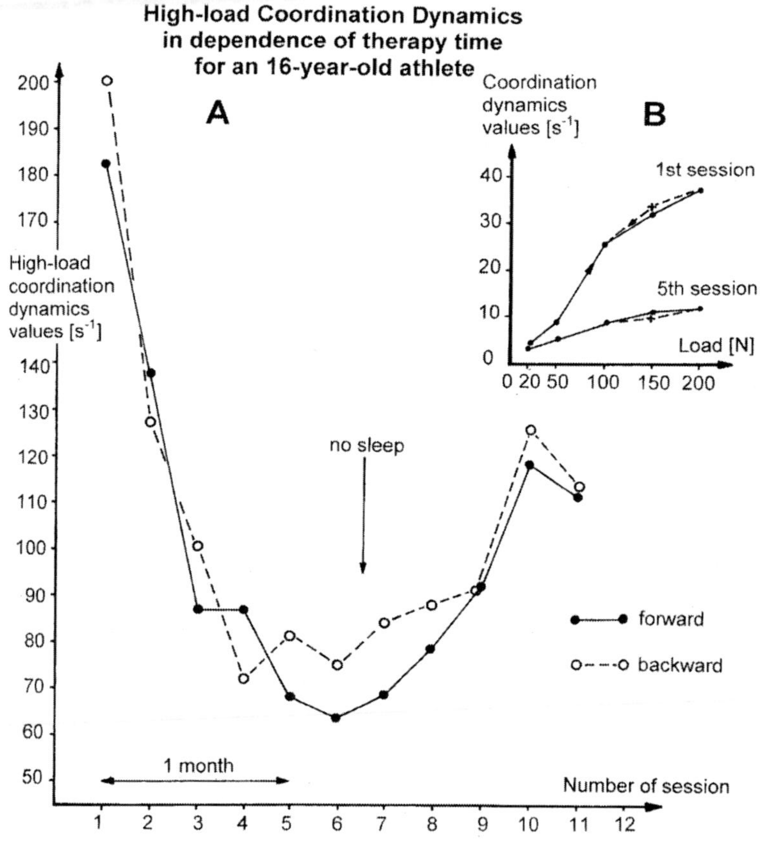

Figure 97. A. Improvement of high-load coordination dynamics (CD) values (for evaluation see B) due to 11 times repeated (low- and) high-load coordination dynamics testing and exercising in the forward and backward direction in a 17-year-old female pupil (athlete). The mean time interval between 2 sessions was 9 days and the number of turns per session was approximately 4200. Note the strong increase again of the high-load CD values following reduced sleep, examination stress, health problems, and lack of motivation, indicated by an arrow and 'no sleep'. B (inset). Coordination dynamics values in dependence on load increase (solid line) and decrease (dashed line) for the 1st and 5th session for exercising in the forward direction. The high-load CD values were obtained by summing up the single CD values.

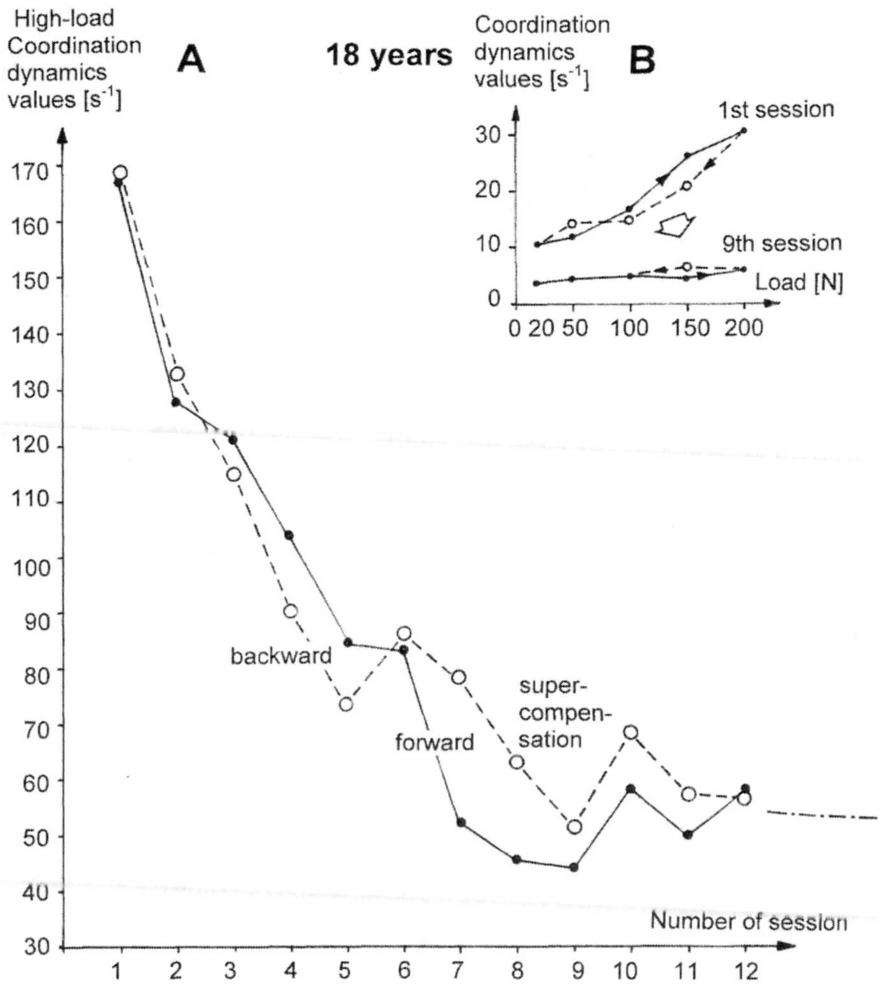

Figure 98. A. Improvement of high-load coordination dynamics (CD) values (for evaluation see B) due to 12 times repeated (low- and) high-load coordination dynamics testing and exercising in the forward and backward direction in an 18-year-old pupil. The mean time interval between 2 sessions was 31 days and the number of turns per session was approximately 4200. Note the transient improved exercising (super compensation) at the 9th session. The dashed-dotted line suggests the further continuation of the high-load CD values. B (inset). Coordination dynamics values in dependence on load increase (solid line) and decrease (dashed line) for the 1st and 9th session for exercising in the forward direction. The hysteresis-like curve for the 10th recording shifted to smaller coordination dynamics values. The high-load CD values were obtained by summing up the single CD values. If the sessions were performed every week or 10 days then the healthy pupil would have learned to get the high-load CD values under $100s^{-1}$ in approximately one month.

5.7. Movement-Based Learning in Short-Term Memory Is Fastest in Young Children

To convince somebody that exercising on the special CDT device is really improving CNS self-organization, this person has to be made to exercise on the special CDT device because they will then experience that this exercising has little in common with a conventional fitness bicycle. This is especially true during backward exercising when the person would feel that his/her CNS has difficulty in generating such movement patterns - and that the movement performance of these patterns improves quickly in the short-term memory.

Time to best value forward (boys and girls)

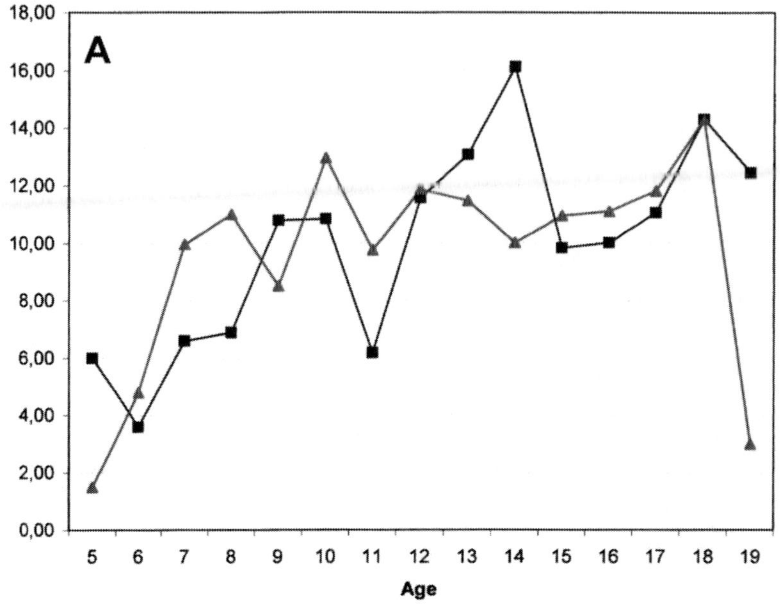

Time to best value backward (boys and girls)

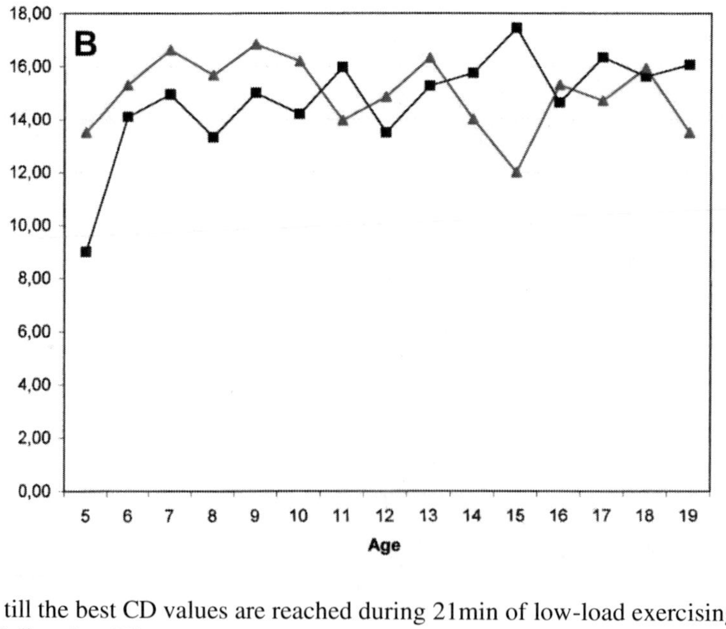

Figure 99. Times till the best CD values are reached during 21min of low-load exercising in the forward (A) and backward direction (B).

It is of interest therefore to measure whether the rate of this movement-based learning in the short-term memory is increasing, decreasing or is the same for all ages. The rate of learning was assessed by the time, which was needed to achieve the best CD value within the 21 minutes of exercising against 20N. As can be seen from Figure 42A, the young children learned fastest. For exercising in the forward direction (Figure 99A), the young children were fastest to reach the best CD value. The time to the best CD value increased (with variations) continuously from five to eighteen years of age. This means that more time is needed for motor learning with increasing age; the rate of motor learning decreased with on-going development.

For backwards exercising (Figure 99B), which is more difficult than exercising in the forward direction (Figure 92), all ages seem to need the entire 21 minute period to reach the best value. It could be expected that younger children learned faster to get their best value for exercising in the forward direction. However, the young pupils had similar problems to the older ones for exercising in the backward direction was to be expected. During development, the CNS prioritizes the functionally more important act of moving forward rather than establishing symmetries. This prioritization can be observed by the ability of small babies to perform the forwards stepping automatism, but not backwards and the ability in small children to walk forwards, but not backwards.

5.8. Movement-based Learning in the Short-term Memory in Spinal Cord and Brain Injury

No real statistics exist for movement-based learning in the short-term memory in the case of an injured CNS. In brain injury, for low-load exercising the best values were mostly achieved after exercising for 15 minutes. The mildly and severely injured neuronal networks need some time for improving function in the short-term memory. In a patient with a complete cervical spinal cord injury, the best values could be achieved at the beginning of the 21-minute period of exercise. Her supraspinal centers were functioning well; she just had to get some information through the injury site to activate the motoneurons physiologically. Also, in incomplete spinal cord injury the learning in the short-term memory was fast, so that the best low-load values were obtained at the beginning.

As will be shown below, the movement-based learning for repair depends strongly on the kind of CNS injury, its severity, and what functions have to be relearned. Low-load exercising on the special CDT device, for example, can be learned faster than the high-load exercising, which is going deeper into the complexity of CNS organization (Figure 100). For further details of the repair by learning in incomplete spinal cord injury, see [2].

5.9. Neural Network Learning Quantified by the Time to Achieve with CDT High-load Test Values under $100s^{-1}$

Now we will turn onto the high-load measurements of patients to meaningfully quantify the rate of movement-based learning in the long-term memory for higher loads, i.e. the rate of repair by learning. It will be shown that there are tremendous differences in the speed of movement-based learning in the healthy CNS and the injured CNS. Repair following therapy of the injured CNS needs much time, depending on the severity of the injury.

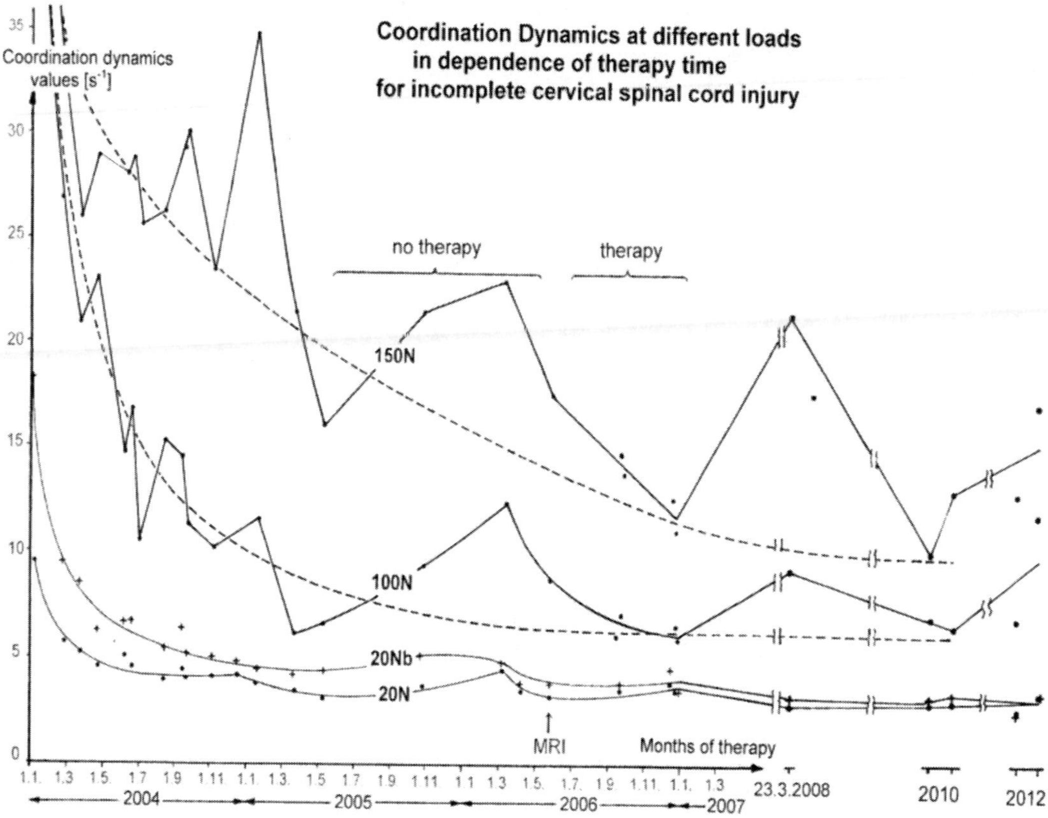

Figure 100. Relation of coordination dynamics values to therapy length for increasing load between 20 and 150N. The loads for forward exercising (dots, 20N, 100N, 150N) are marked at the curves (20Nb = backward exercising (crosses) at 20N). Note that with no therapy the coordination dynamics values got worse and upon administering therapy again, the values improved again, even 2 years after the accident. After stopping therapy, the coordination dynamics values for 100 and 150N increased again (dotted lines).

The high-load coordination dynamics test is used for quantifying the rate of CNS repair by learning. The time required for learning or repair differs strongly between an uninjured and injured nervous system. A healthy athlete learned to get the high-load test value under $100s^{-1}$ in approximately half a month (Figure 97) and a healthy person needed approximately one or two months (Figure 98). Even among athletes, the values can vary significantly. Neural network learning following movement-based learning (CDT) will now be measured in mild and severe CNS injuries or malformations and the rate of learning is measured and later on compared. It is started with mild dysfunctions of the CNS.

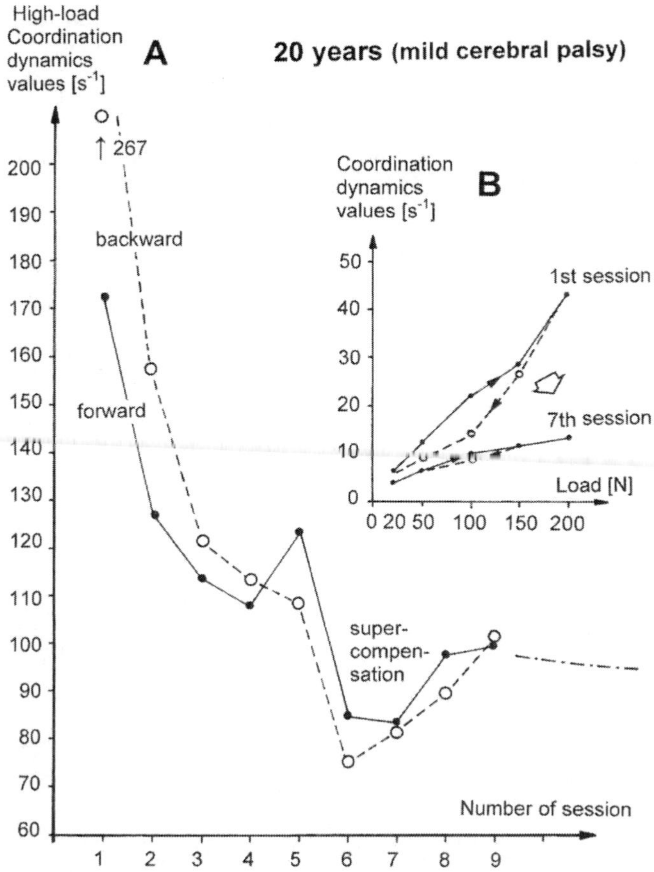

Figure 101. A. Improvement of high-load coordination dynamics (CD) values (for evaluation see B) due to 9 times repeated (low- and) high-load coordination dynamics testing and exercising in the forward and backward direction in a 20-year-old pupil with mild cerebral palsy. The mean time interval between 2 sessions was 5 days and the number of turns per session was approximately 4600. Note the transient improved exercising (super-compensation) at the 7th session. The dashed-dotted line suggests the further continuation of the high-load CD values. B. Coordination dynamics values in dependence on load increase (solid line) and decrease (dashed line) for the 1st and 7th session for exercising in the forward direction.

6. Learning in Mild Cerebral Palsy, Mild Epilepsy and Scoliosis

6.1. Improvement of CNS Functioning in Mild Cerebral Palsy upon Continuous High-load Testing

A 20-year-old pupil with a mild cerebral palsy showed a similar CD value improvement (Figure 101) as the healthy pupils. His super-compensation phase occurred around the 7th session. His plateau value was approximately 100. In comparison to the healthy pupils of the same high school, his values were worse at the beginning and his plateau value was not as good (higher) as in the healthy pupils (Figures 96-98). Still he reached the $100s^{-1}$ in one month. The rate of learning was approximately the same as in the healthy pupils. The cerebral palsy and the healthy pupils were not overweight and liked moving. The pupil with mild cerebral palsy (now studying at university) walked an hour to school every day. This fast walking to school improved his CNS functioning, but the high-load exercise on the special CDT device further improved his CNS functioning by movement-based learning.

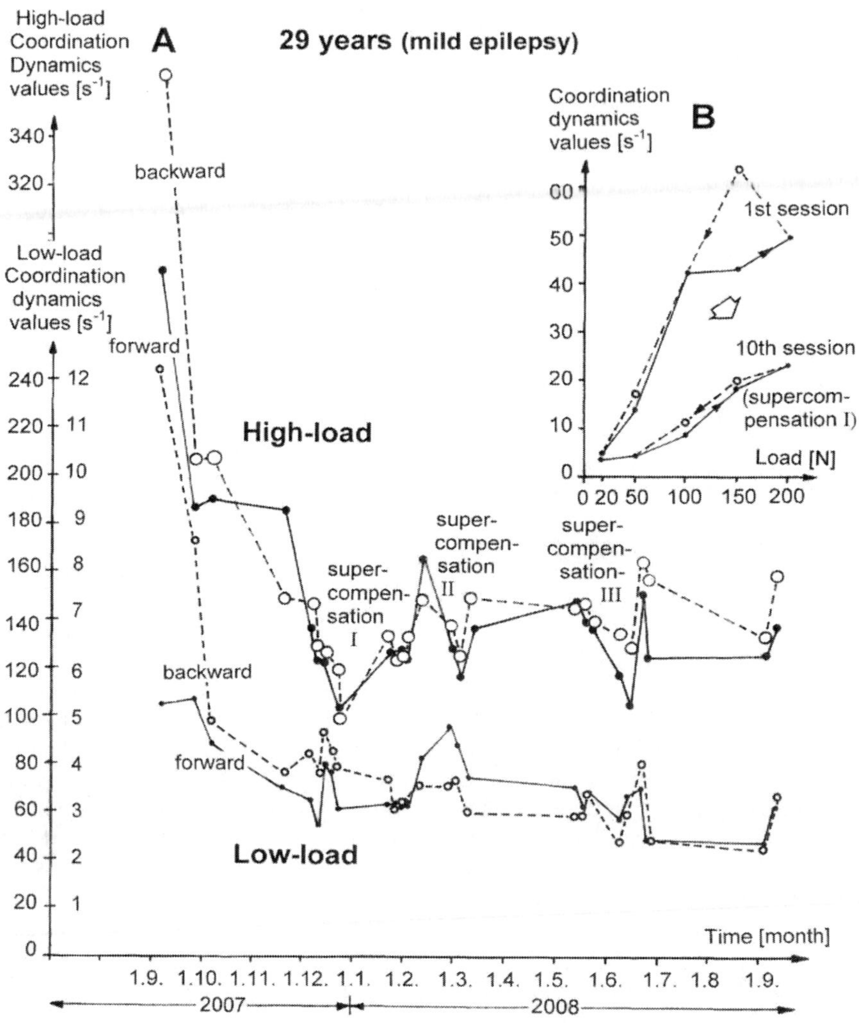

Figure 102. A. Improvement of high-load and low-load coordination dynamics (CD) values upon repeated low- and high-load coordination dynamics testing and exercising in the forward and backward direction in a 29-year-old mother with mild epilepsy. The mean time interval between 2 sessions was 13 days and the number of turns per session was approximately 4300. Note the first transient improved exercising (super-compensation I) at the 10th session; with further exercising sessions, further super-compensations (II and III) occurred. Note further that with these 27 exercising sessions, a plateau with slight decline was reached. The low-load coordination dynamics values declined in a similar way but showed no super-compensation. B. High-load coordination dynamics values in dependence on load increase (solid line) and decrease (dashed line) for the 1st and 10th session for exercising in the forward direction.

6.2. Improvement of CNS Functioning in Mild Epilepsy upon Continuous High-load Testing

A 29-year-old mother of a pupil with mild epilepsy performed the high-load exercising for a longer period (Figure 102). Her super-compensation phase was reached at the 10th session. But because she went on with the high-load exercising to improve her disease further super-compensation periods could be observed (Figure 102A). Her high-load plateau value was around 140; much higher than in the pupils. Also, the single high-load CD values started from much higher values (Figure 102B). Even though her high-load CD values were not as good as the pupils' her disposition to epileptic seizures decreased. Some pupils of the school with mild epilepsy could not be motivated for this demanding training to reduce the frequency of seizures. The organizational effort was too great for them. Movement-based learning was not popular at the high school, even though the head of the school was supporting it. Such CNS and fitness training should be included in school sport lessons.

In Figure 102, the two phases of learning can be seen clearly for high-load testing (the sum of all CD values) and low-load assessment (20N). Initially, the CNS improves rapidly and then slowly reaches a plateau with a small improvement thereafter.

6.3. Rate of Learning (Repair) in Mild Cerebral Palsy and Mild Epilepsy

In the female athlete, fast learning lasted approximately one month (Figure 97) and in a healthy pupil for one or two months (Figure 98). In the 20-year-old pupil with the mild cerebral palsy the fast learning lasted only a month (Figure 101; the period till up to the 6th session was one month), and in the mother with the mild epilepsy approximately two to three months. The improvement of CNS functioning also depends on the frequency of the measuring sessions. Still, the rate of learning in light disturbances of CNS functioning is not or perhaps only slightly reduced. But, as will be shown below, in severe CNS injuries the situation is completely different. The rate of learning will dramatically reduce.

6.4. Learning to Reduce Scoliosis

In 90% of cases, scoliosis is called idiopathic [73]. An important factor is supposed to be a changed innervation of the muscles, which shifts the state of equilibrium [73] dynamically. Assuming that a portion of the idiopathic scoliosis is caused by a slight pathologic organization of the neuronal networks of the spinal cord, then the following can be reasoned: Due to an asymmetric right-left activation of the back muscles asymmetric right-left tension will act on the vertebrae and will draw the spine into a scoliosis curvature. A tight corset reduces the lateral curvature of the spine and stabilizes the posture, but the cause for the scoliosis, namely right-left imbalance of muscle activation has not been resolved. From the mechanical point of view, active and passive three-directional rotational movements of the trunk can be expected to reduce the scoliosis curvature of the spine. Using coordination dynamic therapy, in which primarily rotational movements of the trunk are coordinated with arm and leg movements, the scoliosis curvature of the spine should be reduced actively. Upon an improved self-organization of the neuronal networks of the spinal cord, the too little activated back muscle groups will be activated more strongly, so that the tensions acting at the spine becomes more equal; consequently the scoliosis curvature of the spine should reduce at least during active posture. Rotational movements of the trunk will reduce the lateral curvature of the spine from the mechanical point of view (passive reduction of the lateral curvature), which works in most types of scoliosis and will act to actively reduce scoliosis in cases where scoliosis is caused by a pathologic organization of the CNS. In the latter case, the improvement of the symmetry in the self-organization of the spinal cord neuronal networks (and partly higher centers) will activate the back muscles symmetrically, will make the back muscle mass symmetric, and will make the tensions on the vertebrae symmetric so that the lateral curvature of the spine and the scoliosis form of the thoracic cage will reduce. Irregularity upon jumping on springboard, for example in abduction and adduction, and irregularity in the postural behavior when lying on a big ball support the assumption that in idiopathic scoliosis, there is a slight pathologic organization of the CNS, mainly in the spinal cord. This pathologic organization may be of genetic origin, because monozygotic twins (Figure 103) had both right-left asymmetry (for example, during jumping in abduction - adduction, were switching between jumping modes when exhausted, like jumping in symmetry, jumping in anti-phase and jumping in abduction twice (phase jump by 180°), and both had scoliosis. Even in healthy individuals, one can often observe that the organization of the CNS is not optimal. When normal persons perform fitness training on the special coordination dynamic therapy device, the turning of the levers backwards is often irregular, indicating suboptimal coordination of arms and legs; after exercising a few times a few thousand turns, the backward turning becomes harmonic: the exercised coordination becomes more optimized. It is therefore likely that in some cases of idiopathic scoliosis, asymmetric right-left coordination of the spinal cord neuronal networks is the cause for the lateral curvature of the spine.

A 19-year old female patient with scoliosis was involved in 20 therapy sessions (for approx. 30 min each) on the special coordination dynamic device for 7.5 months. In spite of the low number of therapy sessions, the scoliosis improved. Before treatment, she had pain in the back because of the scoliosis. After the treatment, she still has pain in scoliosis position, but the pain subsides during active improved posture. In cases when the vertebrae are not sticking together and when there are no rigid blocks built in parts of the spine, there may be an active position of the trunk where the spine has little lateral curvature and a more passive posture where there is lateral curvature of the spine. Such more physiologic and pathophysiologic postures may be observed when lying on a ball. Normally, X-ray scans cannot provide information about the curvature of the spine due to the different network states of the CNS. They just show the positioning of the spine when the picture was taken. The scoliosis curvature of the spine may be even related to a graceful positioning of the body. The posture of the sculpture of an ancient Greek (Figure 104) may be a result of scoliosis or of a deliberate graceful posture of the body. The organization of the CNS is influenced by the history of the CNS and by the culture of the society. But in severe cases of scoliosis, often there are also deformations of the whole thoracic cage and other parts of the skeleton.

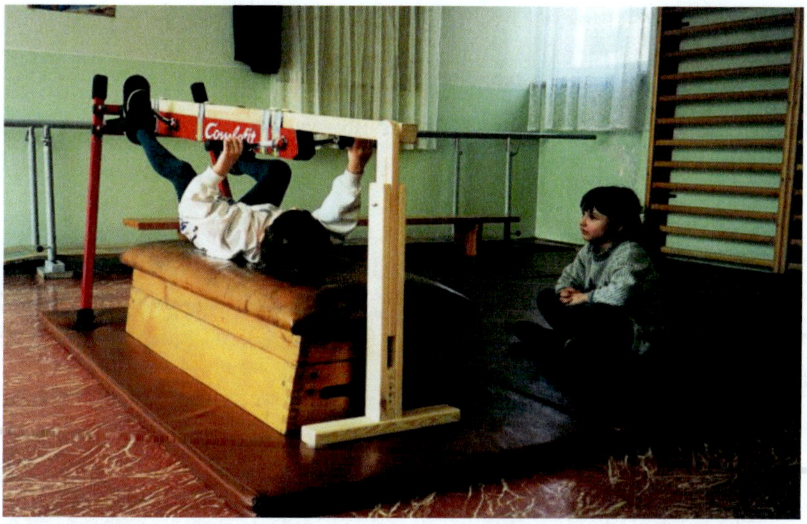

Figure 103. Twins during coordination dynamic therapy. One twin is exercising, the other one is watching. The sitting position of the watching twin is beneficial to organize (and establish on the long-term) a symmetric network state of the CNS, because hands, arms, and legs generate symmetric simultaneous afferent input from touching each other and induce co-movement in the sitting position. The special CDT device is out-of-date.

Figure 104. Sculpture of an ancient Greek. National Museum, Athens, sculpture No. 2585. Note the positioning of the body, which may be due to scoliosis or to the beauty standard of that culture, which may induce scoliosis. Note further that the gracefulness of the woman's body is strongly enhanced by the fold of the thin skirt, only partly reproduced by the picture.

Three female patients between 13 and 15 years of age were in a rehabilitation hospital to treat their severe scoliosis. Intensive therapy was applied including conservative treatment and coordination dynamic therapy. When the therapy was stopped for organizational reasons after 2 to 3 months, the lateral curvature of the spine was reduced in all three patients. In the 15-year old patient, intensive coordination dynamic therapy was continued at home for 3 more months.

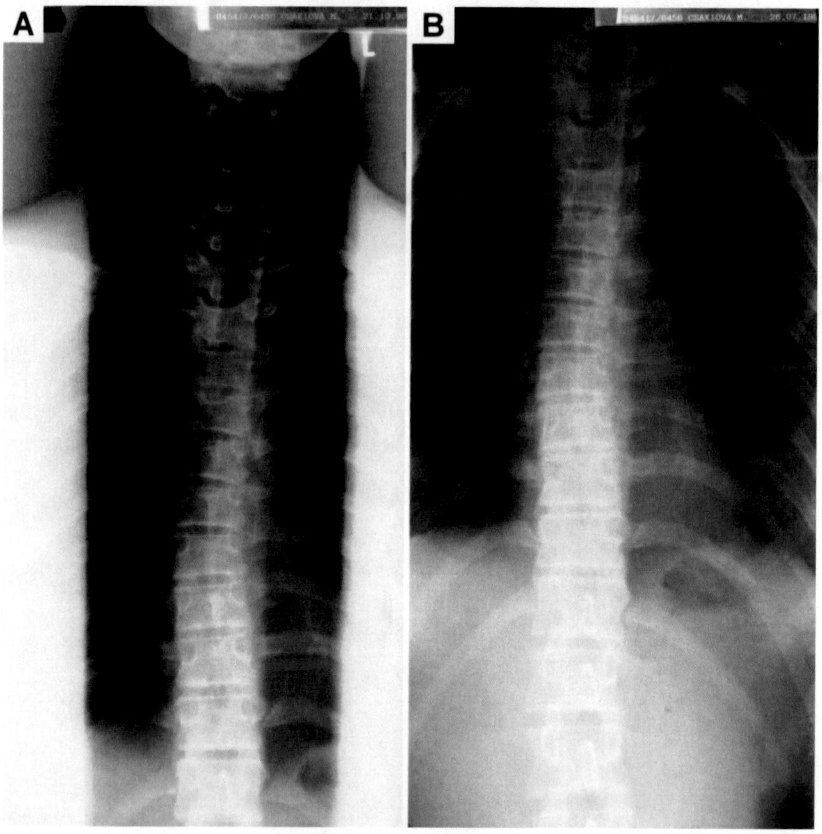

Figure 105. X-ray pictures of scoliosis before (A) and after coordination dynamic therapy (B).

When exercising rotational movements with a stick, the turning angle to the left was smaller than the one to the right. A right-left difference in the rotational movement is typical for scoliotic patients, and is due to the scoliosis deformation of the spine and the costal frame. During the therapy, the turning to the left was emphasized. But when the patient jumped in abduction-adduction, irregularity of these movements occurred. This irregularity in the rhythmic movements cannot be attributed to the scoliosis deformation of the spine and the thorax; it is an indication of a suboptimal organization of the CNS. The argument of suboptimal organization of the CNS is supported by the loss of anxiety before darkness with ongoing therapy. All her life, the patient became afraid when the light was switched off in her room. She could tolerate darkness only when her mother was present. After 6 months of therapy, she lost that anxiety. She learned to get rid of this anxiety pattern. Also, a poliomyelitis patient lost anxiety with ongoing therapy (see poliomyelitis case in [31]). The improvement in two rotational movements is shown in Figure 99 of [31]. With ongoing therapy, the scoliosis improved further. In Figure 105, the improvement of the shape of the spine after six months coordination dynamic therapy can be seen. The scoliosis of the spine and in consequence also the scoliosis shape of the costal frame reduced due to therapy. The curvature of the spine (measured according to Cobb [74]) reduced from 13 to 8°.

A quantification of the rate of learning is not possible because the assessment was not available at that time. But it is shown here that scoliosis can be reduced by movement-based learning. An orthopedic operation to straighten the spine by metal fixation is not necessary.

7. Motor Learning in Severe Traumatic Brain Injury

7.1. Movement-based Learning Has to Start in the Vigilant Coma

Above it was shown that the rate of learning in mild brain injury was not or only a little impaired. As will be shown, the rate of learning reduces dramatically in severe brain injury.

Three boys, aged nine, twelve and fourteen years old (Benjamin, Mario, and Andrej, respectively), suffered severe brain injuries almost at the same time, two in a car accident and one in a bicycle accident. In the nine and fourteen year-old patients, intensive coordination dynamics therapy (CDT) was started approximately five to ten weeks after the trauma in vigilant coma stage. The two patients recovered unexpectedly early from the coma and progressed quickly in their recovery of motor functions. They re-learned running after four months of therapy, even though their locomotor functions were still far from normal. The twelve year-old patient, Mario, did not obtain intensive CDT, only conservative physiotherapy because of many infections and complicated bone fractures, which were not invasively reconstructed. The patient recovered much later from the coma (six months as opposed to six weeks) and had severe extensor spasticity, shortened tendons, problems with several joints, pain, reduced mental functions, grasp reflex in the right hand (reappearance of infant automatism) and nearly no useful motor functions in his legs. Mario received intensive CDT following a delay of five months. The outcome of movement-based learning of the three patients are compared in Figure 107. With respect to the rate of learning and the rate of forgetting it will be concentrated on the patient Benjamin, because optimal learning therapy was administered to him over 15 years and changes of neural network learning assessed. For further details of treatment see [2].

It is common knowledge that in brain injury, and most likely in all CNS injuries, including spinal cord injury, neurotherapy by learning should be started as early as possible following the trauma to avoid pathologic reorganizations. If possible, therapy should be started in the vigilant coma.

7.2. Neural Network Repair by Learning

The two patients Benjamin and Andrej who obtained CDT for six months re-learned walking and running and cognitive functions improved. Mario, who obtained no real CDT for the first seven months following the trauma, could not walk or run and had to be carried from his wheelchair to the training device. He could hardly stand due to extensor spasticity and supination of the feet. Running improved in the two patients Benjamin and Andrej who received CDT. For nine year-old Benjamin, the running times over 22 m reduced over eight months from 14.4s to 5.2s, and for fourteen year-old Andrej, over six months from 18.5s to 6.4s. The jumping in anti-phase in Benjamin increased from 25 good quality jumps in succession to over 250, and in Andrej from 7 to 55. An assessment of the CD was not available at the beginning of therapy.

Benjamin was running whilst playing in the street and was hit by a car driven at 55 km/h. He was thrown several meters through the air. He collided with an iron fence and suffered many injuries, including an open thorax injury on the right side and brain damage of the basal ganglia domain (probably the main cause of impairment of neuronal network functioning) (area = 4 cm x 3 cm) and the frontal lobe. The patient was in a coma for two weeks and subsequently showed some response to stimuli. The intensive care unit medical team forecasted that he would be in a vigilant coma for a further one to two months. After ten days of vigilant coma (≈4 weeks after the accident) CDT was started in the recumbent position. Treadmill walking supported by four persons seemed to frighten the patient (Figure 106A). The patient seemed to have left sided hemiparesis; the left arm and the left hand seemed to show signs of spasticity. Due to a humerus fracture, the shoulder joint could only be moved a little. Even though the movements in the left arm and left hand seemed to be painful for the patient, he was cooperative. Because he could not speak, meaningful communication was not possible. When the Author left the patient after three days of therapy, he was stroked by the patient, which may mean that he was thankful for the therapy, even though the therapy caused him much pain. The CDT was continued by a physiotherapist and the mother, herself a physician. After a further three days, speech returned and the vigilant coma phase ended. The patient said that he had quite a lot of pain in the left arm and hand during the movements. But because he felt that the therapists and the mother wanted to help him, he helped actively during the arm and hand movements as much as possible, in spite of the pain and being frightened (Figure 106A). The recognition of the reality of the trauma was still difficult for the patient. Often he said to his mother: “Mum, please wake me up, this dream is so terrible”. Depressive thoughts were opposed with an intensive therapy (which was anyway part of the neurotherapy he was receiving), not to give the patient time to think his situation over. Further, training using the special CDT device seemed to have a positive effect on his mood. It became apparent when Benjamin emerged from the vigilant coma, that he was indeed hemiparetic on the left (with a slight quadruparesis), including spasticity in the left arm, hand, and fingers and in the left sternocleidomastoids’ muscle, which pulled his head to the left side. The organization of the CNS improved continuously with the on-going intensive therapy, which was performed six times per week, several times per day. On the advice of the Author, the parents refrained from botulin therapy to selectively reduce spasticity (by a blockade of the neuromuscular endplates in the domain of the application of the toxin for approximately three months), which was suggested by a neurologist. With the early beginning of an efficient intensive neuro-rehabilitation therapy, the development of false organization of the CNS (spasticity) can be strongly reduced or abolished; there was therefore no indication for a botulin therapy.

After three months of CDT, the functions of the paretic left arm strongly improved. Thus, the physiologic network states had been strengthened (the corresponding attractor wells had deepened) and the pathologic network states had been weakened (the corresponding attractor wells became more shallow). Only if the left arm can fully be moved in coordination with the other arm and the legs (coupling of the self-organizing network for the right arm and hand movement to the network organization of the other arm and the legs), the spasticity will reduce to an extent that non-rhythmic volitional movements of the left hand and fingers can be fully trained. By oscillator formation therapy, which consists of an entrainment of injured premotor spinal oscillators, achieved for example, by jumping on a springboard, the movements of the paretic arm and leg could be improved.

After six months of CDT (seven months following CNS injury), the paretic arm further improved, the overstretching of the left knee reduced and the patient re-learned running (Figure 106B). It seemed that an essential step forward was achieved in the last three months of therapy, when the patient used the special CDT device in the standing position. At the beginning the patient had great difficulty in keeping his left hand on the lever. Six weeks later, the function of the left hand improved. Several months earlier the patient also had great difficulty in holding the lever of the sky-walker.

After six months of CDT the patient no longer needed support when jumping in anti-phase on the springboard, the left knee was no longer overstretched, the positioning of the left arm, hand, head (reduced spasticity of the sternocleidomastoids' muscle) and trunk were better - hemiparesis had partly subsided. With the improvement of the motor functions, cognitive functions also improved. His short-term memory became better. The short-term memory was still not normal because he often forgot what sentences he should translate, when translating two to three sentences from English to Slovak. He enjoyed attending school for a few hours per day. The improvement of the mental functions corresponded with the healthy looking face, as judged by the parents and the Author.

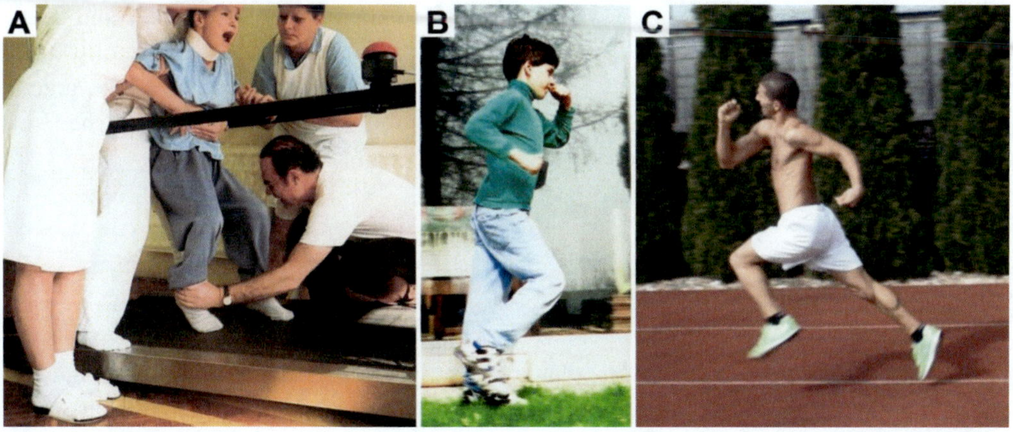

Figure 106. Relearning of running in a patient with severe brain injury. A. Coordination dynamics therapy was started in the vigilant coma stage. The Author is supporting the legs. The patient Benjamin was very afraid when treadmill walking was administered to him. B, C. Running 8 months and 15 years after the car accident. Note, the running after 15 years is faster and better but the spastic of the left hand is worse (C) than 8 months after the accident (B).

One year after the accident, the now 10-year-old Benjamin obtained a further six months of optimal CDT at home. The functions in the left arm and left hand improved further. He could walk and run better after the further treatment. With the left 'bad' hand he could turn over pages of a journal. Benjamin was back to his old school in his old class and was managing well. The strategy to first give priority to the re-learning of the motor functions before going back to school and giving, in this way, priority to the re-learning of the basic structures of the CNS and the attendant improvement in short-term memory was right in this case. With the improvement of the motor functions, the higher mental functions also improved, so that he could better manage at school.

Benjamin was not far from normal again. Even though Benjamin's running was powerful again one year after the accident (Figure 106B), his posture during running was not fully physiologic. It seems therefore that the patient was using the former pathologic posture now at least partly as a habit. The former attractor 'pathologic posture with spasticity' of the injured CNS still somehow existed in the repaired CNS. The patient had the ability to correct this posture, to switch on volition into the attractor state 'physiologic posture'. The possibility to have a bi-stability of the two attractors, pathologic posture and physiologic posture can also be found when tackling idiopathic scoliosis in the case of slight disturbances of CNS functioning. When repairing the bladder two attractors, synergy and dyssynergy of the bladder can both appear transiently (see also Chapter V of [2]). A bi-stability of spasticity and a rather physiologic movement pattern may also exist. Fifteen months after the accident, Benjamin's CNS still had difficulty in organizing the trot gait coordination between arms and legs during walking and during exercise on the special CDT device. When falling, he was still not using protection reaction automatisms. The measured coordination between arms and legs was still not physiologic.

At home, Benjamin further improved, although only a little. He was not able to crawl because of the spasticity of his left hand and arm. When the measuring of low-load coordination dynamics (CD) was introduced, it became possible to follow the evolution of his CNS function. With the introduction of high-load exercising on the special CDT device to activate the CNS more integratively (systemically), Benjamin improved further and became able to crawl. Six years after the accident, his motor functions did not improve any further. The question arose whether Schalow CDT had come to its limits in respect of the potential improvement of CNS functioning by learning, or whether the learning therapy had to be updated and performed more intensively.

7.3. The Rate of Learning Depends on the Efficiency of the Learning Method

It will be shown now that the CNS repair had by no means reached its limits and those measurements of low-load and high-load CD (Figure 110) provided further insight into the reorganization changes of the CNS in this patient.

7.3.1. Improvement of Low-Load CD Following Home Therapy

The low-load coordination dynamics values improved during 5 years (1999-2004) from 12.9 to 11.5 (11%), which was comparably small given the duration of the treatment. The low-load CD values (arrhythmicity of exercising) could not be lowered, despite exercising on the special device for at least twelve hours per week. Even the learning effect of these special movements seemed to be very small or non-existent.

7.3.2. Improvement of High-load CD Following Home Therapy

During home training, the high-load CD for increasing and decreasing the load from 20N to 200N improved within 3 years (2001-2004) with twelve hours of therapy per week. The hysteresis-like curves for increasing and decreasing load improved from a pathologic (Figure 107Aa) to a physiologic hysteresis curve (Figure 107Ab). The values of the high-load CD improved over the three years of home treatment from 229 to 151 (34%) (Figure 107Ca). This is a small improvement taken to the duration of the therapy (three years).

7.3.3. Improvement of the Low-load CD upon Therapy under Professional Supervision

The improvement achieved in low-load CD under optimal conditions within 16 days was high. The value of the CD for exercising in the forward direction decreased (improved) from 9.3 to 5.8 (38%). Initial values taken from uninjured persons are between 5 and 6s^{-1} [75].

7.3.4. Improvement of High-load CD from Therapy under Professional Supervision

High-load CD improved strongly during this short time interval of therapy received at the therapy clinic, even though during this time Benjamin exercised much less on the special CDT device than at home. The hysteresis-like high-load coordination dynamics curves (Figure 107Ba,b) were comparable to those observed for track and field athletes [76]. The high-load CD improved from 175.3 to 97.3 (44%) (Figure 107Cb).

7.3.5. Comparison of Home Therapy and Therapy under Professional Supervision

The values of the low-load CD improved over the five years of home treatment (1800 days) by 11%, whereas an improvement of 38% was achieved during the sixteen days at the therapy clinic. The values of the high-load coordination dynamics improved over the three years (1100 days) by 34% and by 44% during the sixteen days of institutional therapy. The rate of re-learning for low-load coordination dynamics was 0.61%/100days (11%/1800days = 0.61%/100days) for home therapy and 238% /100days (38%/16days = 237.5%/100days) for the institutional therapy place. The efficiency difference in respect to low-load exercising was by a factor of 390 (238%/0.61% = 390). The rate of learning for high-load coordination dynamics was 3.1%/100days (34%/1100days = 3.09%/100days) for home therapy and 275%/100days (44%/16days = 275%/100days) for institutional therapy. The efficiency difference for high-load exercising between home and institutional therapy thus differed by a factor of 89 (275%/3.1%= 89). The efficiency difference, with respect to CNS repairs, as quantified by low-load (390) and high-load CD (89), between home and institutional therapy differed approximately by a factor of 100.

7.3.6. Improvement of Movements upon Treatment at a Professional Therapy Place

Not only did CD improve strongly during professional therapy, but walking also got much better. While walking, Benjamin used the left leg much more actively instead of lifting the hip and swinging the leg. The treatment goal could be approached: at first glance, you could not see that the patient had suffered a severe brain injury.

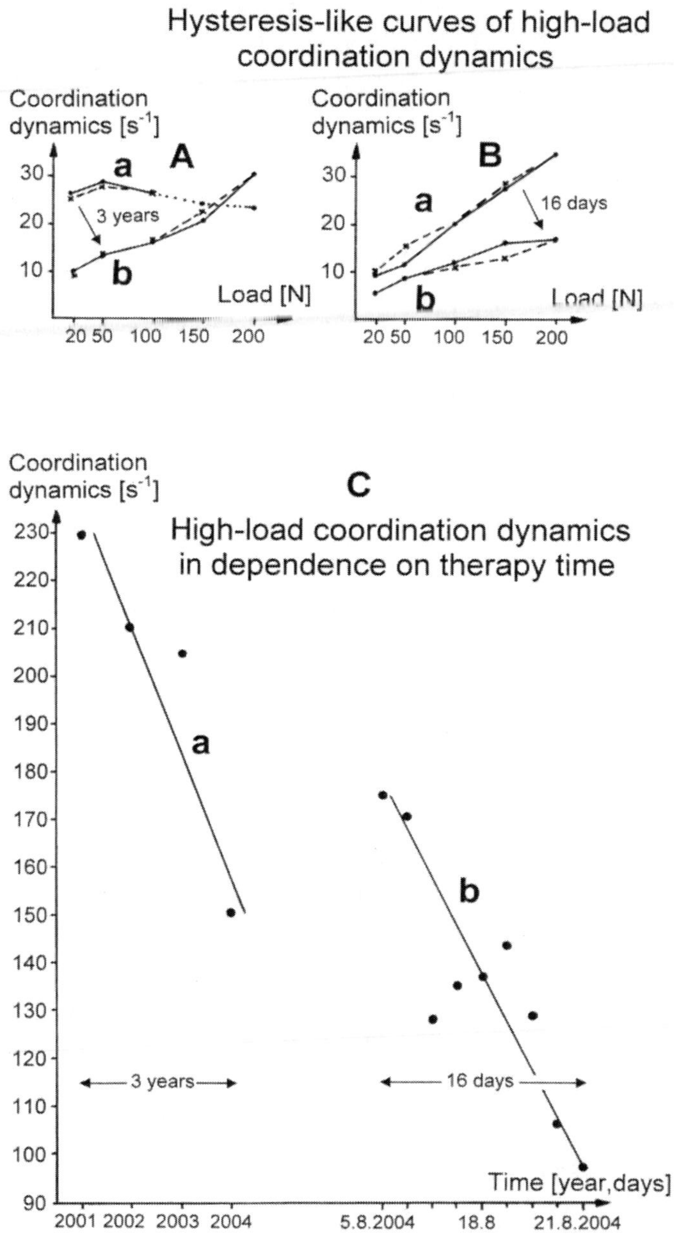

Figure 107. A,B. Improvements (from ‘a’ to ‘b’) of hysteresis-like high-load coordination dynamics values during 3 years of home training (A) and during 16 days of optimal therapy at a coordination dynamics therapy place (B); patient after severe brain injury. C. High-load coordination dynamics values for increasing and decreasing load ($\sum = (\Delta_{20N} + \Delta_{50N} + \Delta_{100N} + \Delta_{150N} + \Delta_{200N} + \Delta_{150N} + \Delta_{100N} + \Delta_{50N} + \Delta_{20N})s^{-1}$) measured during 3 years of home training (a) and during 16 days of institutional therapy (b). Note that the high-load coordination dynamics improved during the 16 days more than during the 3 years before.

7.3.7. Comparison of the Improvement of CNS Functioning after Injury with Physiologic Changes during Individual Development

It was demonstrated that the CNS functioning, as quantified by the low-load CD values, continuously improves during physiologic development within the age range between 3 and 18 years (Figure 92). During the treatment for repair, the values of the coordination dynamics were shown to temporarily worsen, probably as the CNS was changing its self-organization from pathologic to physiologic (Figure 122Ba of [2]). Such temporary worsening of CNS self-organization was predicted by the theory of CD. During the three weeks of intensive therapy, the values of the CD improved in this patient continuously, in similarity with physiologic development.

7.3.8. Oscillatory Firing FF-type Muscle Fibers

In this patient after severe brain injury, the basal ganglia on the left side were probably partly injured. Such partial damage of the basal ganglia may be responsible for the difficulty of functional repair. Rhythmic firing of FF-type muscle fibers was sometimes observed in the left tibialis anterior muscle (Figure 108C), suggesting impaired coordination (including synchronization) between oscillatory firing of motoneurons.

7.3.9. Treatment Judgment according to Surface Electromyography (sEMG)

In this case, the optimal Schalow CDT included exercising on the special CDT device to repair the impaired phase and frequency coordination by means of imposing very coordinated arm, leg, and trunk movements in the lying, sitting and standing position. The innate automatisms creeping, crawling, up-right walking and running were exercised indoors or on beach sand outside for optimal sensory input. Rhythmic dynamic stereotyped movements, like jumping and swinging on the springboard were exercised to entrain premotor spinal oscillators as an entrainment of CNS neuronal networks on the ensemble level. Old learned integrative movements (learned automatisms), like climbing staircases, were also trained. A typical exercise scene at the therapy clinic is shown in Figure 109. All movements were performed with an awareness of the importance of symmetry. To optimally entrain the severely damaged neuronal networks integratively, those integrative movements, which were activated most physiologically, quantified by surface electromyography (sEMG), were exercised most frequently. Surface EMG was performed for different integrative movements to ascertain under what conditions critical muscles showed their best motor program performance (Figure 108).

Due to an excessive palmar flexion of the left hand, the left brachioradialis and extensor carpi radialis muscles were selected as an indication of the quality of CNS function with respect to the arm and hand movements. Because of an excessive plantar flexion of the left foot (and an excessive inward rotation), recordings were taken from the left triceps surae and tibialis anterior muscles to judge CNS function with respect to leg movements. The task was now to get the best sEMG motor programs for all four muscles simultaneously to see which integrative rhythmic, dynamic, stereotyped movements entrain the injured CNS optimally.

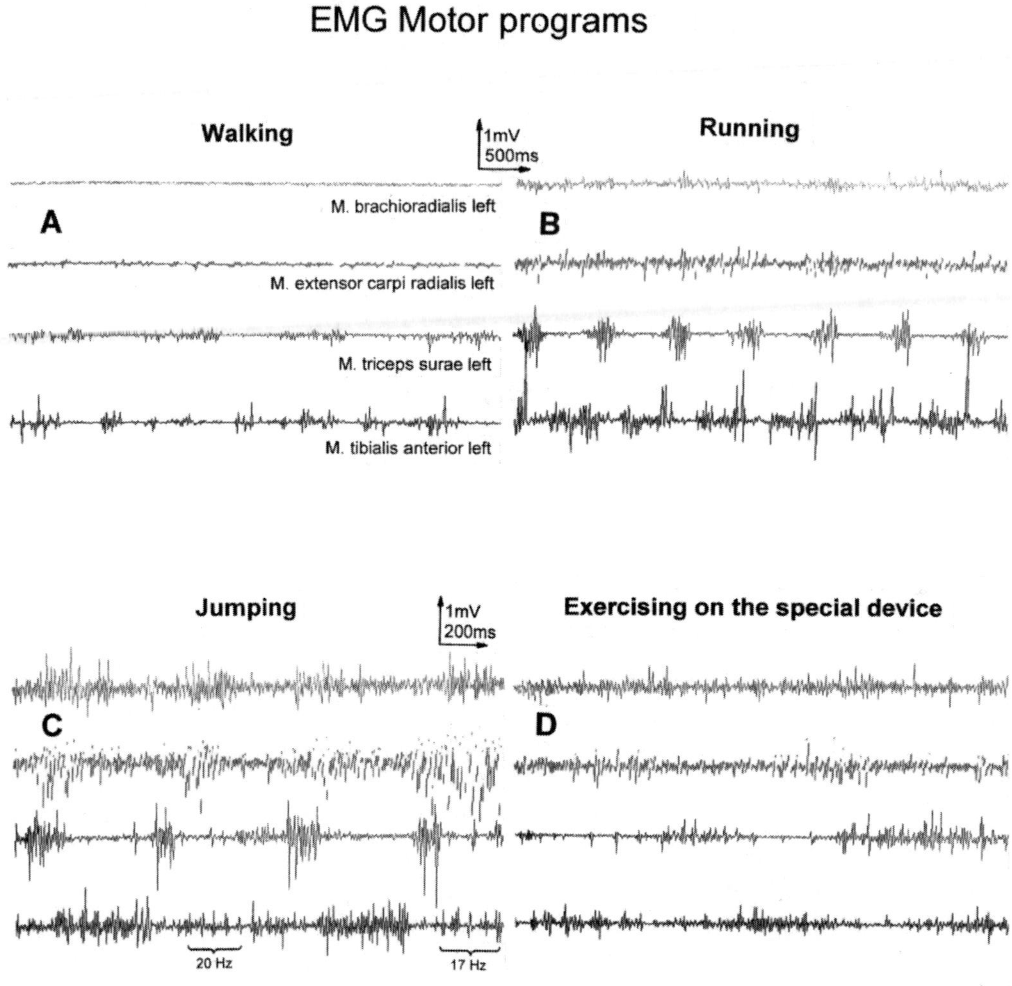

Figure 108. Motor programs registered by surface electromyography (sEMG) during walking (A), running (B), jumping (C), and exercising on the special coordination dynamics therapy and recording device at the load of 100N (D); a patient with formerly very severe brain injury. Note that the critical muscles brachioradialis left and extensor carpi radialis left are not activated during walking (A). On the extensor carpi radialis muscle trace, there may be some artifacts on top of the motor program. On the tibialis anterior trace, rhythmic firing of 20 and 17Hz can be seen, which most likely stems from FF-type motor units innervated by α_1-motoneurons.

It is important to improve CNS functioning integratively (systemically). The human CNS is an open system and if only a small sub-network is being entrained, the pathologic organization escapes the entrainment by shifting to another CNS neuronal network site as could be clearly observed in this patient. When the patient was jumping on the springboard and his left hand was in palmar flexion, he was instructed to concentrate on the left hand, to keep the hand straighter. Then, however, the left foot went into plantar flexion and inward rotation. The patient was then instructed to concentrate on getting the heel down during jumping (to perform more dorsal flexion). He concentrated on the left foot and could improve the position of the foot; but then the left hand went again into pathologic palmar flexion. He was then instructed to concentrate simultaneously on the left hand and the left foot; but then he argued that he was unable to concentrate simultaneously on the arm and the leg. Patients with severe CNS injuries especially have difficulty in concentrating on two things at the same time. It was therefore important to identify complex movements, which would induce good left arm and left foot positioning, measured visually and by good sEMG motor patterns in all the four muscles. Movements, which did not demand very much concentration of the patient, were selected because it allowed the patient to perform the movement many times without getting mentally or physically fatigued.

Figure 109. Some coordinated movements used at the therapy place. Three patients can be seen who had suffered spinal cord injuries and one patient who had suffered a severe brain injury. The exercise is assisted by two physiotherapists. The patient on the left (incomplete cervical spinal cord injury, Sten) is jumping on the springboard to entrain neuronal networks on the assembly level by performing rhythmic, dynamic, stereotyped movements. The patient on the treadmill (Benjamin, sever brain injury) is exercising the automatism walking. Support is provided by the therapist to bring arm movements into coordination with leg movements. The patient in the front (formally complete spinal cord injury Th5/6) is exercising the automatism crawling. Support is provided by a therapist for leg movements. The patient on the right (formerly complete spinal cord injury sub Th10/12) is exercising highly coordinated arm and leg movements on the special coordination dynamics therapy device in the lying position; the coordination between arm and leg movements is imposed by the device.

Figure 108 shows the motor programs of the left brachioradialis, extensor carpi radialis, triceps surae and tibialis anterior muscles during walking (A), running on the treadmill (B), jumping on the springboard (C), and exercising on the special coordination dynamics therapy device (D). It can be seen from Figure 108 that the motor programs of the four muscles for running and jumping on the springboard (B,C) were better than those for walking (A) and exercising on the special device (D). In this patient, running and jumping on the springboard thus seemed to be more efficient with respect to CNS repair than walking and exercising on the special CDT device. Improvements of CNS functioning were quantified clinically, largely by improvement in the performance of movements. Benjamin could walk and run fast after one year of treatment. Only the performance of the left arm and left leg were poor in nearly all movements. The writing performance was poor and his intellectual functions had not fully recovered.

7.3.10. Need for Many Different Movements to Reorganize the Injured CNS

Benjamin did not seem to improve further with home therapy, despite walking and running every day and exercising virtually every day on the special device for one to two hours (approx. 5000 coordinated movements per day). When the patient obtained therapy at the therapy clinic where he exercised many different movements, his coordination dynamics improved strongly, even though he exercised less on the special device than at home. There are two inferences to be drawn from the strong improvement of the CNS functioning during this period. First, after severe CNS injuries, the learning effect of the special movements exercised on the special device is smaller. The special coordination dynamics therapy and measuring device is therefore suitable for the diagnostics of CNS functioning, as improvements of coordinated arm and leg movements are mainly the result of a general improvement in CNS functioning and only little attributable to the learning of the movements themselves. Secondly, single movements, like exercising on the special device or walking on a treadmill will not reorganize the severely injured CNS substantially: Benjamin exercised at home on the special device and was walking and running every day, and he still did not improve further. The therapy has therefore to include many different movements, which have to be adapted to the individual CNS injury of the patient with its unique pathological CNS organization. A specific combination of complex movements and deep knowledge of human CNS organization and repair by learning are needed to achieve CNS repair rather than solely instruments, movements, stem cell therapy or manipulations, broadly targeted electro stimulation, or pure pharmacotherapy.

7.3.11. Need for Optimal Institutional Treatment

The variation in the improvement of CNS organization by a factor of approximately one hundred shows further that an optimal therapy at a therapy clinic is needed for efficient repair of the injured or malfunctioning CNS. Home treatment is helpful but cannot be a substitute for institutional treatment under professional supervision. The difference in the efficiency of the treatment at home and at a clinic, the latter making use of most recent data on CNS repair from human neurophysiologic research is very appreciable indeed. After these three weeks of treatment at the therapy clinic, the Author claimed: "Benjamin, if you come here to be treated for one year, in the end you will not differ very much from the other boys at school any more. You also can exercise at home under my partial supervision, but you may need a hundred years to achieve the same". Benjamin's father, a friend of the Author, asked other friends (neurologists and psychiatrists) for their opinion and did not send Benjamin for a years treatment. The patient got depressed and stopped learning treatment. When Benjamin's CNS functioning got much worse and affected everyday life, he slowly started 4 years later with CDT again.

7.3.12. Stage of Repair 15 Years after the CNS Injury

Benjamin, now 25 years old, obtained CDT, which was still far away from an optimal therapy. As can be seen from the high-load CD values of Figure 110, he progressed nicely up to 2004. When he obtained in 2004, six years after the accident, three weeks of optimal CDT in a therapy clinic, he made significant progress in repairing his CNS (Figure 107). After 2004 he stopped intensive training, because his Father stopped sending him for further intensive therapy. Sometimes Benjamin would exercise on the special CDT device for one hour but at other times he would do nothing for up to a week, apart from the everyday movements. Six years after the intensive optimal therapy he had the feeling that his condition was worse than it had been in 2004. He also felt that his intellect had declined. He therefore started to exercise up to 6,000 movements per day (2 hours). However, because the special CDT device had lost its precision, its beneficial effect on CNS function was very little. In 2006 he started to train slightly more. In 2009, he bought a new special CDT device and he also started to train more intensively. The functioning of his CNS improved again. As can be seen from Figure 110A, Benjamin's CNS functioning got worse between 2004 and 2009 and thereafter started to improve again. He also started to exercise with the help of coaches from university and improved his CNS functioning further. At the end of 2012 however, he still had not reached the level of the professional treatment in 2004.

Then Benjamin made the fundamental decision that he wants to achieve all that he can with his health and to become better than in 2004. By taking part in a three week training camp to improve running for the competition in the world championship for disabled athletes (Figure 106C), his CNS functioning improved and got better than in 2004, with respect to the high-load CD values (Figure 110). A two-week intensive therapy course with the Author improved his high-load CD values substantially. Benjamin achieved high-load CD values normally expected of uninjured people (Figure 110D).

7.3.13. Rate of Repair in Benjamin Quantified by High-load CD Values

Using the high-load coordination dynamics (CD) values, the rate of repair can be compared with those of the healthy school children. In healthy pupils and athletes the decline of the high-load CD values to reach $100s^{-1}$, approximately one month was required (Figures 96-98,110D). Benjamin needed approximately five years to get his high-load coordination dynamics values under 100 in 2004. If one includes the time during which no therapy was performed, then he reached the plateau in thirteen years. If one takes two months as the average for an uninjured person and 60 months (five years) and 156 months (thirteen years) for the repair in Benjamin (Figure 110A), then the repair/improvement in Benjamin needed 60 to 156 times more time. The huge difference of time needed to improve CNS function can clearly be seen by comparing the healthy curve (110D, inset) with Benjamin's improvement curve (Figure 110A). The rapid rate of improvement in Benjamin in 2004 with optimal CDT indicates the rate of repair depends strongly on the efficiency of the treatment.

In conclusion, the rate of learning reduced in Benjamin by approximately a factor of 100 in comparison to the healthy CNS.

7.3.14. Rate of Learning and Forgetting

If in Figure 110A the changes of the coordination dynamics values are split into learning and forgetting periods, then Benjamin needed five years of learning to get his high-load values under 100. According to Figure 110, Benjamin's neural networks forgot some of the improved functioning in five years. After forgetting of quality of CNS functioning, Benjamin learned in four years again to get the high-load values under 100 (Figure 110).

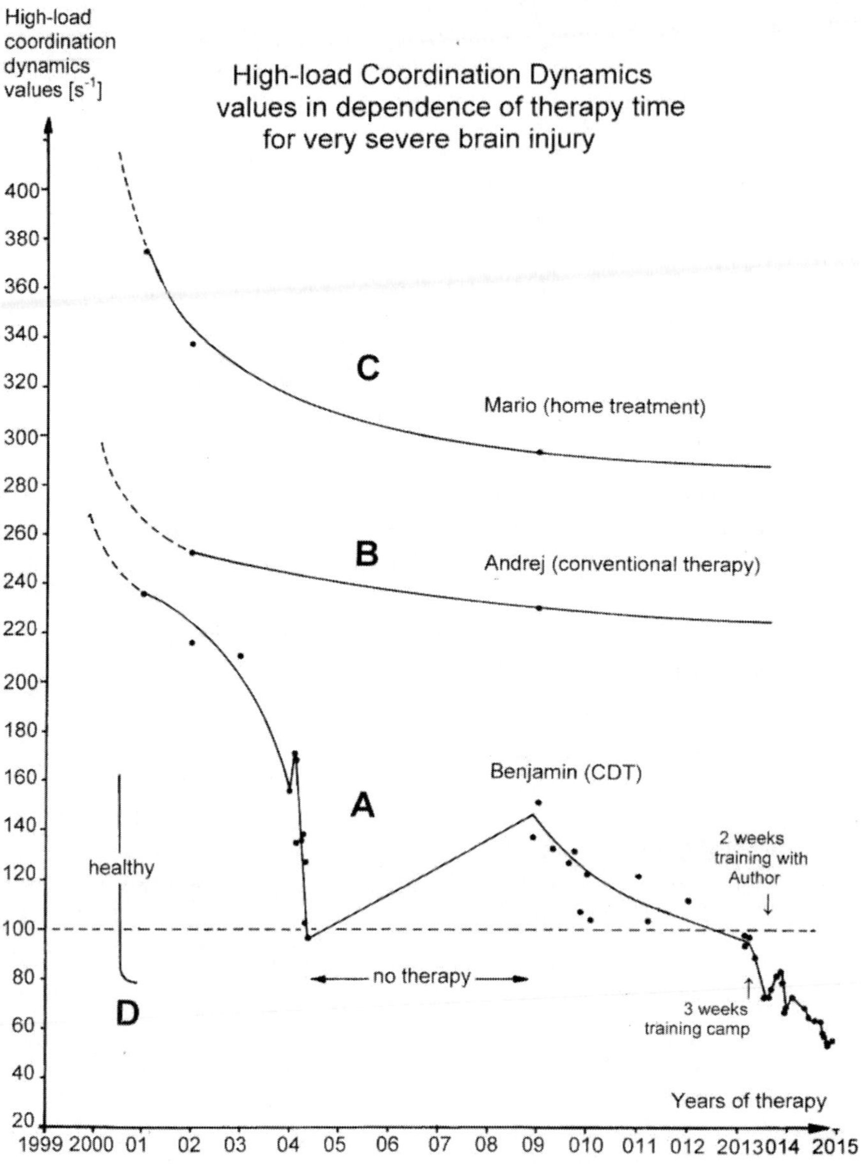

Figure 110. High-load coordination dynamics (CD) values in dependence of therapy in the patients with severe brain injury Benjamin (A) (CDT), Mario (B) (home training) and Andrej (D) (conventional therapy). The high-load CD values of healthy pupils upon repeated exercising on the special CDT device are inserted according to scale (D). Note that the high-load CD values of Mario and Andrej are far away from the healthy case with respect to CD values and time for improvement. The high-load CD values of Benjamin reach the values of healthy pupils but with much more exercise time.

7.3.15. Little Learning with Inefficient Learning

The patients Mario and Andrej never reached a high-load value of 100 (dashed line). The efficiency of their treatments was so low and did not reach the repair goal and they will have to live as disabled persons for the rest of their lives unless optimal CDT is administered to them. Even in adult patients with severe brain injury, a long time after the injury is sustained, substantial progress can be achieved if efficient treatment is administered.

7.3.16. Comparison of Injury and Repair between Benjamin Mario and Andrej

The curves of the high-load CD values in Figure 110 also provide information concerning the severity of the injury and the speed of repair. Mario had at the beginning much higher (worse) high-load values than Benjamin. His injury was probably more severe than that of Benjamin. Andrej, on the other hand, started with values close to that of Benjamin. His CNS was probably similarly damaged to that of Benjamin. But if we compare the curve of Mario with that of Andrej, one can see that Mario started from much higher values than Andrej. His curve declined more than that of Andrej. Even though Mario had a more severe CNS injury than Andrej, the repair of his CNS was more efficient. He obtained home treatment partly supervised by the Author. The home treatment was better than conventional therapy.

The curves of high-load CD values give something of an overall picture of severity of the injury and the speed of repair (Figure 110).

7.3.17. The Society Has to Look for the Health of Their Children

Only CDT could repair Benjamin's severely injured CNS. The partly supervised home treatment could repair Mario's injured CNS better than conventional neuro-rehabilitation could in Andrej. The Author predicted these outcomes at the beginning of therapy. In these cases the neuro-pediatrician of the conventional rehabilitation clinic wanted to apply CDT. However, she was pressured by more senior figures to continue with out-of-date treatment. Benjamin was included in an EU program to make videos about children with severe brain injury to show how they can manage in everyday life. The film about him won the competition. A decision was made not to include treatment, even though proper treatment is essential for managing better later everyday life. Evidently, treatment that can partly repair severe brain injuries was not of interest to the EU administration.

Since the developing (and repairing) CNS has difficulty in generating two patterns simultaneously, such as playing ball in the street and looking out for approaching cars, the necessary complexity is not yet achieved), the society has to take better care for their children in the way that they do not suffer injuries and that the children get adequate therapy if an accident occurs. Some European legal systems still decide that car insurance companies need not pay if a child is hit by a car, and the child is at fault. How can a child be at fault when his or her CNS cannot sufficiently organize two patterns at the same time?

7.3.18. Comparison between Optimal CDT and Sport Training

For practical reasons the Author was not able to offer the patient Benjamin optimal CDT again as in 2004. Benjamin started to train with university sports coaches in 2009. His high-load CD values improved to a similar extent, if not speed, as during optimal CDT treatment in 2004. With optimal CDT however, he needed eighteen days and with the sport training he needed four years for the same improvement. The repair in 2004 was approximately fifty times faster than that in the period 2009 to the end of 2012 (Figure 110). There are two obvious reasons for this. Firstly, optimal CDT is much more efficient in repair than sports training with respect to the repair of the injured CNS. Optimal CDT therapy also includes sEMG measurements to ascertain under which training conditions the best motor program is achieved (Figure 108). The more physiologic the exercise is with most patterns, probably the repair is most efficient. Generally, the coaches had specific knowledge about human neurophysiology for repair, did not quantify the progress with an assessment like the coordination dynamics measurements and were not using sEMG.

Secondly, Benjamin was approximately six years younger when CDT was administered to him. The rate of repair of the injured CNS depends strongly on the age of the patient. Younger patients have a higher rate of repair. In very young patients the rate of repair is probably tremendously higher. Premature born babies probably have the highest rate of repair, even though it is difficult to administer CDT to them. But special CDT devices are made for very small children (Figures 15,86). In premature born babies the rate of neurogenesis (from endogenous stem cells) is most likely much higher than during later development. Newly born nerve cells, once integrated in the existing networks, will have a greater impact on repair and memory than changing more or less the connectivity only. Substantial 'correction en route' during development is possible and can be used for the repair of the injured CNS, caused for example by the hypoxia due to premature birth.

7.3.19. Treatment Started Already in the Vigilant Coma State

The CNS of the patient Benjamin could be partly repaired by CDT. As stated above, the treatment has to start as early as possible. The possibility cannot be excluded that Benjamin's tremendous improvement is partly due to the early start in the vigilant coma stage (Figure 145A). The same may be true of Mario. Mother, physician, physiotherapist (Figure 120 of [2]) and the Author all believed in early treatment.

7.3.20. Rate of Repair Depends on Age and Treatment Quality

The progress in repair depends on the severity of the injury, the quality of treatment and the age of the patient. Benjamin offers the possibility to compare the rates of repair for different treatments. Above it was shown that Benjamin needed between five and thirteen years to improve his high-load CD values to what a healthy person learns in two months at the most. The rate of repair in Benjamin was 60 to 156 times slower.

In 2004, it was shown that the repair efficiency in Benjamin could be improved by a factor of 100, when he was 15 years old.

In 2013, a coach in a training camp was able to improve the high-load CD values in Benjamin, below the values of 2004 in three weeks (Figure 110A). But in 2013, the Author was also able to improve the CD values substantially with twelve days of optimal CDT (Figure 110A). The patient reached the values of trained healthy persons (Figure 128D). A comparison of the efficiency of treatment between the coach and the Author seems possible. But the comparison of treatment is complicated by the possibility that the benefit from treatment may be latent to some extent.

Further, by comparing the progress made by the patient in 2004 and in 2013, it seems that the rate of repair was higher in 2004, most likely because the patient was younger then. In 2004 he improved by 80 in 16 days. This means that Benjamin's CD improved by 5 per day.

In 2013, he improved by 27 in 90 days, which equates to an improvement of 0.3 per day. In 2004 therefore, the rate of learning was approximately seventeen times faster. However, when the nervous system is already functioning quite well, judged by a high-load CD value of under $100s^{-1}$, then it is much more difficult to improve its function further in both the injured and uninjured persons (Figures 110,96).

7.3.21. No Limit of CNS Repair if CDT Is Applied Continuously over Several Years at the Limit

It will be shown now that there seems to be no limit of CNS repair, if CDT is applied continuously over many years at the limit and is supervised by the Author who can upgrade the therapy if needed.

It is predicted by the System Theory of Pattern Formation that the intrinsic dynamics F_{intr} can be changed, i.e. repaired by the behavioral information $F_{inf}(X,t)$ (the training of the different movements) of relative strength (c_{inf}) (volume of a certain movement training) according to the equations of motion of collective variables (the vector X):

$$dX/dt = F_{intr}(X) + \sum c_{inf} F_{inf}(X,t)$$

It was further predicted that when with a certain training the CNS functions cannot be changed any more, then the treatment has to be changed. However, the System Theory provides no information about the specific changes needed to upgrade the therapy and the intensity to be applied.

It seems that when for example the walking performance is no longer improving, the performed movements cannot reach (activate) damaged networks for repair, which are urgently needed for improving walking. But it could also be that in the chain of events for repair, some other networks have to be repaired first before the neural networks before walking can be improved in its function. And last but not least, the benefit from newly built nerve cells when exercising at limits seems to take place only after a year of treatment.

With the patient Benjamin, it was shown in Figure 107 that through an upgraded therapy in a professional setting, his CNS functioning could be improved further, even though Benjamin thought that he had reached the limit of repair in 2004.

It is always good practice to repair the CNS because one never knows what complication may come with aging. For example, patients with mild poliomyelitis are often able to walk. In spite of lost motoneurons, the CNS manages to generate the walking pattern. But with aging and degeneration of the CNS, polio patients may lose the ability to walk and have to use a wheelchair (post-polio syndrome).

The patient Benjamin (traumatic brain injury) also had problems with the improvement of walking, running, and jumping performance. In 2014, he asked the Author for more supervision. And indeed, during one week of intensive supervision two movements could be found, namely certain jumping (Figure 111B) and walking patterns, which reached those damaged neural networks that contribute to the improvement of the walking and running patterns.

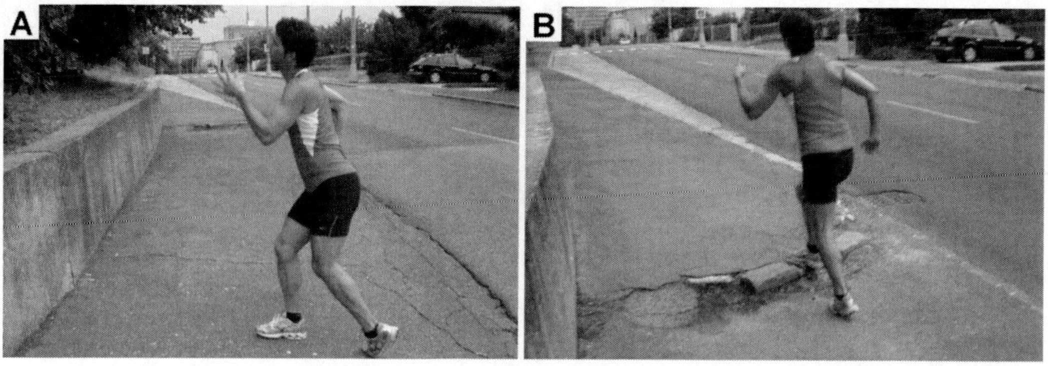

Figure 111. Jumping in anti-phase of a now 25-year-old patient Benjamin who suffered a severe brain injury 15 years ago. A. Left foot is in a pathologic position. B. When jumping against a step, the performance of jumping is much better.

At an age of 24, the main problem in CNS functioning was still the pathologic functioning of the left arm and leg. The free jumping in anti-phase improved the coordinated movement of the left arm. But during this free jumping in anti-phase, the left heel came up (Figure 111A). The physiologic positioning of the left foot is important for all movements and needs to be corrected so that the CNS does not learn a pathologic pattern. Now, by freely jumping against a step, the left foot took a physiologic position and the left arm moved (Figure 111B). Only the left hand and foot were a bit spastic. This kind of upward jumping made the anti-phase jumping somewhat physiologic: The left arm moved and the left foot was in the physiologic position. When jumping a step downwards, the left foot still went into pathologic position.

Interestingly, when climbing staircases upwards, the left foot performed a physiologic movement. When going downstairs the foot positioning was pathologic, in some similarity to the jumping in anti-phase, the step up and downwards.

The backward walking performance was quite good in Benjamin. The only problem for covering a distance is that the eyes are on the wrong side of the body. The normal forward walking performance on the other hand was poor and had to be improved. The forward walking could be improved in the short-term memory by simulating and activating the stepping automatism. The stepping automatism (Figure 18a,89) is mainly located in the neuronal networks of the spinal cord and is stimulated for example by the heel strike. The patient had suffered a traumatic brain injury and the stepping automatism should therefore be only little disturbed. When Benjamin was walking in such a way as to simulate the stepping automatism by exaggerating the lifting of the knees during walking and striking the ground with the heel first, the left leg was not swinging out any more and the pathologic inward rotation of the left foot reduced.

It seems likely that Benjamin can further improve his walking and running performance when training these two movements.

The instruction ‘put the heel down or do not swing the leg outwards during walking’ will not help very much to improve the walking performance, but finding certain movements, which functionally reach the damaged networks, can.

When Benjamin got better and his high-load CD values got better, he wanted to compete with the Author and to surpass his good high-load values. The Author cannot compete with Benjamin’s fast running. However, nobody so far can beat the Author in the high-load test where good nervous system functioning, quantified by the coordination of arm and leg movements, is also needed.

7.3.22. Comparison of High-load CD Values between a Disabled Sportsman and the Author (Aging)

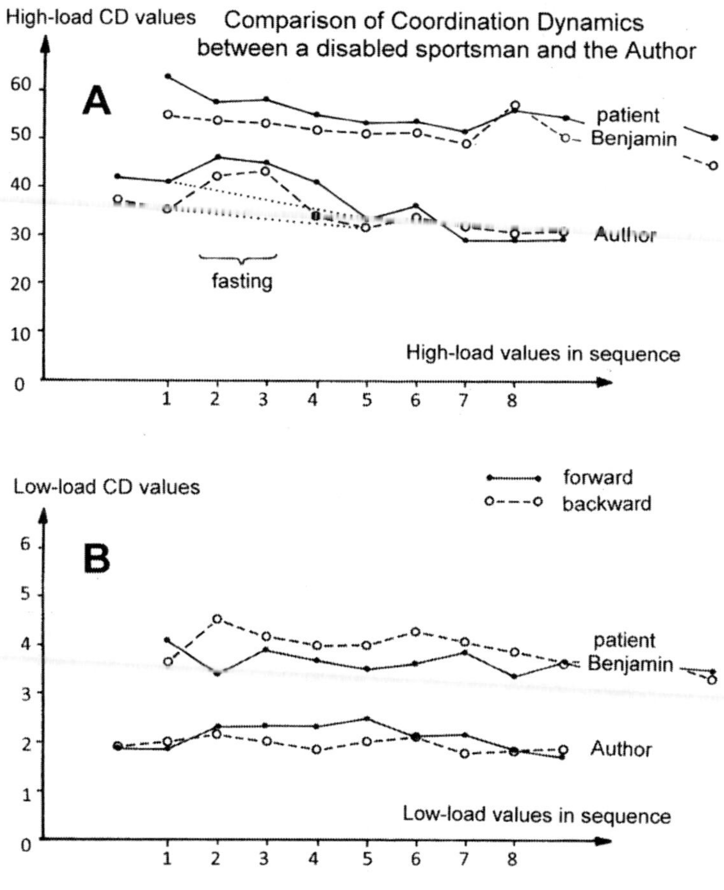

Figure 112. Comparison of high-load (A) and low-load values in sequence (B) of the patient Benjamin, who suffered a severe brain injury 15 years ago, and the Author.

Figure 112 shows the recent high-load (A) and low-load values (B) in sequence of Benjamin (traumatic brain injury 15 years ago) and the Author (over 70). It can be seen that the disabled sportsman Benjamin could reach good (low) high-load CD values, but he is still far away from the Author's values. Recent improvements were obtained in the patient by jumping in anti-phase, including the arms during jumping (Figure 111). With these coordinated arm and leg movements during jumping, the performance of the left arm improved. On the other hand, the low-load CD values were comparably poor (high) in Benjamin in comparison to the Author (B). It seems therefore as if Benjamin is succeeding in getting good high-load values partly by using his muscle strength, whereas the Author needs to put the emphasis on good coordination between arms and legs to compensate for less muscle strength. This means that the Author is taking advantage of better phase and frequency coordination between neuron firing in the CNS.

The competition between Benjamin and the Author can also be seen as a competition between a young person fighting for more repairs of his injured brain, and an older person fighting by use of the nervous system for more rejuvenation of the body. The attempt to reduce the biological age will be picked up below, under aging.

7.3.23. Rate of Learning Quantified by High-load Testing

Learning and repair aspects can be explored by evaluating all high-load measurements. The overall improvement of CNS functioning while performing CDT, can nicely be seen by plotting the high-load values, as was shown in Figure 110. Another example for measuring the rate of learning in the severely injured brain can be seen in Figure 113. The 25-year-old patient of Figure 38 suffered a severe brain injury in a car accident and performed CDT to repair the brain. For details see [2]. It can be seen from Figure 113A that the improvement of the forward-backward symmetry of exercising (and to get the high-load CD values under $100 s^{-1}$) took 40 months. The healthy pupil learned this forward-backward symmetry in half a month (Figure 39). The rate of learning for this forward-backward symmetry was reduced by a factor of 80 in this patient. In stroke patients, the repair of the right-left symmetry is an even greater problem.

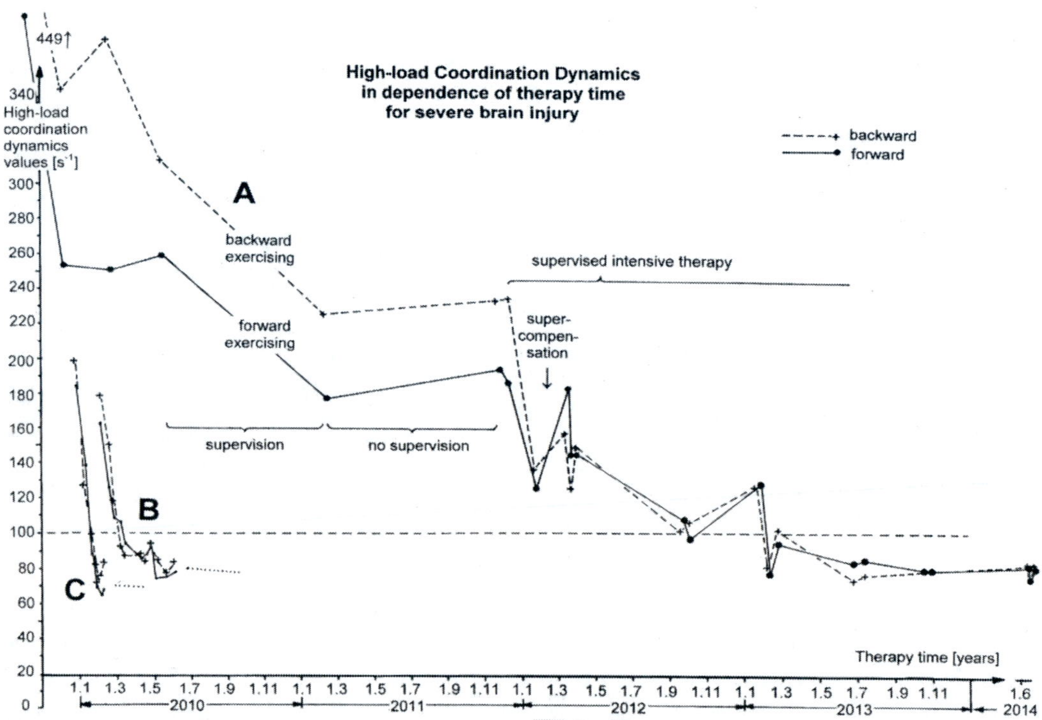

Figure 113. A. Improvement of high-load coordination dynamics (CD) values in a patient with severe brain injury upon coordination dynamics therapy for several years. The high-load CD values were obtained by summing up the single CD values for forward and backward exercising, Δ (high-load CD value) = Δ20N + Δ50N + 100N + Δ150N + 200N + Δ150N + Δ100N + Δ100N + Δ50N + Δ20N + Δ20N). B, C. For comparing the rate of repair, the improvement curves of the high-load coordination dynamics values of an athlete (C) and a healthy pupil (B) are inserted. Note that the brain-injured patient needed much more time to achieve similar good CD values.

From Figure 113A, it can further be seen that supervised intensive therapy is more efficient for repair than training without supervision.

It can also be seen from Figure 113 that the overall CNS repair needed longer than three years. The insets B and C of Figure 113 show the repair curves of the normal pupil (B) (Figure 96) and the athlete pupil (C) (Figure 97) adapted to the time scale. It is obvious that the healthy pupils learned faster. The CD value curves reduced quickly to cross the dashed 100 line, and to reach the plateau, indicated by a dotted line. The overall performances of the movements were also better. The CD values for forward and backward exercising started from around 390 to 450 and those from the pupils at 160 to 200.

7.3.24. Super-compensation

In the normal pupil, the phase of super-compensation occurred approximately after three months (Figure 93) and in the athlete after 1.5 months (Figure 94). The movement-based learning depends on the case of the healthy pupils also on the intensity of treatment, i.e. how frequently the assessment sessions were repeated. Repeating the high-load test every three days may be optimal for a start of a fit person. With shorter repetition times, the CNS gets exhausted but with longer intervals, learning is not optimal. In mild cerebral palsy (Figure 101) and in mild epilepsy (Figure 102) the first phase of super-compensation occurred after one month, with something approaching optimal learning, and 3.5 months where learning is not optimal in repeated assessment sessions. In the patient with severe brain injury, the probable first phase of super-compensation occurred after 26 months of therapy (Figure 113A).

In conclusion, in healthy subjects the first phase of super-compensation occurs after approximately two months of movement-based learning. In mild CNS disturbances, the rate of learning is normal or only slightly reduced in accordance with the knowledge that humans are resilient to minor disturbances within any CNS subsystem [77]. In severe brain injury, the rate of movement-based learning with respect to the occurrence of the first phase of super-compensation is reduced by a factor of 10 to 20 (one to three months as opposed to 26 months). The rate of CNS repair by learning in general is reduced by a factor of 50 to 100 (see above). Obviously, the CNS is not resilient to severe injuries. It will be shown below that in very severe brain injury the rate of movement-based learning is so reduced that certain functions cannot be re-learned at all.

7.3.25. The Rate of Repair by Learning Declines with the Severity of the CNS Injury

The high-load coordination dynamics test is used for quantifying the rate of CNS repair by learning. The time required for learning or repair differs strongly between an uninjured and injured nervous system. A healthy athlete learned to get the high-load test value under $100s^{-1}$ in approximately half a month (Figure 97) and a healthy person needed approximately one or two months. Even among athletes, the values can vary quite significantly. Persons with a mild malfunctioning of the CNS needed three (Figure 101, mild cerebral palsy) and 4 months (Figure 102, mild epilepsy) to get the high-load value under $100s^{-1}$. The young man with mild cerebral palsy was probably comparably good in movement performance because he walked every day 7km to school, which was healthy in general and improved CNS functioning.

Patients with severe traumatic brain injury could reach the high-load test value of $100s^{-1}$ in approximately 32 (Figure 113, Sotiris) and 48 months (Figure 110, Benjamin). A cerebral palsy patient with high intelligence but poor walking performance needed 19 months of therapy of varying intensity to reach the high-load value of 100, whereas a cerebral palsy girl with poor intelligence and moderate walking performance could not manage the high-load test after eight years. Over the years, she turned only with 20 and 50N and never learned to exercise at a load of 150 and 200N. But continence and speech could be learned (Figures 87,88). For further details see [1,2].

A patient with an incomplete C5/6 cervical spinal cord injury (approximately 50% injury) reached the high-load test value of $100s^{-1}$ in approximately 13 months (Figures 51 of [2],100; Sten). A patient with a motoric complete C5/6 cervical spinal cord injury could not reach the value of 100 within eight years. She never could exercise by herself at a load of 100N. Exercising at a load 50N was just possible for her (Figures 19,29; Kadri).

An athlete who suffered a thoracic complete spinal cord injury, succeeded in reaching a value of 100 in a few months (Figure 59 of [2], Avo). But because of his strong upper-body strength and the ability to go over limits, he could manage to reach that value by turning only with the arms, which is not achieving any repair, only an optimization of what is left.

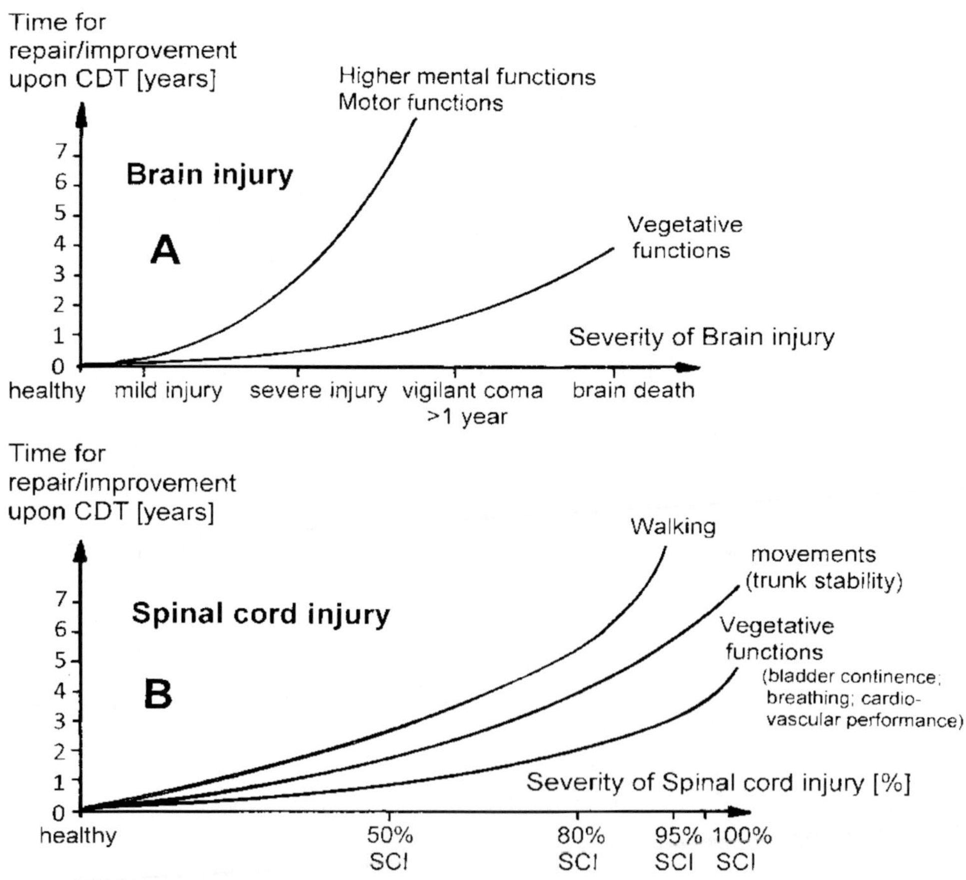

Figure 114. Approximate repair times in dependence on the severity of the brain (A) and spinal cord injury (B).

Two patients with severe traumatic brain injury to whom no CDT was administered for longer times could not reach the high-load test value of 100 within fifteen years (Figure 110; Mario and Andrej) and will probably never reach such quality of CNS organization.

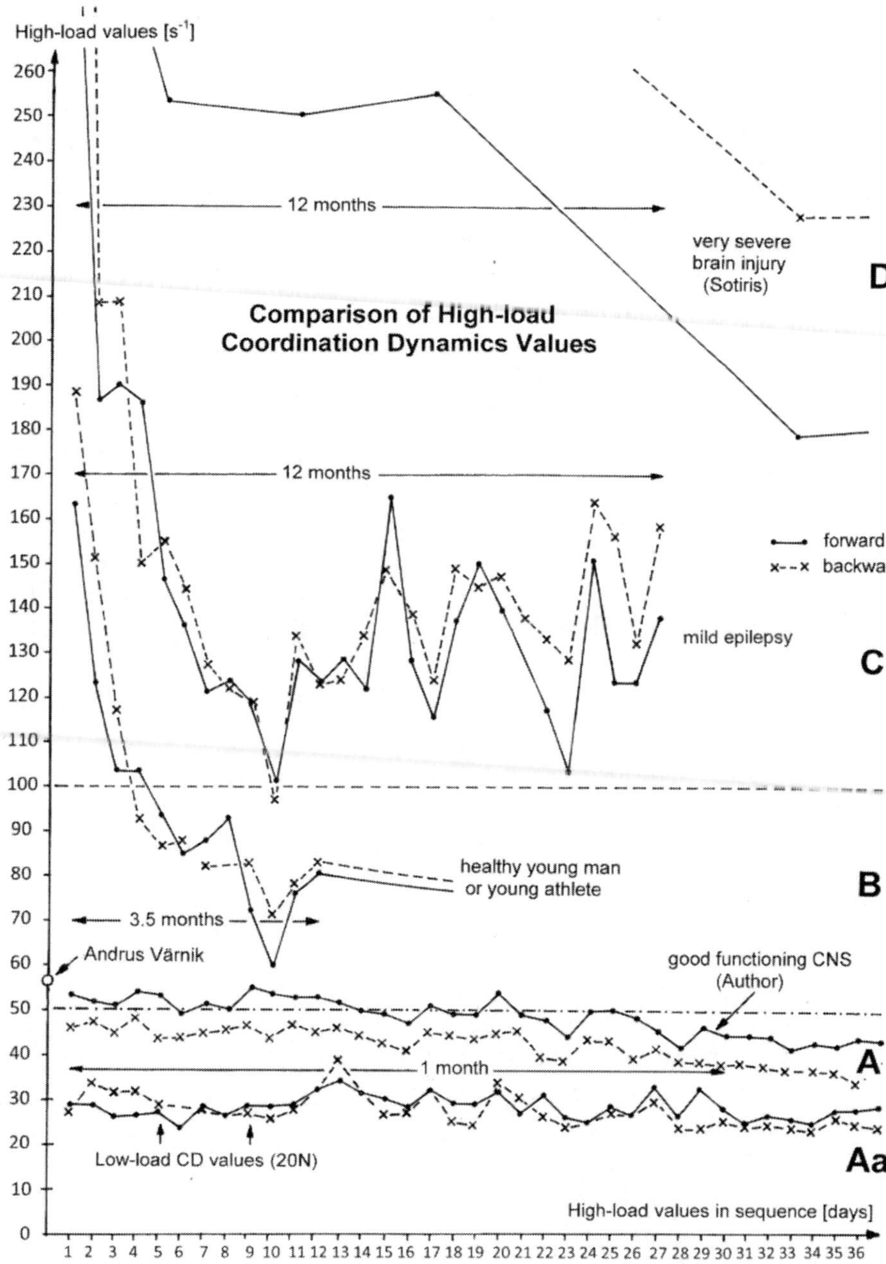

Figure 115. Comparison of the high-load values in sequence of a very well-functioning CNS (A, the Author), a healthy young man or athlete (B), a female patient with mild epilepsy (C) and a young man with a very severe brain injury. "Aa" shows the low-load values in sequence for the Author. The scale for low-load values is that for high-load divided by 10 (for example, 30 → 3). The high-load (HL) CD values are obtained by summing up the single CD values, Δ (HL CD value = Δ20N + Δ50N + 100N + Δ150N + 200N + Δ150N + Δ100N + Δ100N + Δ50N + Δ20N + Δ20N).

Patients with very severe brain injury, and being longer than one year in the vigilant coma state never managed the high-load test so far. Also, brain dead humans cannot perform the high-load test, even though some reflexes can re-appear in some of them if kept ventilated for longer times [2].

These data of movement-based learning (for repair or improving CNS functioning) show that the times for repair by learning strongly increase with the severity of the injury (Figure 114). What a healthy person learns in half or a month, a patient with a severe brain injury needs approximately three years. Or, what a healthy person learns in one year, patients with severe brain injury cannot learn at all.

7.3.26. Repair by Learning Depends on the Kind of Injury

As was shown above, the repair of movements, quantified by the high-load coordination dynamics values, depends also on the kind of CNS injury. Patients with a complete spinal cord injury (SCI) C5/6 cannot relearn walking, but the higher mental functions are not impaired, whereas patients with a severe brain injury can relearn walking, but their higher mental functions may still be impaired. In SCI the outcome of learning is in part determined by the site of the injury (Figure 81) and injury severity. In brain injury, it is more difficult to relate the rate of repair by learning to the site and the severity of the injury. Injury of brain stem nuclei is often not taken into account. Thalamus injury has a strongly negative effect on the rate of repair. In some injuries, for example in hypoxic brain injury, it is difficult to locate the site of injury. Even though it is difficult to clearly relate the rate of repair to the site of brain injury, some important conclusions can be made about the quality of repair by learning.

7.3.27. Comparisons of the Improvements of High-load CDs for the Healthy and Injured CNS

Figure 115 shows the high-load values in sequence of a very well-functioning CNS (A, the Author), a healthy young man or athlete (B), a female patient with mild epilepsy (C) and a young man with a very severe brain injury. The measured values varied between 30 and approximately 300.

Even though the time periods of training vary between one month and a year, it is obvious that the high-load values, characterizing the functioning of the CNS, vary dramatically. That the patient with the severe brain injury and the biggest impairment of phase and frequency coordination had the highest (worst) values is understandable, because his CNS was severely damaged. But that the top athlete Andrus Värnik, World-Championship javelin gold medal winner, does not have similar values to the Author is astonishing. This is because one would think that a top athlete would have a very well-functioning nervous system.

This means firstly that sportsmen can improve their performance further, especially in terms of coordination, by integrating CDT into their training program. Secondly, the CD values characterize only one, albeit an important, aspect of CNS functioning. Football or tennis players could certainly improve their ball skills when including CDT in their training program. By improving the coordination dynamics, there is learning transfer to vegetative and higher mental functions. It is also possible to reduce stress and improve mental disorders, but further research is necessary.

7.3.28. Learning by High-load Training Has a High Impact on CNS Repair

The rationale for the Author to perform many high-load tests in sequence was first to gain an insight into how hard it is to do them. A physician should feel what he expects from his patient. Secondly, it needed to be ascertained whether the test could be done every day for a sustained period. Thirdly, it is of interest to determine just how good (low) high-load values can be achieved in general. As Figure 115 shows, high-load tests can be done every day without overloading the body, as demonstrated by the fact that the Author's values did not get worse. Performing the high-load test is hard and requires mental fortitude. As expected, the values improved continuously with successive test performances.

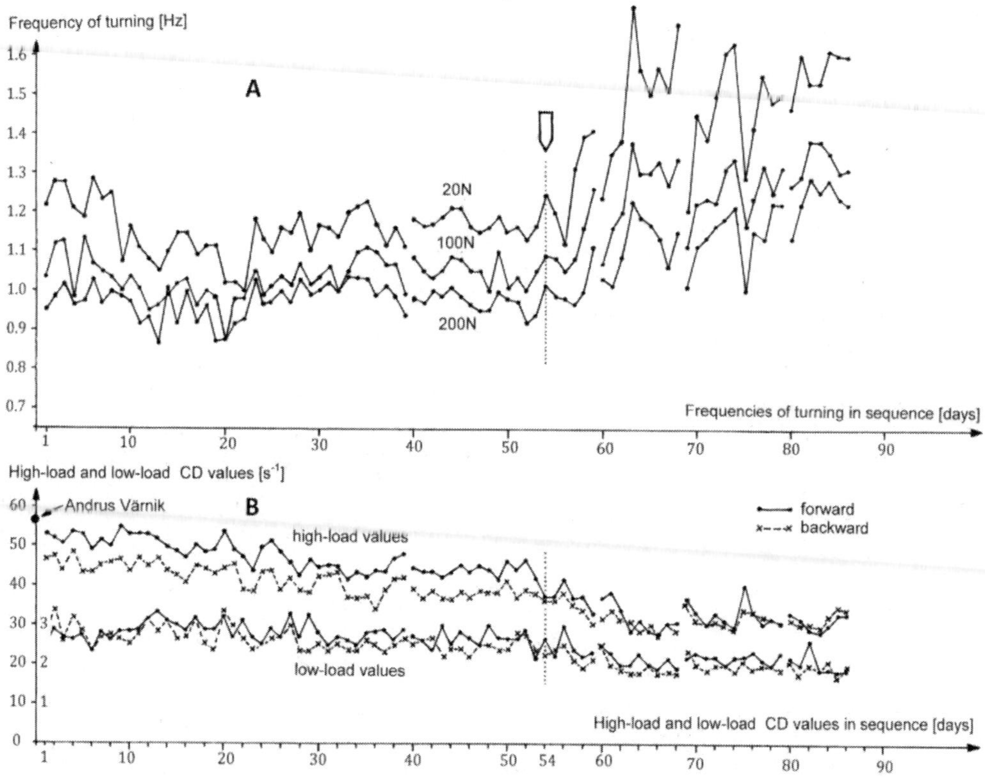

Figure 116. Continuous high-load tests (every day apart from a few breaks) of the Author with a load increase from 20 up to 200N and back to 20N. In "A" the turning frequencies are given for 20, 100 and 200N for exercising in the forward direction. In "B" the high-load test values and the CDs for 20N (lower traces) are given for exercising in the forward and backward direction. Note the strong improvement of the CD values after 54 days of exercising in sequence.

Unexpectedly, after 54 days of continuous high-load tests (Figure 116) the high-load test values got significantly better and the turning frequency got increased, as if some specific changes had taken place in the CNS and improved its function. The improvement of the high-load CD values does not originate in the increased turning frequency f (CD values = df/dt:f), because a person's performance is best if the person turns at his natural frequency. Varying the turning frequency from the inner frequency helps to repair the CNS by increasing motor pattern variability, but the best value for turning is achieved at the own natural frequency and the best values are used to characterize CNS function. That the Author could turn faster with better rhythmicity was therefore due to an improvement in CNS function. The inner frequency increased because of better CNS function. High-load testing in sequence every day up to 200N seems to improve CNS function and helps in repair, provided of course that the patient is able to do it.

Chapter III

Neural Network Learning in Coma Patients

Abstract

Neural network learning is also possible in patients who are not conscious. The learning is un-volitional. It is shown that coordination dynamics therapy (CDT) can be applied to coma patients. If the injury is not too severe, the patients recover earlier from the coma. In a patient with very severe brain injury, first the vegetative functions improved when CDT was administered. After three years, including two years of CDT, the patient was brought out of the coma.

The progress of neural network learning can partly be judged by the impression of the face of the patient and possibly by still working protection reactions. It is emphasized that such patients need efficient learning treatment to survive and to have a chance to be brought out of the coma. The very severely damaged CNS neural networks can often not recover spontaneously. Efficient therapy is needed. Only when the patient becomes conscious again the mental and motoric conditions become visible.

1. Learning and Memory of Neural Networks in Coma Patients

1.1. Repair Strategies at the Neuron Level to Implement Repair in Coma Patients

1. *Repair depends on learning and memory formation, mediated or supported by epigenetic mechanisms.* The epigenetic is the interplay between genes and environment resulting in phenotype and epigenetic landscape.
2. Epigenetic mechanisms like DNA methylation are probably sensors for movement-based learning and memory formation and fine modulators of neurogenesis in the adult CNS with CDT.

3. The *epigenome consists of non-coding RNA and chromatin*, a proteinaceous matrix surrounding DNA. The dynamic interactions of post-translationally modified chromatin proteins, covalently modified cytosine's inside DNA and non-coding RNA define the complex pattern of gene expression beyond the four bases of DNA.
4. The hippocampus plays an essential role in learning and memory. In the hippocampus, there exists a specialized form of neural plasticity, which is the generation of new functional neurons from stem cells occurring throughout life. Adult hippocampal neurogenesis contributes to learning and memory formation.
5. *New neurons are important for learning and memory formation* (besides functional reorganization), i.e. for increasing the rate of repair, for the following reasons:
   a. The insertion of new neurons helps to store the memory of the same activity that led to the creation of the neuron.
   b. *Activity-dependent neurogenesis enhances the learning of new memories and degradation* and clearance of previously stored *unwanted memories* like spasticity, because the synapses, dendrites, and axons can be devoted more fully to the newer memories. The old neurons with large and complex axon and dendritic trees are difficult to change. They can only be changed with sustained effort.
   c. *New neurons seem to improve the accuracy of relearned patterns* (from model study [51]) this means that new neurons help to improve phase and frequency coordination of neuron firing and pattern stability.
   d. The *advantage of new neurons* seems to be dramatically *greater* in networks that had been *more active* and had been *required to store more memories* [51]. The advantage of neurogenesis for memory storage in heavily active networks is that it provides an increased rate of repair if movement-based learning is administered aggressively and if different movements are trained.
6. *Specific natural network activity is required for multiple aspects of repair.*

   Specific activity is essential for correct migration of interneurons and controls the development and repair of their axons and dendrites. During repair, there is a specific requirement of network activity in shaping the cortical integration of specific neuronal subtypes. Newly built neurons are likely electrically active shortly after their birth and participate in the early network activity that contribute to circuit maturation during repair during CDT.
7. *Specific activity is required for migration and maturation at several stages of repair. A break in CDT may invalidate the whole chain of repair events.*

   Specific interneuron subtypes require activity for migration and morphological maturation at two distinct stages of development [51]. Newly build neurons may even require specific activity for migration and maturation at several distinct stages of repair. During a break in CDT, the specific activity, required for neuron migration, maturation and network integration may not be supplied at one of these stages so that the chain of repair events is severed and the whole repair chain has to be started anew.
8. *For optimal repair specific downstream and upstream activity is needed.*

   Since coma patients cannot move by themselves, only the specific movement induced afferent input is available for repair. But since the movement induced afferent input during support by a therapist improved motor programs (Figure 42), it is likely that the specific activity induced by the movement induced afferent input, can to a certain extent also offer the specific activation necessary for proliferation and migration of neurons and network maturation. Further, some automatisms like the blink or swallowing reflex are partly working and activate motor patterns.

9. *Drug application may undermine repair.*

   Altering the level of neuronal excitability within genetically targeted neurons from drug application, for example antiepileptic drugs may have profound consequences on multiple aspects of the repair of select types of neurons within a population of neurons, as well as their associated gene expression. The pain-killer ‘Contergan’, taken during pregnancy, changed gene expression and the babies were born without arms.

10. *Excitation-neurogenesis coupling [51]*

    a. Excitation increases or decreases neuron production directly by excitation-neurogenesis coupling.
    b. The *excitation acts* indirectly on the surrounding mature (hippocampal) cells through *depolarization-induced release of grows factors.*
    c. *Adult neurogenesis* is enhanced by excitatory stimuli and *involves Ca^{2+} channels* and NMDA receptors.
    d. The *Ca^{2+} influx pathways are located on the proliferating stem/progenitor cells* (NPCs), allowing them to directly sense and process excitatory stimuli. The Ca^{2+} signal in NPCs leads to rapid induction of a proneural gene expression pattern.

11. *Integrative coordinated movements* have to be trained to allow functional reorganization and new nerve cell *integration across very large distances.* CDT has to activate injured and uninjured networks to enhance physiologic CNS functioning and learning transfer.

12. *Conclusion for optimal therapy according to the present stage of knowledge.*

    If there is similarity between development and repair, animal (mice) data also hold in humans and the principles of neurogenesis of the hippocampus also hold in other parts of the brain, albeit to a much lesser extent, then the patient has to be trained at his limits (1) to induce substantial building of new nerve cells. The treatment has be continuously administered (2) to support all stages of repair at the progenitor level as migration, maturation and integration and the networks requiring repair have to be activated specifically (3) to generate repair-friendly micro-environmental properties in the networks. No drugs should be administered that change neuron excitability (4).

With the above repair strategies in mind, we shall now examine brain repair by learning in coma patients.

1.2. The Duration of the Coma Depends on the Severity of the Brain Injury and How Early Efficient Treatment Is Started

In even more severe brain injuries than the above-mentioned cases (Benjamin, Sotiris) the limit of repair possible through CDT is reached. However, improvements from CDT, which are possible in patients who have, for example, lost approximately 30% of brain tissue are of interest for medicine in general.

The patients of the Author who suffered a severe brain injury were all in the vigilant coma first. The severity of the brain injury seemed to be partly reflected in the time period necessary to get them out of the coma spontaneously. The therapy should be started in the coma stage to reduce pathologic reorganization and to shorten the coma period.

Benjamin, who suffered a severe brain injury at the age of nine when he was hit by a car, came out of the coma after seven weeks. The treatment was started as early as possible in the vigilant coma stage. Two other boys with similarly severe brain injuries (Mario and Andrej) needed up to six months to emerge from coma mainly because their treatment started either late in the coma phase or shortly after it had ended.

On the other hand, Benjamin's brain injury, even though severe, was not as severe as in the patient Manolis, which will be reported below. In an eighteen year-old patient (Georg) with a very severe brain injury, CDT was unable to get him out of the vigilant coma. A few years after the accident, he died from a lung infection in hospital. However, a fifty year-old man who suffered a very severe brain injury following a bicycle accident came out of the coma. For years, he appeared to be in a non-vigilant coma. It was only later discovered that he was able to communicate through blinking. Objective judgment is necessary to clarify whether a patient is in the coma or not.

1.3. Volitional and Un-volitional Neuronal Network Learning

When the patient is conscious he can move volitionally and initiate automatisms volitionally and can stay in the pattern. However, the coma patient is unconscious and cannot start a movement or automatism volitionally. Some protection automatisms still function in such patients. In the patient Manolis, discussed below, the blink reflex was still working. When something moves suddenly towards his eyes, the eyes closed. The swallowing reflex also partly worked in him. Maybe other reflexes or automatisms are partly activated when moving the patient passively. It is therefore not only the afferent input which activates the neuronal networks for learning and re-learning, the motor pattern from reflexes and automatisms also contribute.

In the final stage of the vigilant coma the patient Benjamin (Figures 121,106A) partly understood that his mother and therapists were trying to help him and he tried to actively participate to activate the movements even though he could not move and speak as he later related when he came out of the coma. His intention activated networks and helped in this way to repair the networks by learning. Communication with the coma patients is therefore helpful for repair, even though there is no certainty as to how much the injured neural network patterns can be activated in the patient. The voice of the mother normally goes most deeply into the networks because the memory was formed deeply in the childhood.

On the other hand, there have been brain-dead humans who moved the legs coordinately when a leg was touched. This means the spinal cord and parts of the brain stem reorganized (learned) and made movements possible when the leg was stimulated by touch. The remaining neural networks of the brain dead human learned but the brain dead human could not become conscious again because of the damaged brain.

Patients who suffered very severe brain injuries and were in a coma for years lay between a healthy subject and a brain dead one. The question is: where is the border of damage between these two cases, which should indicate whether to decide to attempt through efficient treatment to get the patient out of the coma and attempt to bring him back to meaningful life or to let him die in peace? Administering inefficient treatment to a patient who has suffered a very severe brain injury may slowly kill the patient, because the severely injured CNS needs efficient help to survive. It cannot help itself any more. The difficulties in administering efficient learning treatment is discussed in the case below. Cursory decisions made with insufficient knowledge are dangerous for the patient. Even the Author experienced how difficult it is to predict the prognosis of a patient with a very severe brain injury. Only deep knowledge on the functioning of the human neural networks under physiologic and pathologic conditions, combined with a lot of experience can help. Even the regulations units to initiate speech pictured partly in Figure 87 indicate how complex CNS organization is.

Another problem can also be encountered. The possibility cannot be excluded that in some cases physicians do not fight sufficiently for the future of the patient who suffered a very severe brain injury, because it is very difficult to make money with administration of efficient movement-based learning therapy. When, on the other hand, the patient finally suffers brain death because of insufficient treatment, organs can be explanted and re-implanted in other patients. Money can be earned with transplantation of organs.

2. Case Report: Patient Out of Coma by 2 Years of CDT, 3 Years after Car Accident

2.1. Optimal CDT in a Coma Patient

A 21-year-old male patient suffered a very severe brain injury in a car accident. A metal shard went through his skull into the brain. Parts of the brain were removed and a shunt was installed to control brain pressure. An attempt to reconstruct the skull was unsuccessful. After several operations in a university hospital, he was placed in a rehabilitation center and provided with twenty minutes physiotherapy per day. His relatives felt that this was insufficient and paid for an additional hour of therapy per day. This was still insufficient to even maintain his level of CNS function. Prior to the three months of conventional therapy in a rehabilitation center the patient was able to respond 'yes' and 'no' by closing and opening his eyes respectively. After this conventional therapy, he was no longer able to respond with yes or no. The ability to communicate with the patient was lost. The injury was most likely so severe that CNS functioning got continuously worse.

After the three-month rehabilitation period, the patient was taken home and CDT was commenced one year after the accident. There was always the fear that if the treatment is not sufficient the patient may die.

Prince Friso of the Dutch royal family suffered a hypoxic brain injury and died in 2013 after being in a coma for eighteen months.

2.2. Visible Anatomical Damage due to the Car Accident

An MRI was performed one year after the accident at the same time that CDT was started. It showed that many parts of the brain were injured. (1) Figure 117A,B,C seem to show a hydrocephalus, but the ventricle enlargements were secondary to the loss of nervous tissue. If one compares Figure 117 with Figure 88, a cerebral palsy patient but no enlarged ventricles, then it is obvious that the ventricles of the patient in Figure 117 were significantly enlarged. The first, second and third ventricles were enlarged in the coma patient. The fourth ventricle was normal. (2) The corpus callosum was very thin. Impaired communication between the two hemispheres has to be expected. (3) The frontal lobe was injured (Figure 117C) on both sides. (4) The left thalamus was slightly injured. (5) The frontal and parietal lobe showed atrophy. (6) There was great loss of parenchyma of the right hemisphere where the metal shard had entered. (7) Valerian degeneration of the pyramidal tract took place probably due to the injury of the sensory-motor cortex. (8) The cervical spinal cord seemed to be undamaged, but bones seem to touch the spinal cord and may exert some pressure with the consequence of loss of spinal cord functions. (9) No sign of acute ischemia can be seen in the patient's brain. (10) The valve artifact from the shunt can be seen in Figure 117C and D. There may have been no part of the brain left undamaged.

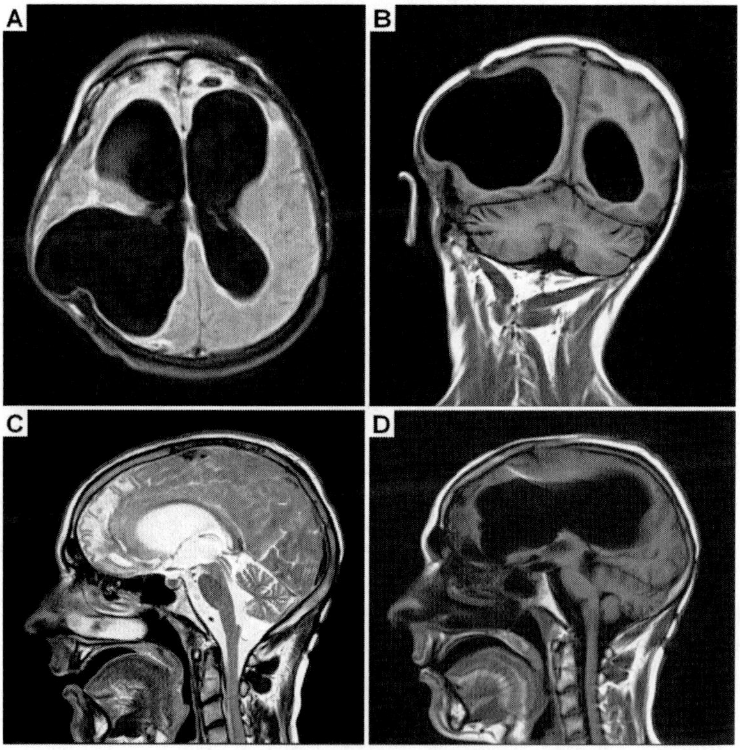

Figure 117. MRI of a 22-year-old male patient with a very severe traumatic brain injury. The pictures were taken one year after the car accident. Since a metal piece went into the brain during the accident, brain tissue had to be removed. A shunt was installed to regulate brain pressure. Because of the loss of nervous tissue, the ventricles became enlarged looking like a hydrocephalus. Many parts of the brain were damaged during the accident. Both frontal lobes were damaged (A, C). The artifact from the shunt can be seen in 'C'.

2.3. Pathologic CNS Organization

The left leg of the coma patient showed extensor spasticity and the right leg flexor spasticity. The right arm showed extensor spasticity and the left arm flexor spasticity. Strong rigor was present, but no tremor at the beginning of therapy. The joints were not working properly due to having layed in bed for one year with insufficient mobilization. A bit of clonus developed later on in the left leg and a bit of tremor developed in the hands. There was some dysregulation of the sympathetic division. One foot was warm, the other one was cold. During exercising, the cold leg was sweating.

When exercising on the special device in the lying position (Figure 118A) the brain pressure was higher, because the liquor transport via the valve was not as good. Upon exercising in the sitting position on a special CDT device or during sky-walking (Figure 118B,C,D) the liquor transport was better. The brain pressure could be felt by touching the skin above the open skull.

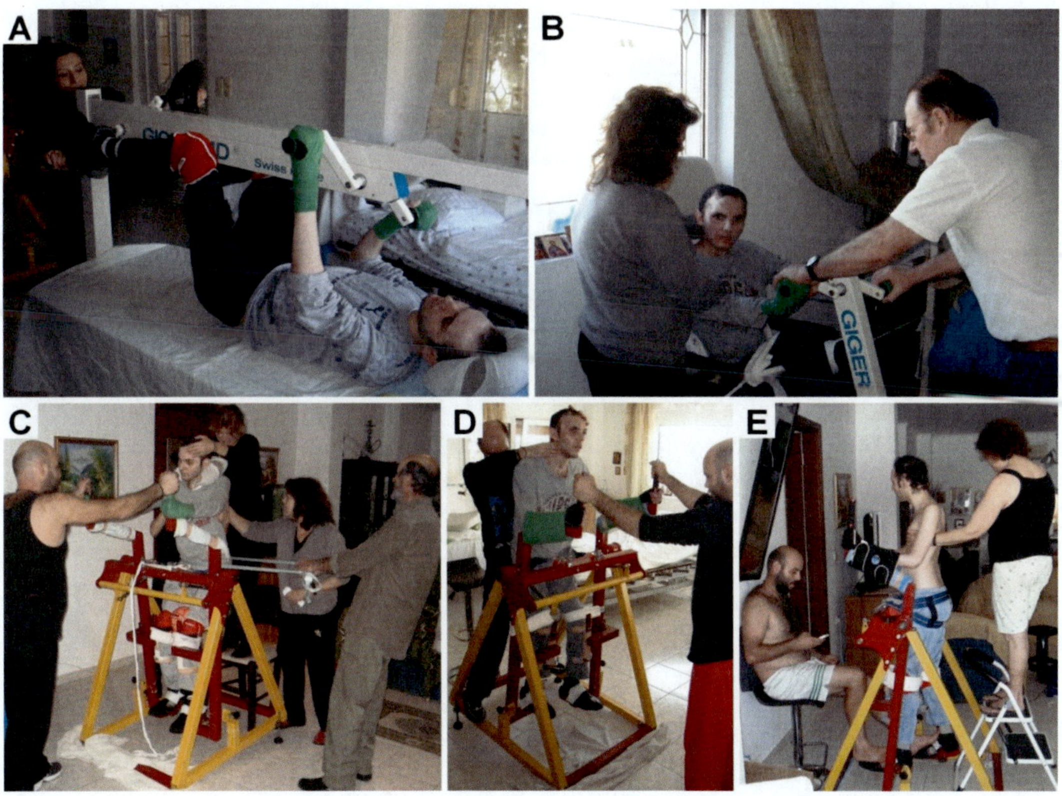

Figure 118. Movements performed with a 22-year-old male patient in the vigilant coma. All movements are passively performed with the patient, because the patient is in the coma and does not move by himself. A,B. Exercising on the special CDT device in the lying position (both hands are fixed in the prone position) and sitting position. C,D,E. Exercising on the sky-walker. Note that the trunk and head control improved from ‘C’ to ‘E’; in ‘E’ no support of the head was needed any more after 5 months of CDT.

2.4. The Pain and Consciousness Problem

Ideally, treatment would have been started slowly and then reach a treatment time of twenty hours per week in approximately three months. However, it was not possible to increase treatment intensity slowly in this patient because the benefit from the treatment may have been insufficient to counter the ongoing pathologic processes, putting the patient's life at risk.

The joints were a significant obstacle to therapy. Because they were not really used for one year, it was difficult to mobilize them. For movement-based learning therapy, the joints obviously need to work.

Pain and other receptors will have informed the damaged CNS about problems in joints, shortened tendons, and stiff muscles. In spite of the damaged thalamus, the thalamus will have attempted to inform the cerebrum of this pain. Whether the hemispheres got the pain messages (natural impulse patterns coding pain) and processed them in the right way is unclear. The natural impulse patterns coding pain (for natural impulse patterns coding touch and pain see pages 292 to 297 of [1]) (Figure 6) could have been changed by the damaged thalamus, so that the cerebrum could not recognize these patterns properly. The other possibility is that the impulse patterns were physiologic, but the hemispheres could not recognize them properly because of their damage. Due to the thalamus damage and severe damage of the hemispheres (both frontal lobes were injured and also one part of the sensory-motor cortex was severely injured), the pain coding impulse patterns will have been altered and the processing in the hemispheres will have been also changed. Still it seemed that the patient's CNS recognized pain and his expression hardened and sometimes even tears were in his eyes when performing certain movements. It is unclear whether the patient really felt pain in the functional deepness of CNS organization where the consciousness is also sited or whether the pain input patterns changed the impression in the face more in the periphery of CNS organization (a vegetative organizational state) in an automatism or reflex like manner.

Pain, or the perception thereof, and the changing impression of the face posed a wider problem. His mother, brothers, and sisters thought that Manolis was communicating with them. The Author could not safely prove that the patient was transiently out of coma. The possibility cannot be excluded that sometimes, when the CNS of the patient was in a good organizational state, achieved by exercising on the special CDT device, the patient responded consciously to his mother.

2.5. Difficulties in Applying the Necessary Efficient Learning Treatment to Coma Patients

In mild brain injuries, the patient's CNS can largely repair itself without treatment. In severe CNS injuries, the patient's nervous system needs treatment so that its functioning does not become too pathologic. In very severe brain injuries, a patient's CNS may destroy itself if the treatment is inefficient; the repair functions are in a sense overtaken by the ongoing pathologic processes. As was shown above, when comparing the rate of learning in a patient with a severe CNS injury with subjects with a mild or no injury, it turned out that the rate of learning was reduced by a factor of 20 to 50 in patients with severe injuries. A patient with a severe CNS injury may need a first approximation of 20 to 50 years to learn what a healthy person can in one year. It must be borne in mind that, in the severely injured nervous system, change takes a long time and the rate of learning depends on the kind of injury.

When the treatment with the coma patient Manolis was started a year after the car accident, the CNS injury was so severe that the nervous system seemed to be destroying itself because the pathologic processes such as high brain and blood pressure, bleeding, and toxins from destroyed nerve cells seemed to be stronger than the repair processes. There was therefore the danger that the patient could die with or without treatment. Another obstacle was his mother. She devoted herself to her son. She did a very thorough job with his care and contributed to the treatment. But she was over-protecting her son. When the treatment increased in intensity it could most likely be seen in his face (see below), then she mother wanted to stop the treatment.

The patient would probably die if he did not get sufficient treatment. With some treatment, the patient may not die and some functions in the functional periphery of the patient's CNS would get better or recover. However, the patient may stay permanently in the vigilant coma. This may not be a meaningful life and in the long-term, the patient may die of a complication like a lung infection. The neurosurgeon, who operated on the patient, stated that, according to his experience, his relatives should be happy if they could maintain the patient's current level. With optimal treatment, as far as that is possible in such a patient however, he might be brought out of the coma, whatever the mental and motoric conditions of the patient would be afterwards.

2.6. Performed Movements

The performed coordinated movements with the coma patient were exercising on the special CDT device in the lying and sitting position and the sky-walking (Figure 118). Additionally the patient obtained eight hours conventional physiotherapy per week. Mobilization and stretching are important, but they contribute only little to repair. All movements had to be performed passively. Due to the flexor and extensor spasticity, rigor and stiff joints with a small angle of motion, it was difficult for the therapists to enable the patient to move his arms, legs, and trunk. The patient altogether received twenty hours movement therapy plus five hours speech therapy per week.

2.7. Progress and Obstacles to Repair after Five Months of CDT

Even though the patient could only move passively, some progress in repair was achieved. At the beginning of therapy several persons were needed to keep the patient in the walking position (Figure 118C), then only two (Figure 118D) and after four months of therapy the patient could sometimes keep his head in the upright position (Figure 118E). The trunk and head control had improved in some similarity to development. The improvement of the trunk control and reduction of the scoliosis like bending of the trunk is astonishing because of the extent of the damage to parts of the sensory-motor cortex. Swallowing also improved and the blood pressure improved (reduced). It seemed that the biggest improvement was with respect to automatisms and vegetative functions.

2.8. Regression of the Coma

The improvement of automatisms and vegetative functions indicates some repair. Can the specific neural network activation with its functional and structural repair also reach the consciousness? Since in coma patients, the coma terminates earlier upon CDT (see above), it is likely that the movement-induced coordinated afferent input from the passive movements helps to get the patient out of the coma. The quality and intensity of the treatment in combination with the severity of the injury will dictate responsibly whether consciousness will be regained.

2.9. Repair after Nine Months of Therapy – First Signs of Regression of Coma

The protection reflexes of the head improved further, including the blink reflex. The blood pressure, a major problem at the beginning, became healthy (after sleep for example, 90/66 mmHg, pp. = 77). The patient learned to swallow: he could eat and drink a bit, but it was reflex activated. The most apparent progress was that the patient seemed to come out of the coma transiently while exercising on the special CDT device in the lying position. After 500 turns on the lying machine, the patient appeared to be more alert and he seemed to be able to communicate with his relatives by opening and closing his eyes. Whether he was in fact indicating saying "yes" and "no" by eye movements was unclear because of the brain damage and especially the damage to the sensory-motor cortex, which seems to have also affected the motor functions of volitional eye movements. In the early stages of repair, early functions are inconsistent.

In patients with somewhat less severe brain injury, there was a clear change from the vigilant coma to the conscious state. When the patient Benjamin came out of the coma, he started to speak. In this very severe brain injury there does not seem to be any clear transition from the coma to the non-coma state. All movements and functions seem to be disturbed or mixed up. Only the protection responses of the eyes seem to be good and precise. Swallowing was also trained and improved consequently.

2.10. Repair after 16 Months of Non-optimal Therapy – Limited Improvements

After 16 months of non-optimal CDT, the condition of Manolis had improved only a little. The trunk stability improved further. Consciousness however, did not. A follow-up MRI showed the same anatomical condition as 16 months before. Figure 119 shows for example the enlarged first and second ventricles, which have roughly the same shape as 16 months before (Figure 117A). There may still have been some pressure from the vertebra onto the spinal cord as in Figure 193A of [2], to be one cause for the high spasticity in the patient. It is not clear whether a further operation would be justified. The anesthetic would put the patient's severely injured brain under severe strain. The display of the blood vessels of the brain via additional contrast medium showed some ischemia. Optimal CDT under this condition could also improve the blood supply of the brain further.

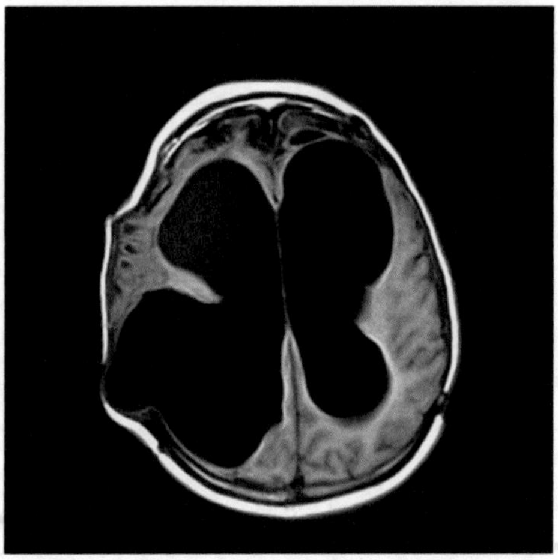

Figure 119. First and second ventricles 16 months after the very severe brain injury of a now 23 years old patient.

2.11. Control of Functional and Structural Repair

Probably the principle that local cellular environments, at the network place of injury and distantly for organization of patterns, are important in controlling functional reorganization and neurogenesis. The repair potential may depend on the kind of local activity (1), access to local activity (2), ability of the local environment to induce activity-sensing competence for repair (3) and factors intrinsic to the resident neurons themselves. Further, functional reorganization and neurogenesis might exist in modes that are switched on and off by the inputs from the movement-induced afferent input. Progenitor cells could be able to distinguish local activity related to new memory storage and repair of (movement) patterns (which should trigger neurogenesis) from activity relating to patterns with no repair load placed on the remaining circuits.

The neuromodulatory inputs available so far are the movement induced coordinated afferent input from the passively performed sky-walking and exercising on the special CDT devices. It is unclear whether this level of therapy will ever be sufficient for the recovery of full-consciousness in an injury of such severity and given the extent of the loss of nervous tissue.

2.12. Judging CNS Organization by the Impression of the Face

Figure 120 shows that the improvement of CNS repair can be seen also in the face of a patient with severe cerebrum and cerebellum injury. The appearance of the face improved over ten years with the CNS repair achieved through CDT (Figure 120).

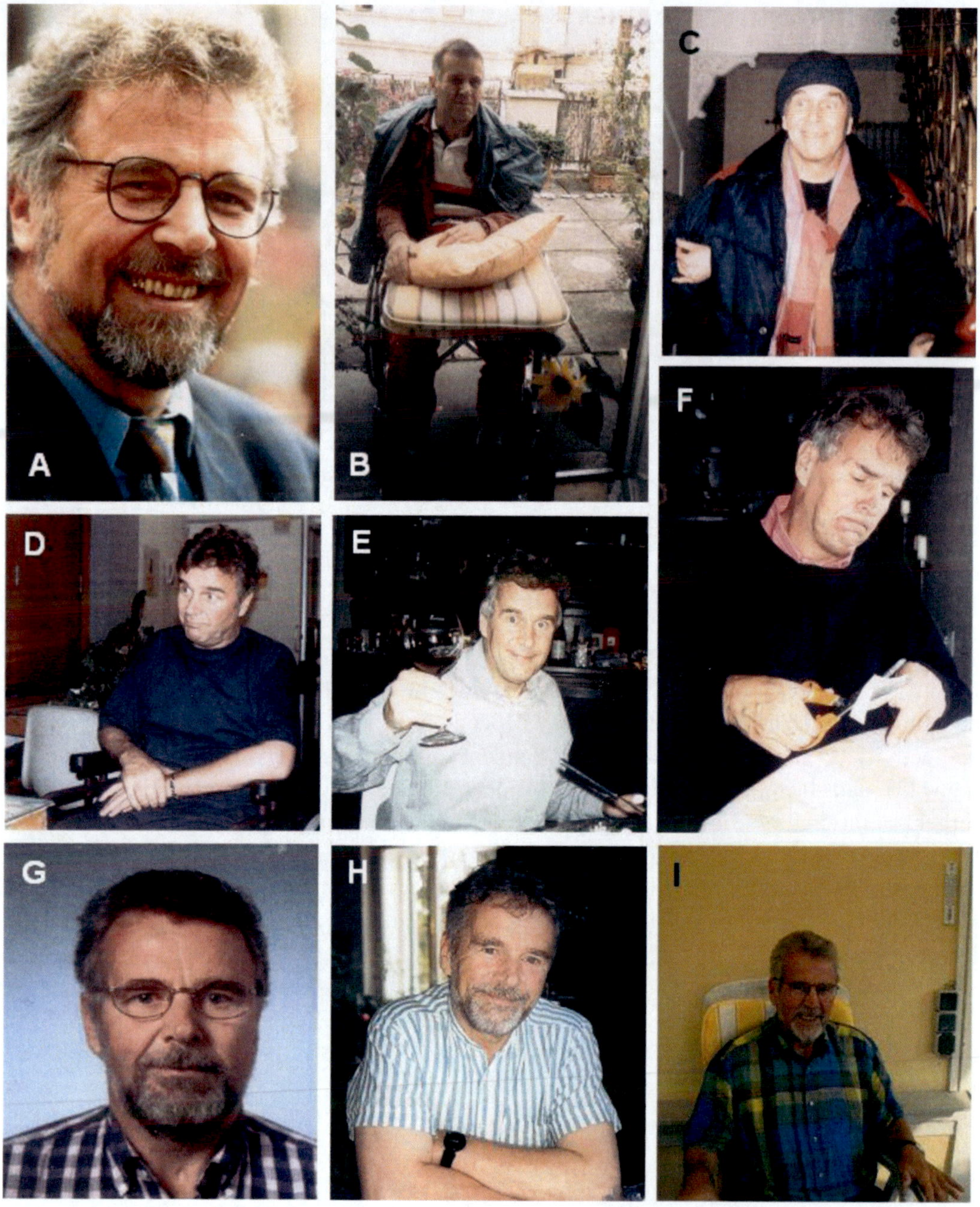

Figure 120. Therapy-related improvement of the impression of the face of the patient who suffered severe cerebellar injury: before the accident (a), after the accident (1995) till 2006 (b-i).

Despite suffering a severe brain injury and being in a coma, the face of the patient Benjamin looks quite healthy (Figure 121). The impression of the face of this patient Manolis, however, seems to indicate a more severe injury (Figure 122). The eyes were veering in one direction, but would react to noise. After exercising on the special CDT device, after approximately 1000 revolutions, the eyes were not veering so much anymore, but were moving more from right to left and backwards. The eyes became livelier, more excited, as if the CNS of Manolis organized itself in a more differentiated manner. The integrative coordinated afferent input improved CNS functioning in the short-term memory. But whether such repeated improvement of CNS functioning is sufficient to get the patient out of the coma was unclear after five months of CDT and subsequently seemed to be more plausible after nine months of CDT. Similar experience was made with another young man who was in a coma for nearly two years due to a very severe brain and spinal cord injury. The more arms and legs were moved, the healthier and livelier the impression of his face became.

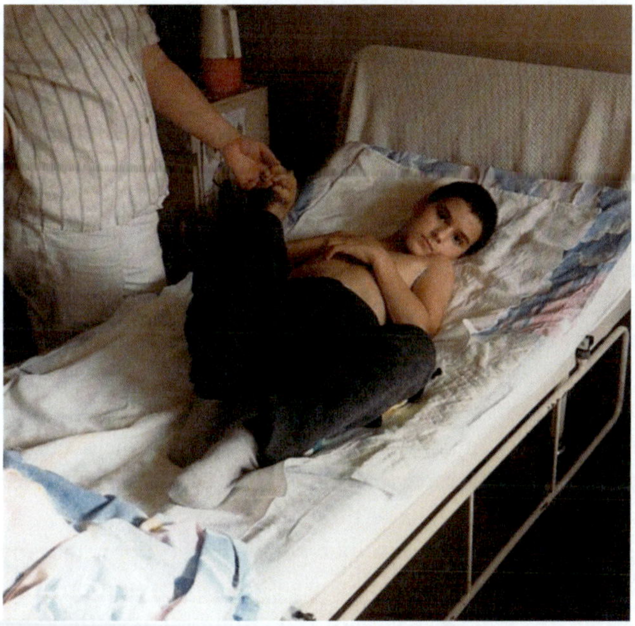

Figure 121. Picture of a 9-year-old patient Benjamin in the vigilant coma following a car accident. The picture was made shortly before he came out of the 7 weeks lasting coma. Note, the impression of the face looks healthy.

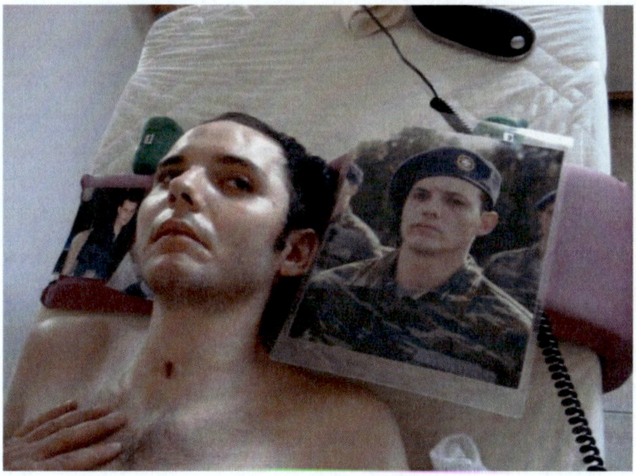

Figure 122. Picture of a 22-year-old patient being 1.5 years in the vigilant coma following a car accident. The impression in his face does not look as healthy as the one of the picture made 2 years before when the patient was in the army.

2.13. Brain Pressure Control

The brain pressure of the patient could be felt when touching the head where the skull was missing. The neurosurgeon checked that the shunt and the valve were working well. According to the neurosurgeon, high brain pressure should be avoided. The brain pressure rose in the lying position. This is likely due to the extended periods spent lying in bed or when exercising on the special device in the lying position (Figure 118A). However, too low brain pressure should also be avoided because of the risk of bleeding in the brain. The brain pressure reduced when exercising in the sitting position on the special CDT device (Figure 118B) or exercising on the sky-walker (Figure 118CDE).

2.14. Normalization of Blood Pressure upon CDT

The blood pressure was very high at the beginning of therapy. The diastolic pressure was often over 90mm Hg. First, the blood pressure increased upon exercising and then the pressure maintained its level during therapy and after five months of therapy, the blood pressure reduced in general and transiently while exercising on the special CDT device in the lying position. The high blood pressure at the beginning of therapy was probably mainly caused by the malfunctioning of the severely damaged CNS. With the improvement in CNS function, the blood pressure became normal, indicating an improvement in the vegetative nervous system. The cardio-vascular performance had improved after five months of therapy and was nearly normal after nine months of therapy.

An average change of the cardio-vascular performance upon therapy in the short-term memory was the following. When lying relaxed in bed, the blood pressure was (110/68; 81) (systolic pressure/diastolic pressure; pulse rate). The impression of the face was similar to that in Figure 122. After being exercised on the special CDT device in the lying position (Figure 118A) for 200 turns (turning frequency below 1 Hz), the pressure was (125/80; 93). The nervous system was more activated by the movement induced afferent input. Some pain receptors in the joints may have been activated. After 900 turns, the blood pressure decreased (123/75; 81). Probably nervous system functioning had improved in the short-term memory with the consequence that the cardio-vascular performance became better. With slightly faster turning, the pressure and the heart rate increased (126/83; 87). Velocity-dependent spasticity may have been activated more and may have induced more pain. Upon reducing the speed of turning, the heart rate reduced again (126/84: 82).

It was shown that in severe cervical spinal cord injury (Kadri; [2]) the cardio-vascular performance could be improved to such an extent that within six months the nutrition of the skin reached the point that below the injury level the skin would no longer bleed when touched firmly and pressure ulcers no longer occurred. Here it is shown that even in very severe brain injury leading to a coma of more than one year's duration, the cardio-vascular performance could also be repaired.

It is likely that in aging (see below) the cardio-vascular performance and microcirculation can be improved through CDT, even though arteriosclerosis will hinder the improvement. The effects of CNS degeneration as nerve cell death and in consequence impaired CNS organization can be partly compensated for.

2.15. Non-optimal CDT

Even though the brain pressure issue was seemingly solved, the cardio-vascular performance was nearly normal, and the administered therapy time was approximately twenty hours per week, the number of performed coordinated movements was still too low. The number of administered turns on the special CDT devices reduced from 23,000 per week at the beginning to fewer than 10,000 after five months to nine months. When pushing the patient closer to his limits and administering more exercising, the blood pressure went up and the impression in the face of the patient became harder as if he was suffering more pain. His mother stopped the therapy to protect her son. It is very important to push the patient to his limits to enhance neurogenesis to get more urgently needed nerve cells for repair. At critical sites of CNS organization of new nerve cells are needed certainly for rebuilding the networks. It is unlikely that with a mere functional re-organization the patients CNS can be improved that much that he comes fully out of the coma. It is not just the loss of perhaps 30% of the brain matter causing the severe functional deficits; in this case, too many parts of the brain were damaged. A neurosurgeon told the Author that he removed nearly one hemisphere of her secretary due to a malignant astrocytoma, but after the operation, she could go on with her work in spite of the loss of brain matter.

2.16. A Lack of a Satisfactory Outcome for Patients who Have Lost Significant Brain Matter

The relatives of the patient were doing a very good job with respect to care, but they were reluctant to push the patient to his limits so as not to cause him discomfort. For these extremely severe CNS injuries, there seems to be no elegant solution. It is important to discover just how much repair is possible in such severe injuries. This knowledge can only come from properly treated patients.

2.17. Out of Coma 3 Years after the Accident

Two years and nine months after the car accident and after 22 months of CDT, the patient Manolis repeatedly could put his finger to his nose when asked. It became clear that the patient had come out of the vigilant coma. How conscious he really is now is unclear. Hopefully his higher mental functions improved and he can tell later on and has not forgotten the present stage of repair by learning. The memory formation also must have been repaired sufficiently to remember. The patient Benjamin ultimately forgot what it was like to be in a coma.

2.18. Meaningful Life

To get a patient out of the coma is only the first step in the repair of the CNS by learning. Now Manolis needs to be brought back to meaningful life. There are patients who got out of the coma, but did not get sufficient learning treatment afterwards to regain a meaningful life. They can communicate via eye blinking but are fully dependent on the care of others. Their mind is entirely trapped in their bodies. Often the family does a very good job with respect to care, but they become worn out over a period of years. No will is left to fight for a better future for the patient. The society must be developed enough to take over the load, apart from proper treatment. This is not the case in Manolis' family for the time being. The same problem occurs with the treatment of cerebral palsy children, especially in autistic children. The parents get worn out from treating the child over the years and the social system society does not adequately assume the burden of care.

Hopefully the patient Manolis will receive further somewhat optimal learning treatment to have a chance to get back some quality of life. It is unclear whether this level of therapy will ever be sufficient for the recovery of full-consciousness in an injury of such severity.

2.19. Learning from the Repair of Severe Injuries for the Mild or Medium Severe Injuries

When performing qualified research and treatment on severely brain-damaged humans, society can learn to treat, with this knowledge, less severely injured patients, and learn from nature to better construct computers and robots. It is also interesting how human neuronal networks create intelligence and consciousness.

2.20. Sexual racism and out-of-date neuro-rehabilitation

A 15-year-old girl in Greece was raped by a man from Pakistan. Afterwards the man tried to kill the girl, Mirto, by hitting her on the head with a stone. The girl did not die; however, she did suffer a very severe brain injury. The man was put to prison for life. Normally such prisoners can leave the prison after 20 years.

Three years later, the author was invited by the parents for a consultation concerning an up-to-date neurotherapy of the now 18-year-old girl. According to the father, Mitro's brain functions had not improved in the last three years by the conventional physiotherapy of Greece and USA. With the excessive administration of drugs in the USA, Mirto's brain functions even seemed to become worse. As argued earlier, drugs cannot repair brain functions. The farther was very worried concerning the patients future, because the severe spasticity seemed to block all movement therapies. Mirto's case was made public in the mass media, especially in Greece and USA. Money was collected from all around the world to help Mirto. Especially people with little money gave money to help the poor girl. Even though the parents urgently need money for efficient treatment and a rebuilding of the house, the collected money did not seem to reach Mirto. Further, Mirto's case is used for the advertisement of an out-of-date neuro-rehabilitation.

The author trained Mirto together with the parents for two days with a special CDT device (Figure 123) to see whether her brain functions could be improved and the spasticity reduced. With the two-day CDT it turned out that Mirto was cooperative during treatment. A communication with eye blinking was clearly possible; she was clearly out of coma. The severe spasticity was reduced, and her motor and higher mental functions improved in the short-term memory, even though the administered movements were only passively. Mirto could not move volitionally. Trunk and head stability were very poor. Still there is indication that her brain functions can be improved.

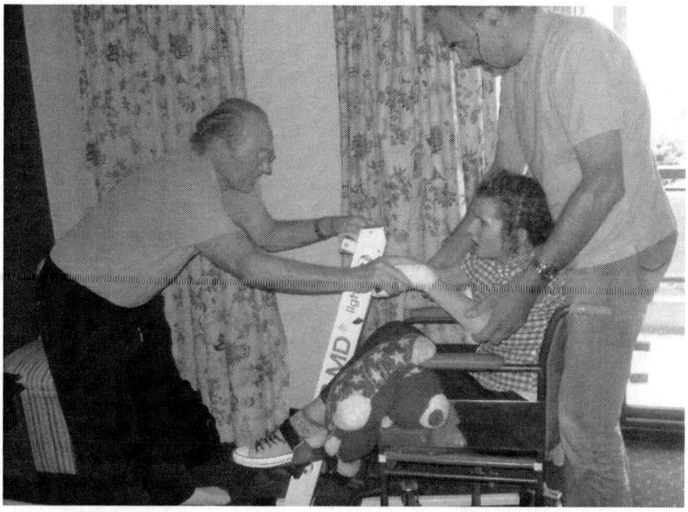

Figure 123. Training on the special CDT device of a patient with a very severe brain injury. Note, the Author (left) is trying to 'catch' the patient's mind during supported exercising and the farther supports head and arms. The left foot is in a pathologic position. The mother makes this picture.

If Mirto does not obtain optimal neurotherapy her prognosis is extremely poor. First raped, then nearly killed and then trapped in her body for the rest of her life. The man who raped and nearly killed her can go to the toilet, Mirto cannot; He can move arms and legs, she cannot. If she does not get optimal treatment in the future, she has to suffer a much worse situation than being in prison for the rest of her life. Only optimal neurotherapy can give a bit of hope for a meaningful life. Handling women as an object and refusing them education and development is a kind of racism. As measured below with the coordination dynamics, the nervous system of a woman functions just as well as that of a man (Figure 130).

Chapter IV

Learning to Improve Health in Aging and Cancer Treatment

Abstract

Movement-based learning is used to improve health in aging and following cancer treatment, i.e. surgery, chemo and radiation therapy. Since the nervous system is involved in nearly all body functions, it could well be that CDT enhances specific physiologic functions and inhibits non-specific pathologic functions. The improvement of cardio-vascular performance by CDT includes the microcirculation and the lymph vessel system. Edema occurring following lymph node resection can be reduced more quickly with the re-building of the lymph vessel system stimulated by CDT. Since repeated prolonged fasting seems to improve the functioning of the immune system by replacing old, damaged and pathologic cells by new healthy cells, a paired CDT and repeated prolonged fasting may improve health and reduce the risk of a tumor recurrence. During CDT and repeated prolonged fasting, performed by the Author himself, fast and transient fast exercising on the special CDT device occurred similarly, as during development. There was some similarity between development and rejuvenation. To live longer with a better quality of life seems to be achievable by movement-based learning and repeated prolonged fasting supported by good nutrition and sleep.

As quantified by the coordination dynamics, women and men have the same quality of CNS organization with respect to the coordination pattern dynamics values.

1. Aging

1.1. To Live Longer with a Better Quality of Life

If only single cells or small groups of cells die in a somewhat distributed manner through parts or all of the CNS, then a repair-friendly microenvironment can be generated for structural and functional repair through CDT. However, if a severe injury is confined to one area, it is then very difficult to create such an environment. Due to the loss of nerve cells, the coordinated firing of neurons becomes impaired and CNS functions become impaired. With CDT, the coordination dynamics can be improved and CNS functions become more physiologic again. In the elderly, it is unlikely that any significant number of new nerve cells can be built throughout the CNS, apart from within some structures like the hippocampus. However, through intensive movement therapy some new nerve cells can probably be built even later in life [78] and structural repair can contribute to functional repair, which is the substantial part. Even if the building of new nerve cells is very limited, functional repair can still optimize what is left of the degenerating CNS.

Many generations dream of a "Jungbrunnen" (fountain of youth) to become young and healthy again. The famous painter Lucas Cranach had pictured such a "Jungbrunnen". In his painting the old people enter the "Jungbrunnen" one side and leave it young and healthy on the other side. Such a dream cannot become reality by using CDT. However, it may be possible to reduce biological age through repair.

In patients with spinal cord (Kadri, Figure 60) and brain injury (Popi) bladder functions could be repaired [2]. Therefore, bladder continence can probably be improved in the elderly through exercising on the special CDT device and (if possible) jumping on a springboard. Repair is however compromised in that the main feeder artery of the conus medullaris is often very arteriosclerotic. The blood supply to the sacral micturition and sexual function center is then strongly limited and the functions become impaired. The cardio-vascular performance can be improved upon exercising on the special CDT device, because CDT can improve the functioning of the vegetative nervous system including the micro-circulation of the blood and lymph vessel system. The functioning of the arterioles is controlled by the vegetative nervous system division, which can be improved in its functions upon CDT. An old saying in medicine is "you are as old as your blood vessels are". In an elderly patient with lung fibrosis (the oxygen is hindered to reach the blood vessels because of the fibrosis due to smoking) the oxygen saturation of the blood could be improved upon exercising on the special CDT device. The patient needed less additional oxygen. The patient could breathe better. The progress was probably achieved upon improved blood supply in the lung. The lung functions were optimized. Therefore, the breathing can also be improved in elderly. If the functioning of the blood vessels can be improved and new cells can be built upon CDT, then also eye functions should become better upon CDT. Operations may be needed only later. Movements can anyway be improved in elderly. By sitting on a chair or lying during exercising, the elderly can even exercise on special CDT devices if they have prostate problems.

A patient with amyotrophic lateral sclerosis could maintain his level of health from four hours of CDT per day over the course of a year. When the patient took a long break in therapy and his doctor changed his medication, the patient died. If challenging patients can no longer move by themselves, supported exercising on the special CDT devices can allow them to live longer.

It is also probable that CDT can improve or delay dementia.

1.2. Exogenous Stem Cell Therapy Is Unlikely to Work

Stem cell therapy to improve CNS functioning has been performed in more than fifty clinics around the world. There are no statistics published on how many stem cell therapies were successful and how many patients suffered complications from the therapy as uncontrolled CNS organization, for example epilepsy or uncontrolled growing, for example cancer. Measurements of any improvements are often insufficient and unscientific. If repair is not measured properly, improvement of CNS functioning cannot be judged. If stem cell therapy is accompanied by a movement-based therapy, some progress in CNS functioning can probably be achieved from the movement therapy alone. It is likely that the stem cell treatment is clinically irrelevant. If a patient has had no therapy before, even movement therapy of low efficiency can bring some progress.

From the theoretical point of view, it is unlikely that stem cell therapy will be successful. Comparably mild, in this case the spinal cord injury in Figure 81 will be taken as an example. If stem/progenitor cells are injected into the cavity of the injury site, the cells cannot be stimulated to proliferate, migrate, and become integrated, because the necessary microenvironment is missing. It is mentioned above that for excitation-neurogenesis coupling Ca^{++} channels have to be opened by membrane depolarization. But the membranes of the injected cells cannot be depolarized, because of missing surrounding activating neuronal networks. If in aging single cells die, then native stem cells can likely be stimulated to divide and proliferate because the necessary microenvironment is present.

A combination of coordination dynamics therapy (CDT) and exogenous stem cell therapy is unlikely to work. As shown in Figure 81, CDT will induce repair around the injury site because of the natural integrative activation of the neural networks. However, this repair-friendly microenvironment will probably not reach the injected cells. Already the distance of communication is a problem. In frog, the distance of communication between two nerve cells is less than 0.1μm (Figure 82). For details, see Chapter I of [1]. Even though new nerve cells are urgently needed for repair (see above), one cannot simply acquire and use them by adding them exogenously. Much more qualified research on adult born and embryonic stem cells is needed, including its relationship to structural repair in humans, before they should be employed in a clinical setting.

The dream of the Early Modern period, namely that it is possible to administer intelligence and wisdom via the funnel of Nürnberg (Nürnberger Trichter), that means improving the complexity and connectivity of CNS functioning, is only a dream. If it were easily possible to change the neuronal networks in humans for improved functioning then humans would have no stable character, because the character is manifested in the complex structure and functioning of the neuronal networks. It is not simply a matter of adding new cells.

But disregarding all the above arguments and assuming that the injected stem/progenitor cells could be integrated into the existing adult networks, than the repair time would have to be considered. It was shown here that the repair times of the brain and spinal cord are in the range of months to several years (Figure 114). If stem cell therapy would work, then one has to expect that the repair would need months to years and not just a few days to a few weeks as was told to patients to whom stem cell therapy was administered.

2. Epigenetic Modification for Anticancer Treatment by Movement-Based Learning

2.1. Activation of Tumor-Suppressor Gene by Exercise

If movement-based learning can suppress genes by modifying the epigenome, it may also be able to activate genes with an anti-carcinogenic effect.

Long term CDT performed at the limit always moved CNS functioning towards more physiologic and more precise functioning. Unspecific pathophysiologic CNS functions like spasticity reduce it. In the framework of the system theory of pattern formation, the stability of physiologic functions increased (the potential well became deeper) and the stability of pathologic states (like spasticity) decreased (the potential well became less deep) (Figures 28,31). Since the nervous system is involved in nearly all body functions, it could well be that CDT enhances specific physiologic functions and inhibits non-specific pathologic functions. Cancer is the uncontrolled unspecific growth of cells. It may therefore be that movement-based learning has an anticancer effect.

It was reported that exercise might exert an anticancer effect by switching on a tumor-suppressor gene [79]. Significant changes in the DNA methylation status of 43 genes occurred in women who engaged, after having completed treatment for breast cancer, in a six-month moderate-intensity aerobic exercise program. An association between gene expression level and overall breast cancer survival was found in three of the women. One of these three was L3MBTL1, a putative tumor-suppressor gene. In the exercising patients, L3MBTL1 was de-methylated, indicating an increase in gene expression. Elevated expression of this gene may be associated with a lower risk of recurrence and improved survival [79].

2.2. Authors Own Experience with Anticancer Effect and Body Function Repair upon Coordination Dynamics Therapy (CDT)

The Author suffered a 'squamous cell carcinoma (epithelioma)' (a malign tumor) in the maxilla (stage between 1 and 2). The tumor was removed by surgery and a neck dissection was performed: Two lymph nodes with formation of metastases were removed and two stages of further lymph nodes and lymph vessels were removed. Radiation therapy and chemotherapy were administered to the tumor area to reduce the risk of tumor recurrence from 30% to 15%. To mitigate the side effects of the anti-cancer treatment, the Author performed CDT for fifteen hours per week, increasing to twenty-five hours per week. Three biopsies of strange growth in the following five years and other diagnostics, including two PET, showed no sign of malignant growth. The exercise on the special CDT device involved coordinated arm, leg and trunk movements and additionally simultaneously coordinated neck, tongue and lip movements.

One year after the tumor extirpation and neck dissection, a reconstruction of the maxilla was performed. The Author was fit by the time he went for the fibula-transplantation operation. He even exercised one hour before the operation on the special CDT device. During the nine-hour operation, pieces of the fibula and the skin with its supplying artery and vein were removed from the right leg. Bone and skin were placed to the former site of the tumor and artery and vein were connected with the temporalis blood vessels. The transplant had to receive blood supply within seven hours in order to avoid rejection. The Author was informed that during the long operation he will get edema and breathing difficulties because of the removed lymph nodes and vessels following a previous neck dissection. However, only minor neck edema occurred during the operation because the CDT had partly repaired the lymph system. One day after the reconstructive operation parts of the transplant were blue, indicating poor blood supply, even though the transplanted artery was open and was supplying blood. The surgeon informed the Author that he would lose some tissue because of poor blood supply.

The Author, having brought the special CDT device into the hospital, exercised on the special device, in spite of the many tubes he was connected to, to improve the blood supply to the head including the transplant site. As the result of the enhancement of blood circulation, the transplant area received enough blood and healed fully. No transplanted tissue was lost. The surgeon was amazed that his complicated work was fully successful. Therefore, CDT had helped to reduce the risk severe neck edema that could occur during the operation and a tracheotomy could be avoided and had improved the blood supply in the head so that the transplant was fully accepted. The Author's hour training before the operation may have helped in this long-lasting and difficult operation. It may have helped to improve the lymph circulation in the short-term memory to reduce edema during the operation. Later on in several small operations performed under local anesthesia parts of the fibula bone had to be removed, because the fibula bone piece was too large for the implantation of teeth, a vestibule had to be reconstructed and implants for the teeth made.

In conclusion, all surgery went well, aided by the continuously performed CDT. CDT helped to repair the body after cancer treatment and may have helped so far to exert anticancer effect. CDT includes not only exercising on the special device but also fast walking and jogging. Since the Author had a hip necrosis on the right side and could only walk with pain, he decided to get an artificial hip to be able to walk well again and to be able to jog a little. The hip operation was successful (see below), so that the Author could again perform the whole movement program of CDT.

Interestingly, the skin from the leg changed into mucosa in the mouth after transplantation and the hair of the skin did not grow any more in the mouth, even though one hair follicle seem to be still there as the Author can feel with the tongue. The dentist argued that nearby mucosa cells migrate into the transplanted skin area, because the changes from skin to mucosa started from the sites of healthy mucosa. But it may also be that the change from skin to mucosa cells was induced by gene expression changes induced by function and/or surrounding changes.

It is very important to note that following the neck dissection (and tumor resection, radiation and chemo therapy) lost or impaired automatisms such as taste, swallowing, coughing, and others, which partly reappeared spontaneously, were enhanced by CDT. I believe that the exercising of the complicated coordinations between pace and trot gait when exercising on the special CDT device, which stimulate the autonomic nervous system, helped to achieve better repair. Even though the eyes are a part of the nervous system and should be reached by CDT, the improvement of the eye functions (wetting of the cornea, opacity of the vitreous body) seems to be most difficult to repair by CDT. Five years after the cancer treatment the wetting of the cornea and the opacity of the vitreous body had improved. Based upon CDT most likely new cells of the eyes were built to improve eye functions. The further repair of mouth, throat and neck functions, and automatisms improved quality of life dramatically. Even though, after five years of CDT not all side effects of the chemo and radiation therapy have disappeared. Sleep is still more disturbed than before the occurrence of the cancer. The chemotherapy probably impaired nervous system functioning and in consequence, many body regulations. It is known that the drug Vinblastine, which was administered to the Author, is nervous system toxic. The patients may get polyneuropathy.

2.3. Difference in the Power of Regeneration

The reconstructive surgery performed one year after the cancer operation was a big load on the Author, and was important for the separation of the nasal and oral cavities. It is also interesting with respect to the power of regeneration of peripheral nerve fibers. With the extirpation of the epithelioma and the surrounding maxilla and teeth, one to two cm of the upper lip became denervated. Since there was no regeneration over the following five years, there was no feeling in the denervated part of the lip and the corresponding lip muscles atrophied. There seem to have been no sprouting of nerve fibers from the healthy lip parts to the denervated ones. On the other hand, the mucosa of the mouth quickly re-innervated, regardless of whether skin from the leg, mucosa from another part of the mouth or pig skin was used for implantation and covering of the transplanted bone.

The area where a transplant was made changed into mucosa and quickly re-innervated. If the healing in the mouth had not been that strong, the Author would have had difficulty with eating and speech. Every change in the oral cavity changes the sounds of speech. The healthy speech centers of the brain have such plasticity that the speech is repaired within two to three weeks. If, on the other hand, the speech centers are impaired by a traumatic brain injury or malformation, then it is very difficult to improve speech. It needs years. It is the CNS, which is the genius and can compensate for many problems in the periphery. Some regulation circuits are pictured in Figure 87.

It is believed that the power of regeneration of nerve fibers is bigger, the closer the injury is to the brain (more rostral). As discussed above, lips and oral cavity mucosa are both close to the brain, but their power of regeneration was very different. High quality reconstructive surgery in combination with movement medicine can show what repair and healing is possible. In cervical C3/C4 spinal cord injuries for example breathing is impaired in addition to the loss of muscle functions. Volitional movement-based learning therapy is not possible any more. But possibly the phrenic nerve can be repaired by passive movement therapy. Only qualified treatment can show which functions can be repaired and which ones cannot. If a patient with a very severe brain injury could recover from coma by administering CDT passively (Figure 118) for 1 year and 9 months, possibly some progress can also be achieved in very rostral spinal cord injury.

2.4. CDT after Hip Replacement

The Author developed a necrotic hip joint on the right side. Exercising on special CDT devices improves normal joint function. However, this hip joint could not be improved in its functioning. There was probably a genetic defect. To reduce the side effects of the cancer treatment the Author performed CDT. However, it is also necessary to perform walking and running. The Author therefore decided to get an artificial hip joint in order to walk again without pain. One day before entering the hospital, the Author brought his special CDT device to the hospital. I exercised before and after the hip operation, apart from the day after the operation when resting in bed was mandatory. When leaving the hospital after a few days, the ward physician said that the healing process was fast.

The surgeon, who operated on the Author, advised me that I should not walk too much in order not to overload the hip. Otherwise, the healing process, including the building of a new joint capsule would take longer. Two orthopedic physicians upon whom he operated trained too much and the healing process took six months longer. At home the Author exercised on the special device, and walked and ran a bit. When he was walked too much the pain increased. The overloading of the transplanted hip set the Author back approximately two weeks in the recovery process. Moving improves the healing process, but the patient has to be careful not to over-train during the healing process. The exercise on the special CDT device was very good for improving the healing process and because very little load is exerted on the hip joint.

Generally exercising on the special CDT device in the sitting or lying position improves joint functioning because the movement stimulus for repair is given; the joint capsule should get enough time to heal and overloading should be avoided.

2.5. CDT after Breast Cancer Treatment to Reduce Edema

As reported above, in order to reduce the risk of a tumor recurrence after the removal of the epithelioma radiation and chemotherapy (Vinblastine) was administered. Lymph nodes and lymph vessels were removed from the neck during the neck dissection. The lymph drainage became impaired and edema occurred in the face and the neck. The Author could successfully reduce the edema within one year when exercising on the special CDT device for at least one hour per day. With the coordinated arm and leg movements, head, lips, and tongue were moved in coordination with the arm movements. During the nine-hour reconstruction of the maxilla, only minimal edema occurred so that a tracheotomy could be avoided. This repair of the lymph system means that the improvement of the cardio-vascular performance upon CDT is repaired. Such a lymph vessel repair will not only be possible for the neck and face by CDT means that not only the blood supply becomes improved but also the lymph vessel system also for the whole body. The lymphatic system inside the body will also be repaired. Further, it is likely that, with a repaired lymphatic system, the immune system is in a better position to fight against remaining cancer cells.

A gynecologist colleague of the Author suffered breast cancer at the age of thirty-four. The cancer and the lymph nodes of the axilla were removed. As usual, edema appeared in the axilla. With exercising on the special CDT device the edema reduced because of a reconstruction of the lymphatic system in the axilla and a better drainage of the lymph due to the movement. The treating physician who was draining the lymph and was astonished that only so little edema was generated. When the gynecologist explained to him that she reduced the edema by exercising on the special CDT device, he did want to listen to her, even though she was a colleague.

2.6. Prolonged Fasting and CDT May Trigger Stem Cell-based Regeneration of Damaged, Old Immune or Other Systems

It was reported that exercise might exert an anticancer effect by switching on a tumor-suppressor gene [79]. Therefore, the Author performed CDT for 15 hours per week to lower the risk of a recurrence of cancer (see above). Vision was also impaired with a delay of one year (cloudiness). While chemotherapy, in this case Vinblastine, and radiation saves lives, it causes significant collateral damage to the immune and other systems. In the Author at least the immune system (occurrence of herpes), the nervous system (poorer sleep, frequency sounds in the ears) and the eye functions were damaged. CDT can hopefully repair the toxic damage of the nervous system over several years. But can the immune system be regenerated efficiently?

It was recently reported that cycles of prolonged fasting not only protect against immune system damage - a major side effect of chemotherapy - but also induce immune system regeneration, shifting stem cells from a dormant state to a state of self-renewal [80].

Because during prolonged fasting organisms minimize energy expenditure in part by rapidly reducing the size of a wide range of tissues, organs, and cellular populations including blood cells, the reversal of this effect during a resumption of eating represents one effect of the most potent strategies to regenerate and/or rejuvenate damaged systems and organs in a coordinated manner. With respect to the immune system that means that when you starve, the system tries to save energy, and one of the things it can do to save energy is to recycle a lot of the immune cells that are not needed, especially those that may be damaged. "PKA is the key gene that needs to shut down in order for these stem cells to switch into regenerative mode. It gives the 'okay' for stem cells to go ahead and begin proliferating and rebuild the entire system" [80].

The body got rid of the parts of the system that might be damaged or old, the inefficient parts, during the fasting. This is the assumption. Now, if you start with a system heavily damaged by chemotherapy or aging, fasting cycles can generate, literally, a new immune system. Reported results indicate that cycles of an extreme dietary intervention represent a powerful means to modulate key regulators of cellular protection and tissue regeneration but also provide a potential therapy to reverse or alleviate the immunosuppression or immunosenescence caused by chemotherapy treatment and aging, respectively, and possibly by a variety of diseases affecting the hematopoietic and immune systems and other systems and organs [80].

The possibility cannot be excluded that prolonged fasting, in combination with CDT may improve CNS functioning after hypoxic or traumatic brain injury. The genetic code is probably not damaged. The state of the epigenome is unclear.

The feasibility and effect of fasting in cancer patients undergoing chemotherapy is quite complicated [81,82]. Practice guidelines for patients undergoing/underwent chemotherapy need to be established. Still, calorie restriction or short-term starvation may be an effective and reproducible intervention for increasing life span, reducing oxidative damage, enhancing stress resistance and delaying/preventing aging and age-associated diseases.

2.7. Prolonged Fasting, CDT and Training at Power Limit to Rejuvenate the Body

The Author decided to fast in addition to performing CDT to regenerate and/or rejuvenate his whole body. By training, best at limits, and resuming eating after fasting one may get new cells and may go in the direction of the fountain of youth. For regeneration and/or rejuvenation, three simultaneous therapies were used by the Author: Firstly, CDT was performed, that means movement-based learning, to repair the nervous system and may switch on tumor-suppressor genes. In the morning, when exercising up to 3000 turns and 130N, the super-compensation was reached in the short-term memory at approximately 2500 turns. This was felt by the Author that he could turn easily and fast for approximately 10 minutes. Secondly, by performing the high-load test and going to the limits by exercising at 200N, new nerve cells and other cells may be building. Thirdly, prolonged fasting should replace damaged cells due to prolonged anesthesia, chemotherapy, radiotherapy, and aging by newly build healthy cells. The fasting periods were three to six days. The macroscopic movements performed during CDT may give the micro-physiologic environment and processes the information needed to achieve the physiologic task of regeneration and/or rejuvenation of the body.

It seems also logic to fast and move strongly at the same time to rejuvenate the body. Our genetics were made or adapted in the last five million years. When prehistoric men were hungry, they had to run first with an empty stomach to catch animals to get food. Food restriction and running and fighting (coordinated movements) therefore also formed genetics by selection. Also nowadays, after or during wars, when people did not have sufficient food and had to walk long distances, they seem to have been healthier than when there is plenty of food and no need to move much.

2.8. Reduction of the Blood Pressure by Prolonged Fasting and CDT

Figure 124 shows the improvements in health, quantified by high-load (HL) CD values and blood pressure, when performing CDT, HL-exercising and repeated prolonged fasting. In "A" the high-load (a) and the low-load (b) CD values are shown, which increase slightly with the seven periods of fasting. After termination of prolonged fasting, the CD values recovered to values of before fasting. The increase and decrease of the HL CD values can be best seen with the continuous five days fasting. The low-load CD values changed only little.

It is important to lower blood pressure. In Figure 124B a blood pressure comparison is given with CDT (check-up) alone, with CDT and HL exercising, and with CDT, HL exercising and repeated fasting together. The blood pressure reduced from performing only CDT (150/90) through performing CDT and high-load exercising (~133/84) to performing CDT, HL exercising and fasting (~120/75). The pulse rate did not change very much. The *reduction in blood pressure* was unexpected and was therefore measured only occasionally at the beginning. In the morning, the blood pressure was mostly measured at rest approximately 30 minutes after exercising on the special CDT device when going with the load up to 130N. In the afternoon, the pressure was measured after exercising up to 200N. It seems that the blood pressure was a bit lower after exercising up to 200N. Anyhow, it seemed that the prolonged fasting and CDT, including high-load exercising, have a natural impact on lowering of the blood pressure. When the Author visited a practitioner, she said that I might have to use β-blocker in the future because of the high diastolic blood pressure. But the blood pressures reduced with fasting and CDT, and were after treatment in the range of 120/75 at rest, which is healthy for men. It has to be seen how much the blood pressure will rise after terminating repeated prolonged fasting and high-load exercising.

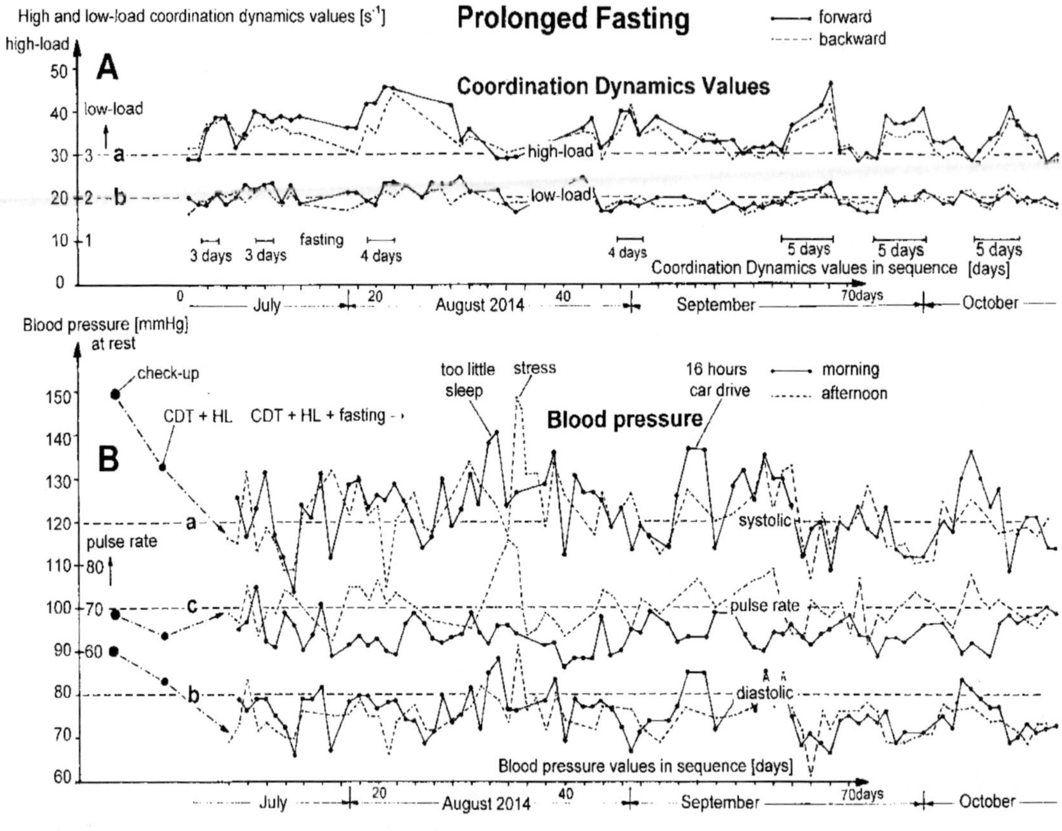

Figure 124. High (a) and low-load (b) coordination dynamics (CD) values (A) and blood pressure and pulse rate (B) in dependence of prolonged fasting during CDT. Note that the high-load CD values increase transiently following fasting. Note further, the systolic (Ba) and diastolic blood pressure (Bb) decrease transiently with prolonged fasting especially with 5 days fasting.

With respect to the lowering of the diastolic blood pressure, it seems that the repeated prolonged fasting for five days and the eating again for five days was most efficient to reduce the blood pressure. But it can also be seen from Figure 124B that too little sleep, long car journeys (16 hours), and especially unpleasant noises (stress) increased the blood pressure strongly. The reduction of the blood pressure by CDT and fasting only makes sense if dangerous blood pressure increases can be avoided.

2.9. Can the Administration of CDT, HL Exercising and Prolonged Fasting Replace in Some Cases β-blocker Administration?

Perhaps in some cases, the performed simultaneous therapies (prolonged fasting and CDT) are an alternative to the usage of β-blocker. But to exercise on the special CDT device during fasting up to a load of 200N is hard and needs a lot of mental discipline and the cardio-vascular performance also must be good to not endanger the heart.

2.10. Reduction of the Biological Age during a Period of 4 Months by Repeated Prolonged Fasting up to 5 Days and Exercising

The Author (180cm) feels good after the repeated prolonged fasting, probably mainly because of the reduction of the body weight from 91kg (slightly overweight) to 80kg (normal weight). The *weight reduction* by 11kg for altogether 29 days of fasting results in approximately 0.4kg per fasting day. The Author ate approximately 15 cranberries each day in order to take necessary vitamins. To digest all the vitamins he was eating a bit, something like 0.1 liter milk with 2% fat. Traditional knowledge in Finland is that five cranberries were sufficient to get all vitamins per day if they are not frozen and were kept in cold water.

Whether the blood pressure stays down after prolonged fasting and HL exercising remains to be seen. Even during 5 days fasting and exercising at high load (200N) on the special CDT device no heart problems could be felt. Stress on the other hand the Author feels immediately on the heart. Obviously, the stress is the risk factor for a heart attack and not the exercising. It even seems that the exercising on the special CDT device at low to medium loads seem to calm the nervous system down. Future research has to show what exercise conditions on the special CDT device can reduce the over-excitation of the sympathetic nervous system division.

In patients with brain or spinal cord injury, we expected structural repair of the neuronal networks after one year or more due to the building of new cells. Therefore, we can expect rejuvenation due to exercising and prolonged fasting up to one year or more. Such long times are supported by the recovery time from a hip transplantation. Even though the Author is exercising quite a lot, he has still not fully recovered from the implantation of an artificial hip on the right side, two years ago. The gluteus maximus and medius are still smaller than on the healthy side as the Author can see easily in the mirror. But he does not feel any impairment of moving due to the reduced muscles. Jumping on springboard may help the partly atrophied muscles recover more quickly in that it provides a means of non-weight bearing exercise. But vigorous jumping may overload the surrounding tissue of the artificial hip. It is interesting and disappointing that walking with the majority of the weight on the left healthy side to avoid pain for more than five years has damaged nervous system a little. Even though the Author is very good at exercising for high loads on the special CDT device, still his CNS has difficulty in generating the pace and trot gate coordination patterns (Figure 125b, lower trace, 137N). The pathologic walking due to pain trained the CNS to generate pathologic movement patterns, which takes more than two years to get rid of. Therefore, for a hip transplantation one may not wait too long for the operation.

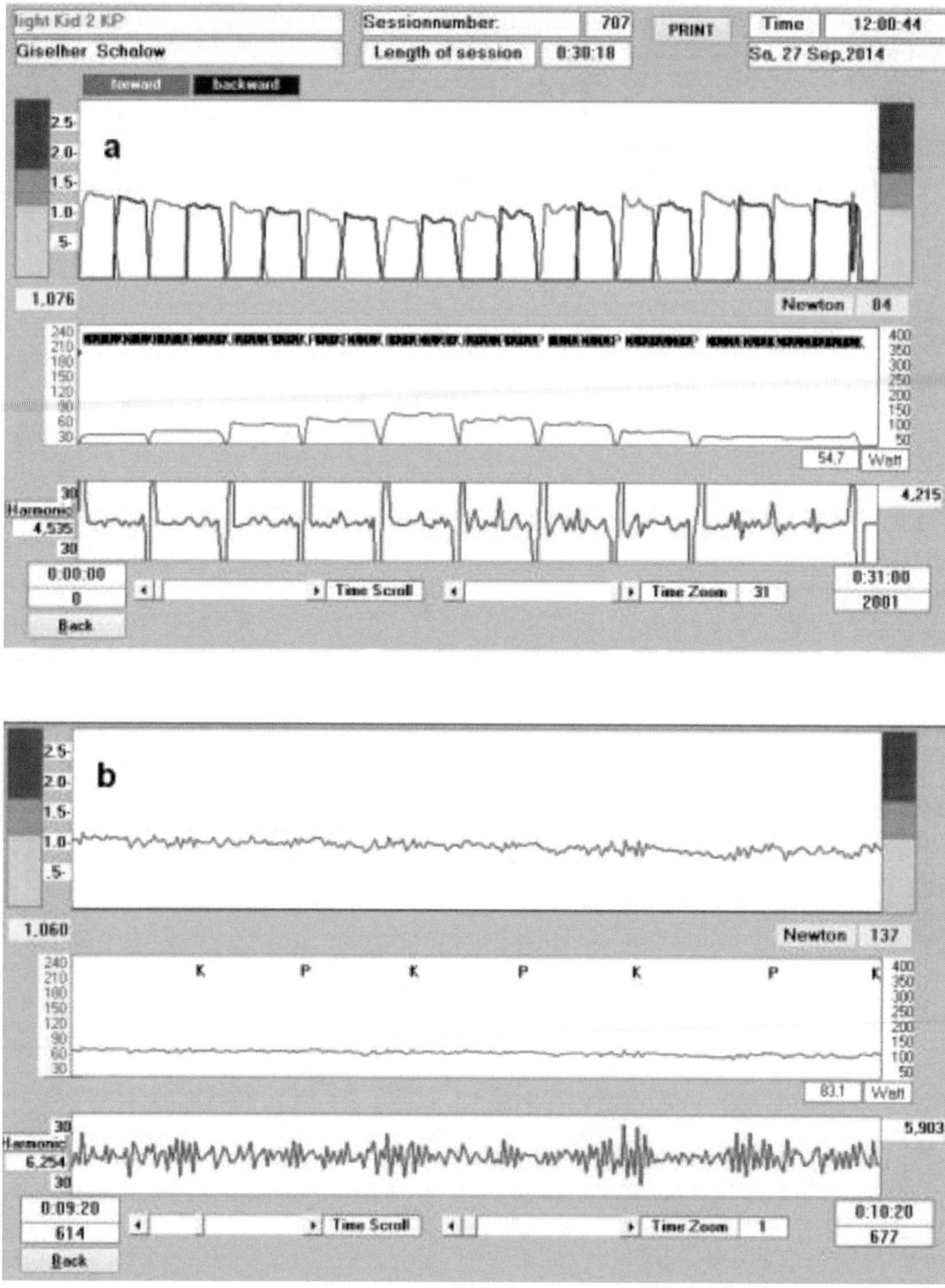

Figure 125. High-load (HL) exercising of the Author. a. Overall test. Note the nice stepwise function of the 'Watt' trace. The interruptions are present, because the Author had to leave the device for changing the load. b. One minute exercising. Note the repeated increase of the arrhythmicity of exercising (lower trace) for the pace and trot gait coordination.

In patients with CNS injury, structural repair was best when treatment was applied continuously for longer than one year. Performing exercise and repeated prolonged fasting is quite hard to do and takes a lot of time. The Author has undertaken treatment in the range of 6 to 8 months. Perhaps it would be worthwhile to go on for some time (see below).

Is the nervous system significantly involved in the process of rejuvenation of the body? It seems so. First, the nervous system is involved in nearly all functions of the body. Second, when the Author felt uncomfortable during fasting, he trained on the special device a bit and felt better. Third, it was analyzed above that in membranes of stem/progenitor cells the Ca^{2+} channels are only opened to modulate gene expression and stimulate excitation-neurogenesis coupling if they are depolarized which means that activity is needed for rejuvenation.

The increase of the high-load CD values in Figure 124A is understandable, because with fasting the power reduced and the Author turned slower to escape the high load (load escape). The CD value equals df/dt.f. When exercising at a lower frequency f, the CD value will increase.

But the low-load CD values also increased a bit (got worse) which cannot be explained by a transient weakness of the body. It seems therefore that the fasting also impaired the coordination dynamics. The Author exercised therefore more often on the special CDT device to compensate for the disturbed CNS organization. The impaired CNS functioning was also felt by the Author when he was driving a car on a narrow road. He felt unsafe and had to concentrate more to stay safely on the road. When eating again, this mild coordination problem terminated and he could easily drive safely on the road again. There seems to be a correlation between the fasting and the coordinated firing of the nerve cells in the nervous system.

2.10.1. Improvement of the Immune System

As a first feeling of the Author, with repeated prolonged fasting the immune system got weaker and with eating again the immune system recovered and worked even better than before fasting. Especially the repeated five-day fasting seemed to improve the functioning of the immune system. A skin disease, appearing approximately two years after radiation and chemotherapies, disappeared.

2.10.2. Repair of the Eye Functions

The impairment of vision, caused by chemo and radiation therapy, reduced. The drying of the eyes at night improved only little. When the Author wakes up at night, it is not clear whether he wakes up because of the impaired sleep pattern or because the eyes got too dry. By being woken up, the eyes become quickly wetted again. As if this eye automatism does not work sufficiently during sleep. It remains to be seen what the long-term outcome will be.

2.10.3. Hip Joint Repair

Most functions of the human body can be improved by CDT including some diseases caused by genetics. Some genetically caused diseases can likely be partially repaired by changing gene expression via the epigenome and other ones not. Here an example. The genetic disease Down syndrome can be improved by CDT. An overused or poorly-used hip joint can be improved in its functioning by CDT. A genetically determined hip joint necrosis cannot be repaired by CDT. The Author suffered a one-sided hip necrosis with aging, which he could not repair with CDT by more than five-year treatments. The not affected healthy hip of the other side is functioning well.

In conclusion, repeated prolonged fasting, in combination with CDT, seems to rejuvenate the whole human body to a certain extent and not only the immune system. After reconstructive surgery, fasting and CDT may help to improve the outcome. One question is, how much can the gene expression be changed in the transplanted cells? Further research has to show how much rejuvenation is possible and what program is optimal to live longer with a better quality of life.

2.11. Coordination Dynamics during a Period of Additional Repeated 6-days and 4-day Prolonged Fasting

The question arises whether a rejuvenation of the body is beneficial if the repeated fasting periods are prolonged up to 6 days (144 hours). First, the Author wants to live longer with a better quality of live. Second, if he wants to push medicine treatment further, he needs further years of good health for having the mental power to accomplish it. Thirdly, it would be difficult to find aging patients after cancer treatment being able or willing to undergo such a treatment, because exercise on the special CDT device during fasting up to a load of 200N is hard and needs a lot of mental discipline, and the cardio-vascular performance has also to be good. Therefore, the Author decided to go on with CDT and prolonged fasting. He first prolonged the fasting periods up to 6 days (Figure 126). To feel good in the body he increased CDT of at least up to 20 hours per week. Approximately 12 000 coordinated movements he performed every day with a load up to 200N (high-load assessment) at different devices and different positions. Jumping in anti-phase and in abduction-adduction was included in the program. Fourthly, is it possible to establish practice guidelines for patients who underwent radiation and chemotherapy?

Figure 126 shows the improvements in health, quantified by high-load (HL) CD values and blood pressure, when performing CDT, HL-exercising and repeated prolonged fasting up to 6 days. In “A” the high-load (a) and the low-load (b) CD values are shown, which increase and decrease with the three periods of 6-day-fasting. The high-load CD values increased more strongly with the repeated fasting. The peaks of highest CD values (poorest CNS functioning) are marked with “1.”, “2.” and “3.”. With ongoing fasting the peaks increased in size, indicating some kind of potentiation appeared during the repeated prolonged fasting. As can be seen from the third period of fasting, the recovery from fasting needed 16 days (Figure 126A). It seems therefore that the period between the two 6-day fasting periods should be at least 16 days. With shorter periods of fasting the recovery period is shorter (Figure 124). After termination of prolonged fasting and recovery, the CD values recovered to values of before fasting, which means values of 30 for the Author were reached again. The low-load CD values also increased but only little. Their increases indicate that the increase of the CD values (worsening of CNS functioning) is not only a lack of missing power due to the fasting, because exercising at 20N is easy to perform from the point of power needed.

Important is also to know how much time the body needs to recover from the prolonged fasting. According to the high-load coordination dynamics values of Figure 126, the recovery time to reach the value 30 again was 17 days after the third 6-day fasting. Other functions may need much more time to recover or improve. The time between two 6-day fasting's should be in the range of 15 days. For shorter fasting periods the recovery time was shorter (Figure 124A) and with the repetition of fasting the recovery time increased (Figure 126).

Because of the long recovery time from 6-day fasting and the load on the body, the Author changed the practice guidelines of repeated prolonged fasting in January and February 2015. He was fasting and eating during periods of four days. As can be seen from Figure 126A, the 4-day-fasting and 4-day eating periods were optimal for the Author from the point of view of high-load CD values. The recovery times were two to three days shorter.

As turns out, to recover from the side effects of the cancer treatment at stage 1.5 the Author needed more than 5 years of using CDT and repeated prolonged fasting for faster recovery. Without these treatments, an aging patient may never recover from the side effects.

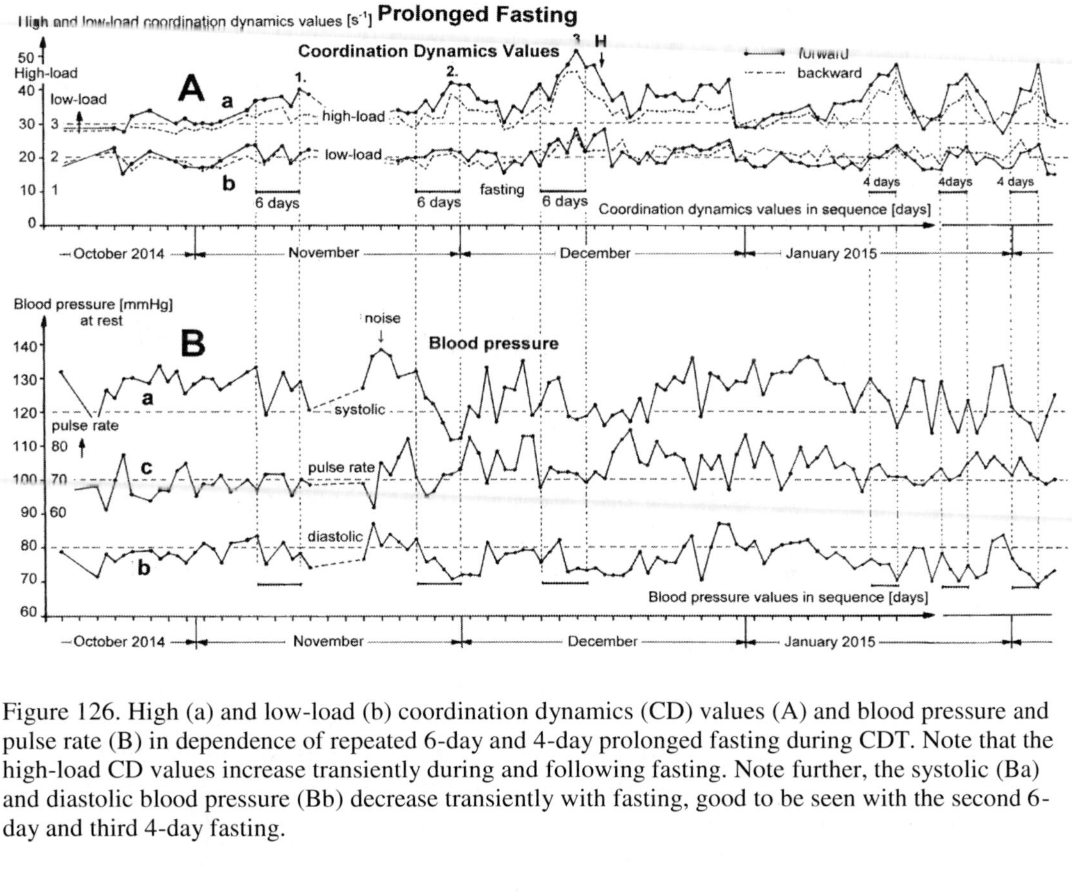

Figure 126. High (a) and low-load (b) coordination dynamics (CD) values (A) and blood pressure and pulse rate (B) in dependence of repeated 6-day and 4-day prolonged fasting during CDT. Note that the high-load CD values increase transiently during and following fasting. Note further, the systolic (Ba) and diastolic blood pressure (Bb) decrease transiently with fasting, good to be seen with the second 6-day and third 4-day fasting.

2.12. Blood Pressure Reduction due to Prolonged Repeated Fasting

The time course of increase in systolic and diastolic blood pressure following reduction due to fasting shows similarity to the CDT value changes. The natural impact on lowering the blood pressure of prolonged fasting and CDT is only transient. Anyhow, it is still good to know that during crisis of hypertension one can reduce the blood pressure also by fasting if, for example, no β-blockers are available and the exercising on the special CDT device needs not to be stopped.

2.13. Beneficial Effects of Prolonged Fasting with Periods up to 6 and 4 Days

a. The mucosa of the mouth in the transplanted area, transformed from the skin of the right leg, started to grow under the tooth prosthesis to adapt fully to the prosthesis in similarity to the normal mucosa adapting to the natural teeth. The Author was first worried a bit by that growing, because such mucosa growing was similar to the former cancer growing. But the main difference to the cancer was that the growing was controlled and not uncontrolled.

   The reduction of the blood pressure, even though mainly transient, was appreciated by the Author, because his blood pressure could be a bit lower.
b. The CNS functioning improved in the way that the hearing of sound became less strong and then disappeared completely.
c. The rejuvenation of the skin was very limited. Wrinkles did not reduce significantly. The scar from the hip transplantation reduced a bit in size especially at the place where the underwear touched the skin. During exercising on the special CDT device this skin site was continuously touched due the movement.
d. When exercising at high loads the Author had the feeling that with the enhanced microcirculation the micro-physiologic machinery found the places where repair was needed. The nose and the upper lip were often itching.
e. Since neurogenesis seems to contribute only after a year of therapy, further improvements can be expected to occur in the future.
f. The general body feeling improved. The Author felt younger.

2.14. Negative Effects of Prolonged Fasting up to 6 and 4 Days

The functioning of the immune system got transiently impaired during the period of 6-day-fasting. After the third 6-day-fasting, Herpes appeared at the lower lip. Further, with the 5th day of fasting a skin problem re-appeared. With eating again, these immune system deficits disappeared. The transient weakening of the immune system with prolonged fasting could be dangerous especially with respect to cancer reoccurrence. With the growing of the cancer, metastases were sent out. Two lymph nodes with metastasis were removed during a neck dissection and additional two further stages of lymph nodes. But it may happen that some cancer cells are not filtered in the first three lymph node stages and get stuck in the next nodes and stay there in a dormant stage. With a weakening of the immune system these cancer cells may become active, but they are not destroyed by the immune system, because of the impaired functioning of the immune system. In the Author's case, the fasting periods should not be longer than four days. Patients who suffered cancer and survived are afraid of a reoccurrence of cancer. This holds also for the Author, even though he had no reoccurrence in the last 6 years.

2.15. Practice Guidelines for Repeated Prolonged Fasting and CDT

Because of the negative effects of 6-day fasting, the Author tried the 4-day-fasting periods again. No Herpes and no skin problems occurred and the recovery times were shorter. No immune system functioning impairment seemed to have been occurred. If these practice guidelines hold also for other persons, then in three weeks a bit of rejuvenation of the body can be achieved by alternating fasting and eating every four days.

2.16. Not or Only Little Repaired Functions by Fasting, CDT and Spontaneous Recovery

a. The nose has still not fully recovered from surgery, radiation, and chemotherapy. The radiated area was situated below the nose and the upper lip.
b. Some scar tissue in the vestibule did not (or only little) reduced with CDT and prolonged fasting. The gene expression and/or modulation of the epigenome of the scar tissue cells could not be changed, even though the mucosa of the mouth was brilliant with respect to rejuvenation. The surgeon argued that only in a quite big operation, in which big parts of the mucosa are taken from other places, there is a chance to reduce the scar tissue. With small operations, the scar will reappear. The Author hoped that with adapted CDT and prolonged fasting the epigenome could be modulated to change gene expression of the cells in at around the scar to replace the scar by mucosa. This did not happen so far.
c. The eye functioning had improved in the 6 years following the cancer treatment, but was still not the same as before.
d. The good sleep of before the cancer could not be re-established again so far, even though some improvements occurred.
e. The side effects of cancer and cancer treatment are very difficult to repair by movement-based learning. But without CDT over the years and prolonged fasting the health condition would be much worse. Surviving the cancer is only the first step in repair or rejuvenation. The second step is to improve quality of life.

2.17. Suggested Practice Guideline for Patients Who Underwent Cancer Treatment Surgery and/or Chemo and Radiation Therapy

CDT combined with prolonged fasting is beneficial in after-cancer treatment, because CDT probably hinders the reoccurrence of cancer. Only the starvation periods should not be too long. In the case of the Author, the fasting periods should probably not be longer than 4 days.

Therefore, the Author had changed the program of repeated prolonged fasting. He fasted now 3 times for 4 days (January and February 2015, Figure 132). The recovery times from 4-day fasting were with respect to the high-load coordination dynamics values with two to three days much shorter than for 6-day fasting. Whether all body functions recovered in that time is unclear. With respect to the training of the cell renewal regulation, the repeated 4-day-fasting is probably more efficient.

The 9-year-old girl of Figure 30 suffered with the cancer operation, a lumbar-sacral spinal cord injury. She is not continent anymore and the leg movements are poor. CDT can be administered to her to repair the spinal cord without being afraid that the movement-based learning therapy enhances the probability of cancer reoccurrence, because movement-based learning therapy reduces the risk of a reoccurrence of cancer by switching on tumor-suppressor genes [79]. However, she needs intensive CDT immediately, because now she is a lovely girl in spite of the incontinence and the movement problems. But in ten years, she is a young lady who is incontinent and cannot walk properly; her quality of life will be low. Actually, the incontinence would then be the main problem; but urinary bladder functioning can be repaired by CDT (Figure 62, Chapter VII of [2]). Unclear is, whether repeated fasting would be helpful for her in addition to CDT. CDT has been started (Figures 30 and 32).

3. Comparison of Neural Network Learning during Development and CDT Plus Prolonged Fasting in Aging

3.1. Overlap of Different Movement Patterns

Even though the sleep pattern could not sufficiently be repaired so far with CDT and prolonged fasting, neural network learning processes took place.

In Figure 125a, it can be seen from the upper trace that the frequency of exercising transiently increased when changing the direction of turning. Figure 116 shows that when the frequency of turning increased, the high-load coordination dynamics values got smaller (better). This means that the neural network organization improved transiently. But why is CNS organization transiently improved when changing the direction of turning? An explanation is that the two movement patterns "exercising in the forward direction" and "exercising in the backward direction" are both activated around the change of the direction of turning. During this transient simultaneous organization of both patterns (overlap of patterns), the general neuronal networks are more integratively activated and by learning (repetition of movements) the network organization becomes better. There is therefore an overlap of movement patterns, even though only one pattern is executed. An overlap of movement patterns is also plausible, because there cannot be chaos in between two movement patterns. All the Author learned from the structure and function of the human CNS is that everything seems to be highly coordinated, exact and tremendously complex.

3.2. Improvement of Neural Network Organization in the Short-term Memory

From Figure 133 it can be seen that, especially when reducing the load from 200N, the frequency of exercising also increased before the change of the direction of turning. Such movement performance improvement is not achieved because the Author prepared himself for changing the direction of exercising. The increase of the frequency started earlier. This performance improvement is due to the improvement of CNS organization in the short-term memory. The Author had the feeling that it was easier to turn and he felt good to turn faster. One is somehow addicted to turn faster.

An opposite feeling the Author had during the periods of fasting. In these periods, he could not turn fast. It was just not possible. When driving a car he felt unsafe because the nervous system was not working well enough. He could not control the car well during prolonged fasting, even though the higher mental functions were working properly in difference to the situation when drinking alcohol.

3.3. Improvement of Neural Network Organization in the Short-term Memory Following CDT and 6-day-Fasting

By comparing Figure 125a with 133 it can be seen that the improvement of CNS functioning in the short-term memory was highest following 6-day-fasting, especially when reducing the load from 200N.

It seems that the longer fasting has a bigger effect on CNS functioning improvement. With longer fasting periods more damaged cells could be replaced by better functioning ones and/or the regulation of cell replacement could be improved by learning. The learning impact on the improvement of neural network functioning is higher when the periods of fasting are longer, because the transient increases of the frequency of exercising were higher. It seems that the advantage of longer fasting is that there is more time for rejuvenation and the disadvantage is that the load on the body is higher. Especially the immune system functions become impaired.

In young patients without cancer, longer periods of fasting may be more efficient in rejuvenating of the body, whereas in aging longer periods become more dangerous, because of too high load on the body.

3.4. Synaptic Potentiation as One Possible Reason for the Improvement of Movement Performance

Above it was argued that neural network organization improved with movement-based learning. Especially coordinated movements are beneficial because neurons work as coincidence detectors (Figure 40).

But the threshold to generate an action potential at the axon hillock can also be reached earlier if each post-synaptic potential is higher. The sub-synaptic potential of a single synapse can be made bigger by repeated activation, which is called potentiation. The timely coordinated potentiation of different synapses will contribute to the improvement of movement performance in the short-term memory.

Above, in connection with the improvement of learning through increasing the integrativity of neural network activation by including vision, it was indicated that presynaptic depolarization facilitates neurotrophin-induced synaptic potentiation [96].

Neurotrophins therefore, participate in activity-dependent modification of neuronal connectivity and synaptic efficiency. Further, changes in NMDA receptor expression provide a mechanism by which brief sensory experience can regulate the properties of NMDA receptor-dependent plasticity. Already one hour of visual experience can alter the complement of postsynaptic glutamate receptors in the visual cortex in vivo [99]. Such rapid modification can contribute to middle-term supercompensation when exercising on the special CDT device, which is appearing after approximately one hour of exercising. Explanations for supercompensation during movement performance will be picked up again below.

3.5. Sub-synaptic Potentials and Active and Passive Impulse Conduction in a Frog Model

To get a better understanding of neuronal network learning sub-synaptic or post-synaptic potentials and their conduction are explained in a frog model. It will be shown that neural network learning involves not only synaptic weights but also changing of membrane properties and passive conduction of sub-synaptic potentials from the dendrites to the axon hillock were action potentials (APs) are generated and active conduction of APs along the axon or axon trees.

A neuron is on average connected to 4000 other neurons in the CNS. In the frog model for simplicity a cell (muscle fiber) is considered, which is innervated by a single neuron. Parts of the membrane conduct potential changes actively like in an axon, and passively like in the soma and dendrites. In Figure 127, a slow muscle fiber is shown in which in one part a sub-synaptic potential is generated (like in the neuron soma) and in the other part, the action potential is generated (like the axon hillock).

Slow muscle fibers of frogs normally do not generate APs to depolarize the muscle fiber to increase Ca^{++} concentration for contraction. The slow muscle fibers are depolarized by the endplate potentials of several distributed endplates connected to one or two thin motor axons. When impaling a voltage and a current microelectrode into the slow muscle fiber (C) and applying constant current, the membrane resting potential can be held at -90 or to -100 mV. Upon applying an additional current pulse for 100 ms, the muscle fiber can be transiently depolarized or hyperpolarized (A,B). When hyperpolarizing the muscle fiber, the slow membrane potential deflection indicates high membrane resistance ($\approx$ 10 MΩ) typical for slow muscle fibers. Twitch muscle fibers (not shown here) show a faster time course, have a lower membrane resistance ($\approx$ 1 MΩ), and much shorter endplate potential. When depolarizing a normal slow muscle fiber by a depolarizing current pulse, no AP is generated (similarly to voltage electrode impaling site 5 in C). When depolarizing a denervated slow muscle fiber (cut nerve supply to the muscle), after 10 to 15 days the frog slow muscle fibers generate APs similarly as in A and B. Upon partial denervation of a slow muscle fiber (C), which is innervated by two motor axons one of which is cut, the innervated muscle fiber part responds with a motor endplate potential (and no AP) and the denervated muscle fiber part with an AP (and no endplate potential). The AP and the endplate potential will spread electrotonically (passively) to other parts of the muscle fiber (lower part of C). It seems therefore that the different parts of the muscle fiber are genetically differently controlled by the different muscle fiber nuclei.

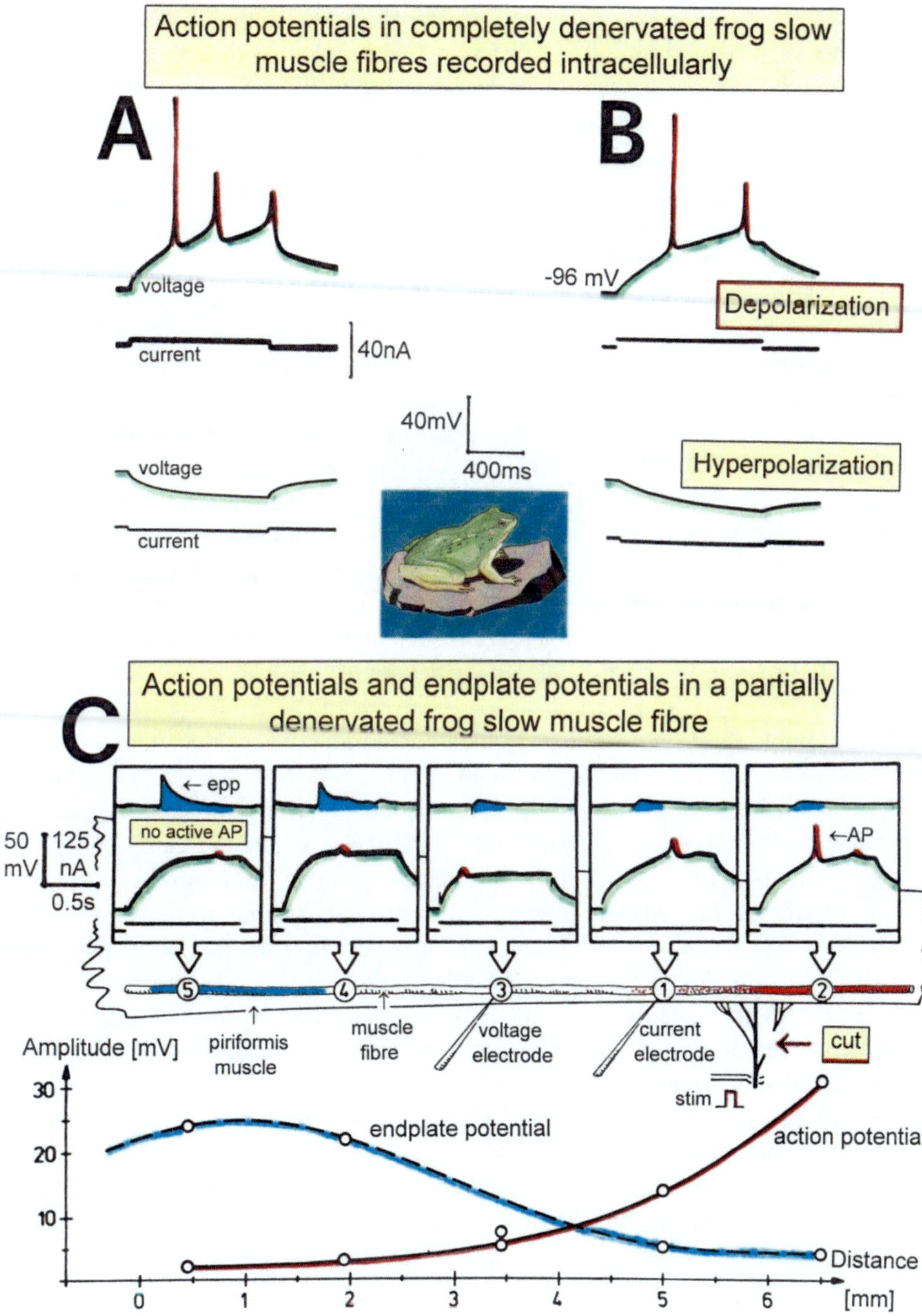

Figure 127. Simultaneous occurrence of action potentials (APs) and endplate potentials in a slow muscle fiber of a frog (rana temporaria).

APs are generated in denervated slow muscle fibers of frog piriformis muscles upon depolarization by transient current pulses of 100 ms duration (A,B). The slow time course of the voltage upon hyperpolarizing the fiber shows that the microelectrodes were impaled into slow muscle fibers. With increasing depolarizing current pulses (and fully developed AP mechanism) first one AP is mostly generated (C, impaling site 2), then 2 APs (B) and then 3 APs (A) (in this case only shown for 3 different fibers). Note with respect to the relative length of the interspike intervals that the 2 AP impulse train in B and the 3 AP impulse train in A bears similarity to the impulse trains recorded from oscillatory firing human α_2-motoneuron axons (Figures 5, 13).

In "C" APs and endplate potentials are recorded along a partially denervated slow muscle fiber by keeping the current microelectrode at site 1 (close to the nerve entrance to the piriformis muscle) and impaling the voltage electrode successively along the muscle fiber from site 1 to site 5. At each site of the voltage electrode impalement, the muscle fiber is directly stimulated by the current pulse from the current electrode (the reference bath-electrode is not shown, muscle in frog Ringer solution) to elicit an AP, and indirectly stimulated, by applying to the nerve supplying the muscle a voltage pulse of 0.5 ms duration and 1 to 10 V amplitude, to elicit an endplate potential. The results of both stimulations are shown in the pictures related to the sites of successive impalements of the voltage electrode.

In the partially denervated slow muscle fiber (C), the endplate potential did not elicit an AP generated in the other denervated part of the muscle fiber. But often the endplate potential evoked in the still innervated muscle fiber part (here left side) is still high enough in the denervated muscle fiber part (here right side) to elicit an AP there, which is then actively conducted in the denervated area of the muscle fiber and spreads electrotonically (passively) into the innervated muscle fiber part. For further details of this frog model, see [85-88] and Chapter I of [1].

This frog example shows how delicately membrane properties can change at the cellular level with functional consequences, which cannot be understood with molecular, genetical or pharmacological methods alone. Avoiding regeneration, it seems possible to increase by training the size of one or two left synapses in the innervated area of the muscle fiber, so that the endplate potentials can generally reach the threshold for AP generation in the denervated muscle fiber segment. It is conceivable that similar mechanisms are working in the human CNS, e.g. with respect to the integrated functions of dendrites (part for electrotonic potential spread), cell soma (summing point) and axon hillock (AP generation for active conduction along the axon), when changing neuronal network properties by re-preformation of the neuronal network structures due to changes in efficacy and properties of synapses and changes of the excitability of membranes.

Ca^{++} ions are involved in many functions and not only in excitation-contraction coupling in muscle fibers or neurogenesis-excitation coupling (see above). As shown in this frog model also membrane properties change following innervation changes and nerve tissue damage with partial denervation of neurons. The changes following nervous tissue damage or neural network learning are tremendously complex also on the neuron level.

The repetitive stimulation of a neuron axon, giving rise to an adding up of sub-synaptic potentials and potentiation is simulated in Figure 127A,B by the transient constant current depolarization pulse. The kind and form of impulse patterns for the communication between neurons is very important. An example of communication patterns, among neurons in the human CNS, are given in Figure 45.

As reported above, presynaptic inhibition will also modulate the sub-synaptic potential. Potentiation of synaptic efficacy by brain-derived neurotrophic factor (BDNF) is greatly facilitated by presynaptic depolarization. This potentiation depends on the relative timing of depolarization and reflects an enhancement of transmitter secretion from the presynaptic neuron [96].

With this complexity of neuronal network organization in mind, neural network learning during development and during movement-based learning with prolonged fasting are compared.

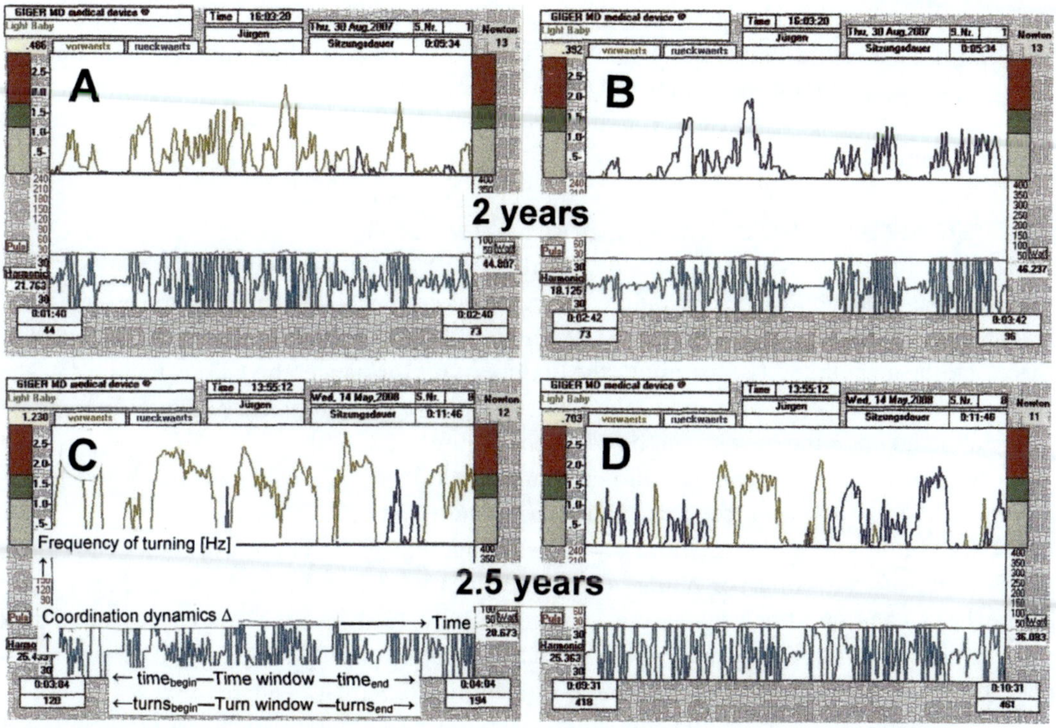

Figure 128. Development of CNS complexity, quantified by coordination dynamics measurements (very early stage of coordination dynamics (CD) assessment during development). Upper traces = exercising frequency; lower traces = arrhythmicity of exercising (CD). A,C = forward exercising and B,D = backward exercising. A,B. At an age of 2 years healthy Jürgen cannot turn continuously on the baby CDT device: the frequency varies very much and the turning shows many stops. C,D. With 2.5 years, the young child can turn more continuously on the baby device and the frequency of exercising is much higher.

3.6. Neural Network Learning During Development

3.6.1. Neuronal Networks Need Complexity to Generate Complex Patterns

At an age of 6 months, the infant Jürgen (Figure 18) was too small and not able to turn on the special coordination dynamics therapy (CDT) device. At an age of 1 year, the infant was made familiar with the device and he liked it. With 2 years, Jürgen was tall enough to exercise on a baby CDT device and he wanted it strongly. But he was not able to exercise continuously (Figure 128). He got stuck again and again after 1 to 3 turns. He solved the problem by himself of not being able to turn with arms and legs continuously, by taking the hands or the feet from the device and avoiding in this way the complicated coordination patterns of arms and legs (with respect to coordination) for which his CNS had not developed so far the needed complexity. Jürgen thus nicely demonstrates that high complexity of neuronal network organization is needed to be able to perform coordinated arm and leg movements with changing coordination between arm and leg movements (pattern change). For turning only with the legs (fitness bicycle) no high network complexity is needed and not learned. At an age of 3 years, he became able to exercise on a special CDT baby device (Figure 129); even though the rhythmicity of turning was poor, especially in the backward direction. With 5 years, Jürgen was able to turn on the larger children's CDT device in the forward and backward direction. The exercising in the backward direction was less smooth. The assessed coordination dynamics (CD) will give further insight into the maturing of the CNS during development by structural changes, including higher connectivity and movement-based learning.

3.6.2. CD Assessment between 2 and 5 Years of Age

The generation (increase) of CNS neuronal network complexity during development can be quantified, if the child is performing complicated coordinated arm and leg movements and the quality of performance measured. The child is made to exercise on the special CDT device and the arrhythmicity of turning (the quality of performance, the CD) is measured. The improvement of the performance and the reduction of the arrhythmicity of exercising can be nicely followed up in the developing boy Jürgen (Figures 128,129).

With 2 years of development, Jürgen could not turn continuously on the baby device (Figure 128A,B). The mean frequency was low (Figure 128A,B; upper trace = frequency; no continuous line) and the arrhythmicity of turning (lower trace) was high. The mean arrhythmicity of exercising per minute (the CD value) was with 44.8 for forward exercising (Figure 128A) and 46.2 for backward exercising (Figure 128B) very high. The values would have been even higher if the infant would not have got stuck during the turns. At an age of 2.5 years, the frequency of exercising became higher for forward (mean frequency = 1.24Hz, Figure 128C) and backward exercising ($f_{mean,back}$ = 0.71Hz); the corresponding frequencies of turning at an age of 2 years were $f_{mean,forward}$ = 0.49Hz and $f_{mean,back}$ = 0.39Hz. At an age of 2.5 years the CD values had decreased (became better) to $\Delta_{forward}$ = 20.8 and $\Delta_{backward}$ = 36.1. As can be seen from Figure 128A through C, the child could not turn continuously in the forward or backward direction. He made mistakes in the direction of forward and backward exercising; he mixed them up a bit. The continuous intention to exercise in one direction was not fully developed; the symmetries of CNS organization were also not fully matured. The exercising in the backward direction was worse in comparison to the exercising in the forward direction.

At an age of 3 years, Jürgen was taller and became able to turn on the special CDT device for children (Figure 129). This is the size of the device, which is used for assessing all later developmental stages. Exercising on the larger CDT device is more difficult and the coordination dynamics values cannot directly be compared with those obtained from the baby instrument. The CD curves obtained from the children's device (Figure 129A,B) show clearly that it was more difficult to turn on the larger device. The frequency of exercising was smaller ($f_{mean,kid,forward}$ = 0.42Hz (against $f_{mean,baby,forward}$ 1.24Hz); $f_{mean,kid,back}$ = 0.24 (against $f_{mean,baby,forward}$ 0.71Hz) and the coordination dynamics values higher ($\Delta_{forward,kid}$ = 33.6

(against $\Delta_{forward,baby}$ = 20.8); $\Delta_{backward,kid}$ = 51.2 (against $\Delta_{backward,baby}$ = 36.1)). Further, the rhythmicity of turning was still so poor, that no preferential turning patterns could be seen (Figure 1291A). The exercising in the backward direction (Figure 129B) was worse than the exercising in the forward direction (Figure 129A), because the CD values were bigger (worse) and the child changed un-volitionally more often into the wrong direction of turning. At an age of 3.1 years the CD improved further. In the forward direction of exercising, Jürgen no longer made mistakes in the direction of exercising and also the backward exercising became more stable. In the forward direction of exercising the difficult and easy coordination patterns became visible. For the easy coordination patterns around the pace gait (P) and trot gait coordination (K), Jürgen could turn quite fast (Figure 129C) and among the difficult coordinations between pace and trot gait he got stuck.

At an age of 5.5 years, the coordination patterns of arm and leg movements became much better. The arrhythmicity of exercising became smaller ($\Delta_{forward,kid}$ = 21.3, $\Delta_{backward,kid}$ = 27.8) and the frequency of exercising higher ($f_{mean,kid,forward}$ = 1.07Hz, $f_{mean,kid,back}$ = 0.78Hz) (Figure 21E,F). But these values were still far away from those of the adult mother (Figure 129G,H). Her arrhythmicity of exercising was much lower and her frequency of exercising higher ($\Delta_{forward,kid}$ = 4.1, $\Delta_{backward,kid}$ = 3.6; $f_{mean,kid,forward}$ = 1.35Hz, $f_{mean,kid,back}$ = 1.32Hz). The motor maturation from 5 years on up to the adult stage was assessed and is shown in Figure 92.

3.6.3. Preference of Function to Symmetry Learning

In general, the frequency of exercising increased with age (Figure 130) and the coordination dynamics values (the arrhythmicity of exercising) decreased (Figure 92). At the beginning during the measuring period between 2 and 3 years, upon exercising on the baby device, the performance of the forward exercising of the infant Jürgen became better than those exercising in the backward direction. Then later on, in the age period between 3 and 5 years, the performance of exercising in the backward direction improved more than those in the forward direction. That means, the exercising frequency in the backward direction improved more than that in the forward direction and the CD values for backward exercising improved more than those for forward exercising. This means, the developing CNS gave first priority to establish the forward moving, probably because the forward moving is more important. Then, at a developmental stage, when this movement pattern was somehow established, the development took care of the symmetry of moving. It seems therefore that during development function was given priority to symmetry. This movement principle is in accordance with the learning (establishing) of walking. The developing child often learns first the forward walking in the age range between 12 and 18 months, and then, 3 to 6 months later, also the backward walking. The stepping automatism probably becomes also first operational in the forward direction, because the newborn can only step automatically in the forward direction (Figure 44). When partly re-appearing, following spinal cord injury, the stepping automatism can be induced (during supported treadmill walking) in the forward and backward direction. During development, forward crawling becomes operational before backward crawling.

It could be that the backward movements are stored differently in the CNS, in similarity to the storing of languages. The mother tongue is stored differently than later learned languages. Following brain injury, it happened that the patient could speak first in a foreign language and only later on in the mother tongue.

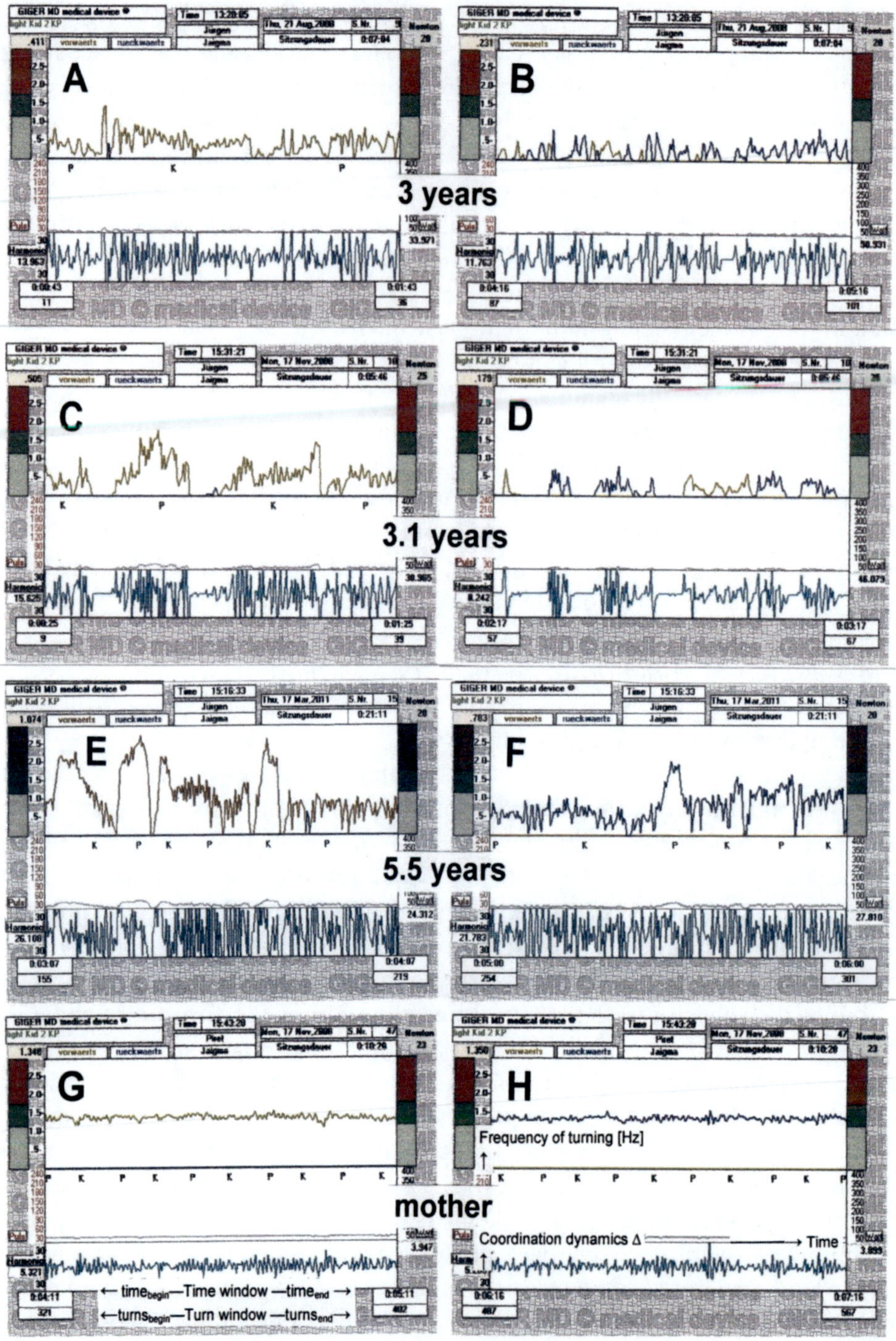

Figure 129. Development of CNS complexity, quantified by coordination dynamics measurements (early stage of coordination dynamics (CD) assessment). Upper traces = exercising frequency; lower traces = arrhythmicity of exercising (CD); A,C,E,G = forward exercising and B,D,F,H backward exercising. A,B. With 3 years the child Jürgen becomes able to exercise on the children's device. The frequency of exercising is low. The child still makes mistakes in the direction of exercising, especially upon turning in the backward direction (B). C,D. Two months later he can turn at higher frequency.

Now the difficult coordination's between pace (P) and trot gate (K), for which a higher neuronal network complexity is needed, becomes visible (C): he gets stuck or nearly stuck between the patterns pace and trot gait. The backward exercising is still very poor (D); mistakes in the direction of turning still occur. E,F. At an age of 5.5 years, Jürgen can turn more continuously at higher frequencies. The difficult coordinations between pace and trot gate are still visible (reduction of the frequency). For forward exercising, transient high frequencies of turning (4 peaks) occurred (E). For exercising in the backward direction one period of exercising with high frequency occurred (F). G,H. Coordination dynamics of the mother, for comparison. Note, the frequency (f) of exercising is nearly a line (constant frequency) and the arrhythmicity of turning (df/dt /f) is small for forward (G) and backward (H) exercising. Note further, there is a large difference in performance of exercising on the special device of the 5-year-old healthy Jürgen and the mother, exercising on the same children's device.

3.6.4. Functions, Which Need Less Network Complexity Are Learned and Repaired First

As can further be seen from Figure 129, the infant could perform the exercise better on the smaller baby device and better for lower load on both the baby and children's device. Therefore, for smaller movements and exercising with less load (less Newton), less muscle fiber activity is needed, and in turn less integrative and complex network activity. The less integrative functions first become operational, and could be measured during the repair in a patient who suffered an incomplete cervical spinal cord injury. With CDT, the patient re-learned first, after 6 months, the turning for low load (20N) in the forward direction on the special CDT device. For exercising against a load of 100N, he needed 24 months and against 150N more than 36 months (Figure 100). As re-learning period, the time was taken when the CD values had reached a plateau of low value and the CD values were in the range of those of healthy persons.

3.6.5. The Quality, Complexity, and Stability of CNS Neuronal Networks of a 5-Year-old Child Is Far Away from Those of the Adult Mother

In Figure 129, it can be seen that the arrhythmicity of exercising (the CD values) reduced (improved) and the frequency of exercising increased with development. The neuronal network complexity and quality of organization of the CNS had increased strongly during these 3 years of development. The intention for fast exercising (to improve CNS) was no longer hindered by the undeveloped variability in patterns organization. But the quality, complexity, and stability of CNS functioning of the child, as judged by the CD, was still far away from those of the adult stage of development, as can be seen when comparing the infants CD values with those of the mother (Figure 129). The mother could turn smoothly (with little frequency variation) for all patterns between pace and trot gait and the CD values for forward and backward exercising were small, indicating that her CNS neuronal network complexity was good enough to generate all the necessary patterns for smooth exercising. Good enough network complexity means that the network connectivity was high and that the phase and frequency coordination of neuron firing was high and allowing the generation of difficult movement coordination patterns.

It is shown below in a cross-sectional study that with increasing age the coordination dynamics values continuously decrease (Figure 92) till they have reached the adult values, whereas the frequency of exercising shows an increase and decrease (Figure 130).

3.6.6. Transient Very Fast Exercising

An interesting movement performance phenomenon for learning was already observed in the child Jürgen. When he became able to exercise sufficiently fast, he also sometimes turned transiently very fast (Figure 129E,F). As if somebody was telling him to turn very fast for a short time. It will be analyzed below that such transient fast exercising of children may be a strategy of the CNS to improve its coordinated firing of neurons additionally. During repair by learning, the patient can exercise on a special CDT device to improve the injury or malformation impaired phase and frequency coordination of neuron firing. During normal development, the CNS may use the fast performance of different movements, as for example fast running or crawling, to improve continuously the phase and frequency coordination of neuron firing, which is necessary, because the developing network (increase of connectivity for example) is changing all the time. Following severe cervical spinal cord injury, it was shown that with every bit of regeneration of the spinal cord (re-appearance of function), the CD values got transiently worse (Figure 84). Upon exercising on the special CDT device, the transient fast moving during CNS development becomes nicely visible (see below).

3.6.7. Development of Coordination Dynamics (CD) Values between 5 and 19 Years of Age for Girls and Boys

Figure 130 shows the CD values of boys and girls upon exercising in the forward direction (A) and backward direction (B). As can be seen from the curves, the values decreased continuously from 5 to 19 years of age. There may have been a transient increase of CD values at an age of 10 to 11 years. The size of the age groups varied between 5 to 25 pupils per group. The frequency of exercising increased first strongly and then decreased slowly (Figure 130C,D). The boys turned fastest in the forward direction ($f_{average,forward}$ = 2.02Hz) at an age of 9 years. Exercise in the backward direction was fastest also at 9 years of age, but with a lower frequency (1.59Hz).

The values and group sizes for girls were similar to those of the boys. As in the case of boys, the girls turned on average fastest in the forward direction (f = 1.82Hz) at an age of 9 years. For exercising in the backward direction the girls were fastest at an age of 13 (1.48Hz).

Plotting the coordination dynamics values of male and female pupils together, it turns out that Since the coordination dynamics values of boys and girls were similar (Figure 130), the data of boys and girls were gathered together to give the coordination dynamics values of Figure 92. The most group sizes for each age range were now between 30 and 50 (Figure 92). Even though the group sizes were on average very large, details of the development seem only partly to become visible; the longitudinal study of the young child between 0.5 and 5 years of age offered more information. The transient increase of coordination dynamics values for exercising in the backward direction (worsening of CNS organization) at an age of 11 years (Figure 130B) may have significance. An expected increase of the CD values during puberty (in the age range of 14 years) cannot be seen. But because these pupils were not very cooperative at that age (as during school lessons), the group size was comparably small. The frequency of exercising was highest at an age of 9 years.

At an age of 5 and 6 years boys and girls were not very cooperative. They often were afraid of the assessment situation, even though the measurements were performed in their familiar surroundings.

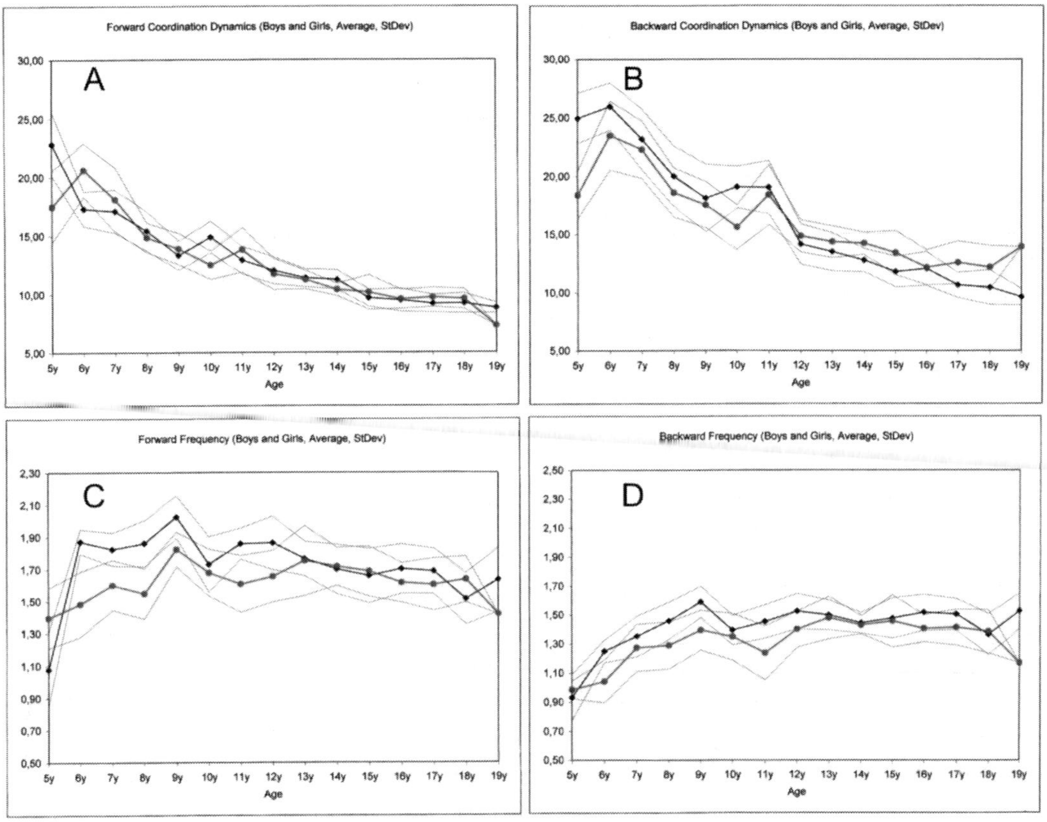

Figure 130. Coordination dynamics values and frequency of exercising of boys and girls in dependence on age. "■" are boys and "●" are girls. Note that the average group values are very similar. The standard deviations overlap much.

In conclusion, the CD values decreased with age, which means with on-going development. The frequency of exercising in the forward direction first increased strongly between 5 and 9 years of age and then deceased slowly. This can be understood upon watching the pupils during exercising. The young children wanted to turn fast (sometimes they were even saying it) but they got stuck again and again. The continuous exercising was hindered by the missing network maturation. Especially the complicated coordination patterns of arm and leg movements between pace and trot gait could not be sufficiently generated by their CNS neuronal networks. The network complexity and accuracy were not developed well enough in that developmental period. When Jürgen was older, he even wanted to have the on-line measurements, seen on the screen of the computer, explained (Figure 90F).

3.6.8. Equality between Women and Men

An important observation, with respect to equality between women and men, the girls' CNS was functioning as good as that of the boys with respect to low-load coordination dynamics (CD) values. Also for high load, the CD values of women and men are similar. The female director of the high school managed better with the high-load assessment than the male sports teacher! But with respect to a high-load assessment, the composition of muscle fibre types (fast and slow) and their upstream fast and slow neuronal sub-networks have to be taken into consideration (Figure 13).

3.6.9. Fast Movements to Improve Phase and Frequency Coordination among Neuron Firing for Improving CNS Self-Organization

From Figure 130C it can be seen that the frequency of exercising first increased and then decreased. The slow decrease of the exercising frequency with increasing age, after the fast increase, can be understood. The high frequency of exercising was used by the CNS for improving its self-organization. The increased movement frequency improves the quality of performance because the kinetics is smoothing the movements and the improved movement-induced afferent input improves CNS organization by movement-based learning. With on-going maturing during development this kind of movement-based learning becomes less important and the frequency of exercising decreased. Fast moving can be observed in infants and children in everyday life. Generally, walking and running (especially when performed fast), jumping, and training balance seem to be necessary for a physiologic development of the nervous system.

3.6.10. Occurrence of CNS Instabilities due to Lack of Sufficient Movement-based Learning

In reverse, if pupils are not moving sufficiently, they may not get only overweight, but we have to expect that variations, instabilities and small mistakes in CNS development can no longer sufficiently be repaired by movement-based learning. A higher incidence of diseases in children caused by suboptimal CNS development has to be expected. The occurrence of narcolepsy already in children (which normally occurs at an age between 20 and 30 years) may be such a disease caused by suboptimal CNS development.

3.6.11. Transient Fast Moving

The learning from fast movements to improve CNS functioning (most likely by improving phase and frequency coordination) seems to be necessary during development, because of increasing network complexity. The intention to move fast can be seen by the transient fast exercising of children on the special CDT device, frequently observed in the age range between 6 and 9 years. For a few seconds the young pupils were turning very fast (Figure 129E,F; Figure 131C,D), just because they liked the feeling, as an older person with a slightly malfunctioning CNS explained. In this way the CNS is getting the input it needs for proper development. Such transiently fast moving can also be observed in patients with spinal cord injury, cerebral palsy, epilepsy and other CNS diseases.

A 6-year-old boy with mild cerebral palsy turned comparably slowly in the forward and backward direction (around 0.5Hz) but turned transiently very fast (up to 2.5Hz) (Figure 131A,B). His healthy and sporty 10-year-old sister turned generally faster (in accordance with the typical age values) but the transient fast turning was comparably not that fast (Figure 131C,D). It seems that the injured or malformed CNS needs especially the fast exercising for improving CNS self-organization, even though this fast exercising is hindered by the impaired self-organization.

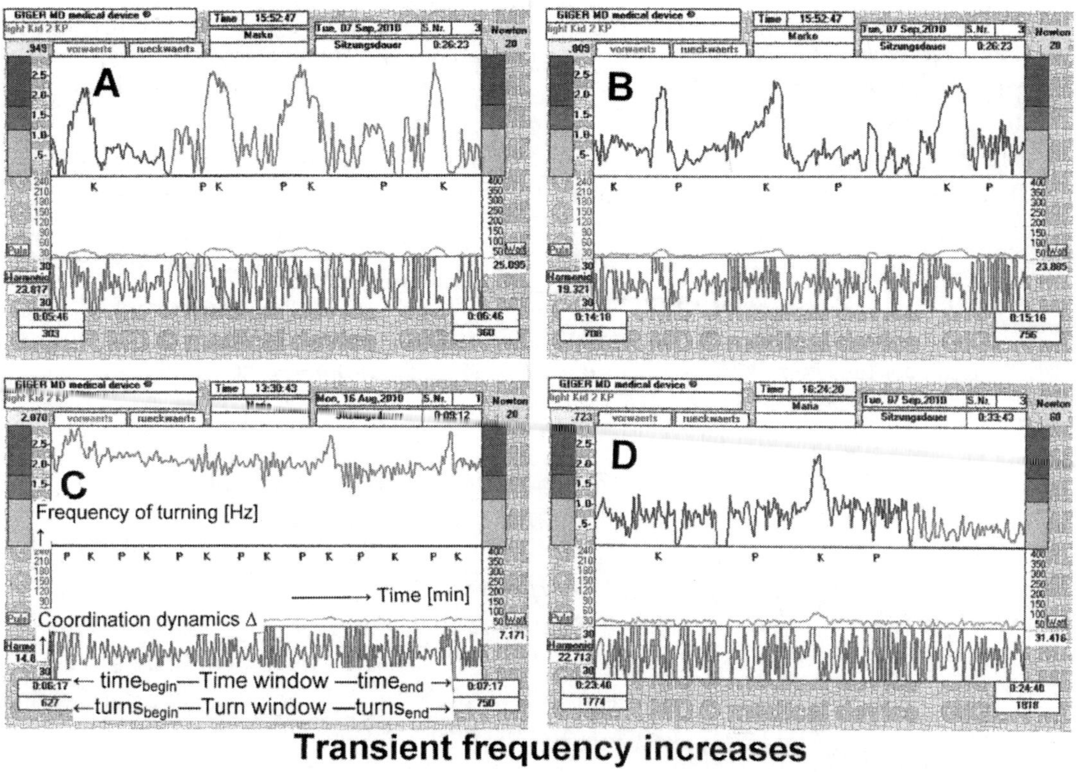

Figure 131. Transient frequency increase of exercising in a 6-year-old boy with cerebral palsy (A,B,) and in a healthy 10-year-old girl (sister) (C,D). Note that the boy with a mild cerebral palsy is exercising at a low frequency, but the transient frequency increases are very high in comparison to those of his healthy sister.

Adults with a healthy CNS turn mostly at a frequency in the range between 1 and 2Hz. Athletes turn at higher frequencies, partly determined by the sport they perform. But healthy adults do not turn un-volitionally transiently fast. This transiently fast exercising seems therefore to have something to do with the improvement of the suboptimal functioning CNS. It is probably used by the CNS for improving its phase and frequency coordination for improved self-organization and for a functional CNS repair. If the fast exercising and the transient fast exercising are related to the improvement of CNS functioning during development, then it is interesting to see how it is with the fast exercising in aging when performing CDT or in aging when performing CDT and prolonged fasting. It will be shown that during exercising in aging and especially with additional prolonged fasting the CNS uses the fast exercising and transient fast exercising to improve CNS functioning.

3.7. Neural Network Learning in Aging when Performing CDT with Respect to Fast Exercising

In Figure 116 it was shown that the nervous system functioning improved when performing coordination dynamics therapy (CDT). The improvement of CNS functioning was quantified by lowering of the CD values especially for high load. Also the frequency of exercising increased (Figure 116). The decrease of CD values and the increase of the frequency of exercising is a measure for improved CNS functioning. The un-volitional increased frequency of exercising is caused by the improved of CNS functioning und is used also to improve CNS functioning.

The un-volitional transient fast exercising was not observed when performing CDT alone. Since the Author made this experiment on himself, he can safely say that transient fast exercising is un-volitional. There is similarity to development because fast exercising is improving CNS functioning. An interesting question is now, how it is with the fast and transient fast exercising when performing CDT and repeated prolonged fasting. Are CDT and prolonged fasting also related to fast and transient fast exercising? It will be shown below that fast and transient fast exercising do occur during CDT and additional prolonged fasting.

3.8. Neural Network Learning in Aging when Performing CDT Plus Repeated Prolonged Fasting with Respect to Fast Exercising

3.8.1. Increase of Fast Exercising in Aging

When performing CDT in aging, the coordinated firing of neurons improved and the assessed coordination dynamics improved. The high-load CD values got smaller (Figure 116B) and the frequency of exercising increased (116A). But transient fast exercising, which occurs in healthy development (Figure 131CD), did not occur. The crossing of network excitation as for example via synfire chains [49,94] is only trained partly. Exercising at higher frequency is only possible if the coordination of excitation crossing pathways is improved. Also the traffic at street or road crossings needs to be more coordinated if more cars at higher speed approach the crossing.

3.8.2. Occurrence of Transient Fast Exercising When Performing CDT and Repeated Prolonged Fasting

When performing in aging CDT and repeated long prolonged fasting together, ‘transient fast exercising’ occurred in addition to ‘fast exercising’. The frequencies of exercising are shown in Figure 132C. Exercising frequencies of up to 1.96Hz were reached. When performing CDT alone, only frequency values of up to 1.7Hz were reached (Figure 116). With repeated fasting up to six days the frequencies of fast exercising were higher. The repeated prolonged fasting for four days in January and February 2015 shows nicely again the increase and decrease of the turning frequency. Also, in the high-load assessment additional increase and decrease of the turning frequency can be seen, especially before and after the change of the direction of exercising (Figure 133). By comparing the frequency values of Figure 125a with those of Figure 133, it can be seen that with CDT and fasting the frequencies were higher in general and were higher before and after the change of the direction of exercising then with CDT alone.

In addition to higher turning frequencies in general with fasting also ‘transient fast exercising’ occurred quite often (Figure 134), even though their amplitude was much smaller than during healthy development (Figure 131C,D).

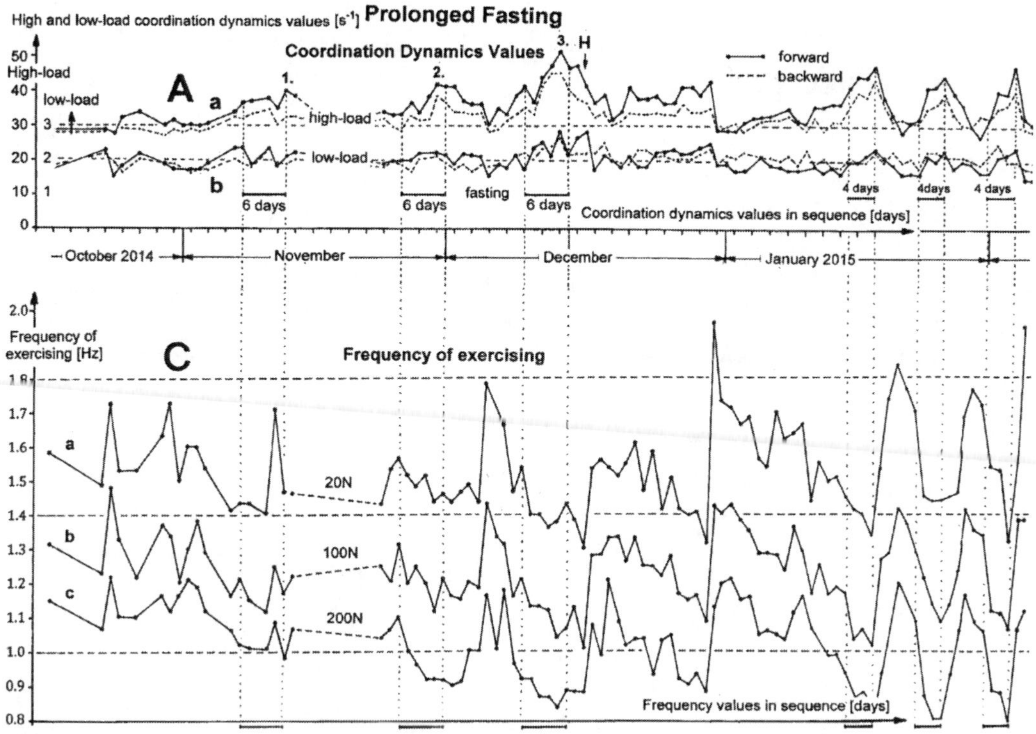

Figure 132. Continuous high-load and low-load tests in relation to the frequencies of exercising during repeated 6-day and 4-day fasting. A. High- and low-load coordination dynamics values (same as in Figure 126A). B. Turning frequencies for 20, 100 and 200N for exercising in the forward direction. Note the strong frequency increase following each fasting for four days in 2015.

The increase of the frequency of 'fast exercising' and the occurrence of 'transient fast exercising' indicates that more repair processes by movement-based learning took place when CDT and prolonged fasting were performed together. Since the higher turning frequencies and the occurrence of transient fast exercising indicate better movement-based learning, CDT and simultaneous prolonged fasting seems to activate some kind of rejuvenation of CNS functioning. Because the CNS is involved in nearly all functions of the human body probably also other functions of the human body improve or will improve in its functioning. One can expect especially improvements in those functions, which are related to movements as for example the cardio-vascular performance. However, as the plasticity of and the regeneration in the mucosa of the mouth following cancer treatment indicate, there were other functions, which improved in the Author. The problem remains that the rejuvenation needs months till years and some rejuvenation occurs spontaneously. On the other hand, functions get worse with aging. It is therefore difficult to safely relate rejuvenation to CDT and prolonged fasting.

3.8.3. Worsening of CNS Functioning during PROLONGED Fasting

As can be seen from Figure 132 during fasting the high-load coordination dynamics values increased and the frequency of exercising decreased. It became more difficult to exercise on the device. The general changes of CNS functioning in comparison to not fasting were the following:

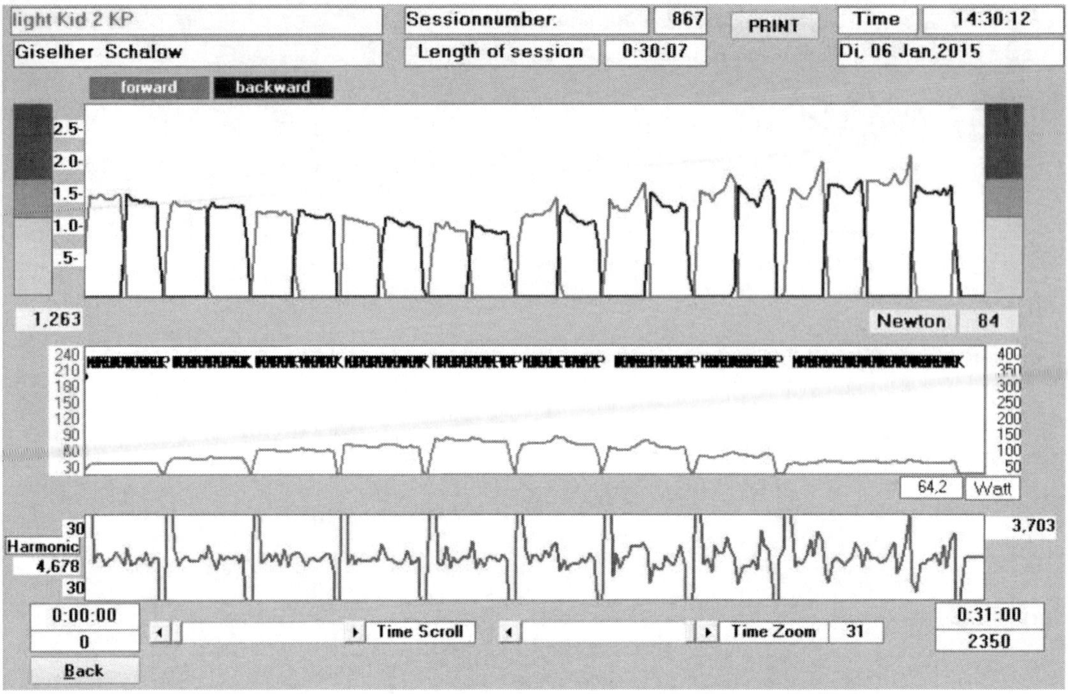

Figure 133. Assessed high-load test. Upper trace = frequency; middle trace load in Watt; lower trace = coordination dynamics. Note the increases of the turning frequencies before and after the change of the direction of turning.

(1) After 3 days fasting, it felt as if the load of exercising had increased by approximately 30N for all loads, as if there was an additional resistance in the joints like rigor.
(2) The frequency of turning was limited. The Author could just not turn faster.
(3) The sweating during the exercising at higher loads was reduced because of the reduced load due to the lower turning frequency.
(4) The Author made more mistakes with the directions of turning, like the young infants did during development.
(5) It became more difficult to do two things at the same time, namely turning and thinking of other things.
(6) Car driving was less safe. The coordination in steering worsened as if the automatic performance was impaired.
(7) It seemed that the exercising in the forward and backward directions was impaired differently.
(8) It seemed as if the CNS in general worked worse.
(9) No tremor occurred.
(10) The exercising with opened or closed eyes did not change the coordination dynamics. With open eyes the visual neuro-feedback helped to exercise more rhythmically and with closed eyes the neuro-feedback (the feelings) from the moving arms and legs was more used for turning more smoothly.

3.8.4. Short-term Supercompensation Changes during Prolonged Fasting

When exercising on the special device during fasting, improvement of the performance could be felt in the short-term memory and could partly be seen on the frequency curve. But the improvement was reduced with respect to not fasting. When changing the direction of exercising, transient faster exercising occurred, but to a lesser extent. Short-term supercompensation, occurring approximately after one hour of exercising, did not occur or only little occurred during fasting. The reduced power of exercising during fasting can only partly explain the worsening of movement performance, because also the low-load CD values got worse. There are some neural network organization changes during fasting.

Long-term supercompensation (Figures 96,101,102), occurring after many days of exercising, are probably related to other mechanisms.

3.8.5. High-load Performance after Four Day Fasting with Eating Again

During normal eating the exercising with increasing load was more difficult than during reducing the load from 200N. After the high-load testing, one is feeling better because of improved CNS functioning. This is the time when the short-term supercompensation occurs. The effect of short-term supercompensation was more evident with the first high-load test after having eaten again. When starting the high-load test, there was resistance in the system, something like rigor. But when reducing the load from 200N to 20N, I felt that the resistance was somehow 'washed' out with ongoing exercising.

3.8.6. Possible Explanation for Reduced Short-term Supercompensation and Rigor by Reduced Neurotrophin-Induced Facilitation of Synaptic Potentiation during Fasting

"Neurotrophins have been proposed to participate in activity-dependent modification of neuronal connectivity and synaptic efficacy. Preferential strengthening of active inputs requires restriction of putative neurotrophin-mediated synaptic potentiation to active synapses. It was reported that potentiation of synaptic efficacy by the brain-derived neurotrophic factor (BDNF) is greatly facilitated by presynaptic depolarization at developing synapses. Brief depolarization in the presence of low-level BNDF results in a marked potentiation of both evoked and spontaneous synaptic transmission, whereas exposure to either BDNF or depolarization alone is without effect. This potentiation depends on the relative timing of depolarization and reflects an enhancement of transmitter secretion from the presynaptic neuron. Thus synapses made by active inputs may be selectively strengthened by secreted neurotrophins as part of activity refinement of developing connections or of mature synapses" [96].

In Figure 82, the competition of two kinds of neurons for the innervation of a target muscle cell (slow muscle fiber) is shown during frog development. The synapse of the fast motoneuron (Figure 82b, profile 1), in some similarity to the α_1-motoneuron of Figure 13, is replaced by the growing synapse of the slowly conducting motoneuron (Figure 82a, profiles 2,3,4), in similarity to the α_3-motoneuron of Figure 13. The competition of the two kinds of motoneurons for innervation takes place probably by neurotrophins and later on additionally by the activity of the neurons. The neurotrophins are probably stored in dense core vesicles. Dense core vesicles can be seen on the right side of Figure 82b. When the dense core vesicles fuse with the membrane, the vesicles open and the neurotrophins are excreted into the space between the cells and will reach the other neuron or muscle cell by diffusion in some similarity to the excretion of acetylcholine from the acetylcholine vesicles, which have no dense core. Plenty of acetylcholine vesicles can be seen in the endplate of Figure 82b. The neurotrophins are probably transported in micro tubes from the cell soma to the synapse, marked in Figure 82d by arrows. For further details of the electrophysiology and morphology of this synapse formation changes during frog development see Chapter I of [1]. The same mechanism takes place during repair in the adult frog following denervation by cutting the nerve supply [85-88].

It is conceivable that during fasting there is a lack of neurotrophins to modulate optimally neurotrophin-induced synaptic potentiation in the CNS, which is felt during exercising as rigor. With eating again and normal activation of the synapses, the neurotrophins are produced sufficiently again and are brought to the places of action. The rigor terminates.

3.8.7. Rigor in Parkinson's Disease

If neurotrophins are sufficiently needed for activity-dependent modification of neuronal connectivity and synaptic efficacy, then in patients with Parkinson's disease the rigor could be explained by a lack of BDNF. Because the Author had some kind of rigor during prolonged fasting and no tremor, the lack of BDNF may not be related to the lack of inhibition in patients with Parkinson's disease, which leads to tremor (Figures 65 and 66). There seems to be more pathology in the functioning of the CNS in Parkinson's disease than just a lack of inhibitions.

Still, by movement-based learning when exercising on the special CDT device, the physiologic movement, given by the device, can 'catch' the pathologic patterns and reduce tremor and improve the motor programs (Figure 76). Further, CDT in general can optimize CNS functioning by modulating efficacies of synapses in the short and long-term memory by coordinated activity-dependent modification and partly compensate for reduced inhibition and reduced neurotrophin supply and other pathologic changes. The deficit in anticipatory postural adjustments can be counteracted by changes in the direction of movements and training pace and trot gait coordination during crawling and walking. With the general improvement of CNS functioning, the increased latency of reflexes will be reduced.

3.9. Comparison between Neural Network Learning of the Healthy Development and during CDT and Prolonged Fasting in Aging

During healthy or normal development, movement-based learning is used to improve continuous CNS functioning by a 'correction en route'. The small mistakes, done by the genetics, are corrected by learning. One strategy of correction is 'fast exercising' and the 'transient fast exercising'. With CDT and fasting CNS, improvement is achieved by a 'corrected en route' when poor functioning proteins and/or cells are replaced by better ones.

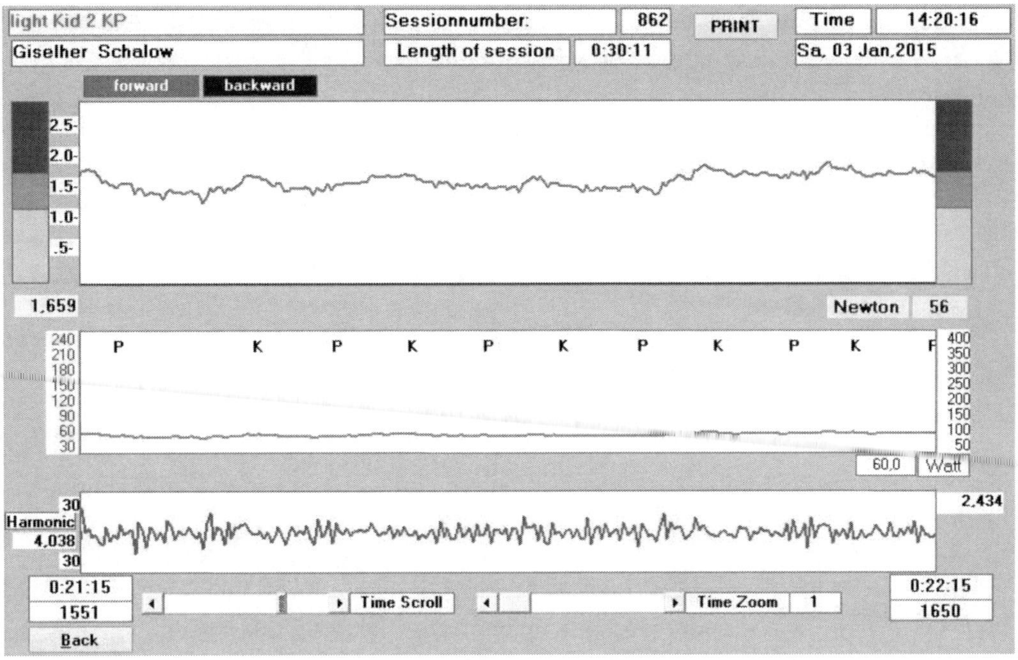

Figure 134. Transient fast exercising. The episodes of fast exercising have an approximate duration of 5s (upper trace) and an amplitude of approximately 0.2Hz.

In aging the transient fast exercising periods have an amplitude of approximately 0.2Hz (Figure 134) and during normal development an amplitude of around 1Hz (Figure 131C,D). Since during normal development, the infants can turn at higher frequencies (Figure 131C; 2Hz for forward exercising) than in aging, and the transient fast exercising has higher frequency amplitude, the possibilities of rejuvenation seem to be very limited in comparison to improvements of CNS functioning during normal development as judged by the amplitude of transient fast exercising. However, with CDT and prolonged fasting aging people have a tool at hand to help in a natural way to live longer with a better quality of life.

It is conceivable that CDT in connection with repeated prolonged fasting also helps in cerebral palsy or traumatic brain injury to increase the efficiency of treatment.

4. Neural Network Learning during Deviant Development and Severe Impairments in Aging

4.1. Retarded, Accelerated or Deviant Development of Motor Functions

Many fluctuations occur during normal development. There are accelerations and decelerations in the developmental rate of different functions within one infant. Characteristic for normal development is *variability in motor performance*. But the impaired nervous system is not able to attain such variability in motor performance because the structure of the CNS is deficient in enabling the infant to use various modes of operation for single performance. When there is *serious damage* to the CNS, it is even possible that the *usual modes cannot be developed at all; the brain has to resort to unusual modes*. Distorted motor patterns arise, which fall outside the range of the normal variability and which are *stereotyped, due to the deficiency of the 'hardware'*.

Due to the reduced variability in motor performance of the impaired nervous system, patients with CNS injury or malformation have serious problems exercising on the special CDT devices because the device imposes complex motor pattern variability. But motor pattern variability can partly be learned from the device. Due to the deficiency of the neuronal structure, the development of motor functions may be retarded, accelerated, or deviant. Some functions may not develop at all, while others show only a decrease in variability.

4.2. Learning for Repair by Recapitulating Development

The mechanisms, or strategies, developed by the newborn, infant, or young child are not peculiarities or movement patterns that occur only in the neonatal or later period. They belong to one phase of the developmental course of the nervous system. These patterns are elements which have been tested and trained and which can be modified for use in more differentiated and complex motor behavior during later development. Their integrity must be regarded as a prerequisite for normal development during infancy. Sufficient crawling must be trained in order for the infant to have no coordination problems between arms and legs in later life. The stepping movement of the newborn (Figure 44) (characterized by the emphasized lifting of the knees) is partly re-appearing in severe cervical spinal cord injury. Upon partial regeneration, induced upon CDT, the stepping automatism is partly changing again into volitional walking. The patient with severe brain injury of Figure 111 can run quite well (Figure 106C) but the forward walking is very pathologic (not shown). But by the emphasized lifting of the knees during walking, he partly activated the stepping automatism of the newborn, which is located mainly in the spinal cord and was not damaged by the traumatic CNS injury, and the performance of walking improved in the short-term memory.

The aim of learning is not only to equip the deviant infant or child with just one efficient mode of motion for each motor performance, but to lay down a foundation in the CNS for adaptive responses in specific circumstances and to induce learning transfer to other (movement) patterns, which cannot be trained. Jumping on springboard and exercising on the special CDT device improves continence in severe cerebral palsy (see repair in cerebral palsy) and spinal cord injury by learning transfer. Exercising walking in the backward direction on treadmill improves the forward walking in stroke patients by learning transfer.

The *learning process may be hampered by a deficiency of the neuronal structure on which they are dependent*. Stimulating the patient to use uncommon modes of operation may enable him to achieve results, which he cannot achieve spontaneously. If an adult patient or child cannot jump in anti-phase on springboard (Figure 109) continuously, he may learn it by including, for example, a swing in between the jumps. Elements of function must also be learned, which are necessary for more complex motor patterns. If one recognizes which learning processes are deficient, one may then be able to offer the infant a cue to start other learning processes or help him to eliminate errors and to progress in the right direction. If treatment strategies become more personalized to the needs of individual infants or children, it may become more difficult to evaluate the effectiveness of any single treatment program. But if CDT would be 10 times more efficient than conventional treatments, differences in effectiveness can be found. Proper treatment of the handicapped infant or child consists of close co-operation between an active parent and an active therapist.

4.3. Learning for Repair

Pathology: Because of experienced adverse perinatal events, such as hypoxic ischemic encephalopathy, bronchopulmonary dysplasia, or bacterial meningitis we may find retarded, accelerated, or deviant development of motor and other functions. Some functions may not develop at all, while others show only a decrease in variability. Combinations of these possibilities may occur within single infants and result in individual motor and other patterns. The deviant patterns are characterized by a lack of variability and assume a thoroughly stereotyped appearance.

In most cases, some parts of the brain are more damaged than others, and some parts are not damaged at all. The undamaged parts will try to become operational. In correspondence with the localization, extent, and type of the damage, it may happen that processes which in themselves are normal, cannot bring about good results, because some areas of the brain which are also necessary for the accomplishment of the particular motor function are deficient. For example, certain functions that have developed normally cannot be integrated into larger patterns because other elements of these larger patterns are missing or have not developed sufficiently. In general, both impaired and healthy parts of the brain mature over time and thus lead to increased complexity, which has direct repercussions on the quality of the learning (trial and error-elimination [83], see below) processes.

The well-known symptoms and signs of cerebral palsy in the first year of life (poverty of movements, stereotypy of posture and motility, inability to 'discover' new motor possibilities, neglect of one extremity, stereotyped extension of the legs during vertical suspension, head-lag during the traction test or during sitting), can all be traced back to a lack of trial and error-elimination processes (learning) as a consequence of deficient brain structure. In the case of an inability to 'discover' new motor possibilities, there may be a disturbance in the chain of events because errors are not recognized (or not eliminated), with the result that the processes stop prematurely [84].

4.4. Problem Solving Therapy by Learning during Deviant Development

From the therapeutic point of view, trial and error-elimination in abnormal infantile development can be of great value in that it may indicate which type of treatment is needed. It is of crucial importance to recognize the deficiency or absence of particular functions and learning processes. If one recognizes which processes are deficient, one may then be able to offer the infant a cue to start other processes or help him to eliminate errors and to progress in the right direction. It may be sensible to 'teach' the infant specific modes of operation to avoid wrong postures or wrong movement patterns, or to stimulate him to use uncommon modes, which may enable him to achieve results that he cannot achieve spontaneously. It has to be assessed, which functions are left intact and the ways in which the child can avail himself of these functions to promote further development [84].

4.5. Motor Learning and Problem Solving Therapy by Learning

Motor learning, such as learning to ride a bicycle or play piano, is partly different to learning for CNS repair or optimizing development. The memory of motor skills (nondeclarative memory) such as riding a bicycle is called procedural memory. Nondeclarative memory is formed in brain regions other than the hippocampus (probably in the sensory-motor cortex and brain stem nuclei). The problem solving learning tries to repair sub-networks, which are necessary for functioning and learning. By inducing trial and error-elimination processes in subunits of the normal developing nervous system an optimal development is achieved [83].

4.6. Development and Repair of Neuronal Networks by Movement-Based Learning

One way to understand the complexity of neuronal networks is to understand how their connectivity and how their impulse patterns in circuits emerge during development. The traditional model of brain development includes two phases; an early phase during which a coarse wiring of the nervous system is laid out, and a later phase during which the coarse connections are refined. In this model, the developmental events that underlie the coarse wiring are the result of predetermined genetic programs and occur independent of neural activity, whereas the refinement is a result of interactions between the nervous system and the outside (and inside, for example urinary bladder function) world. In the newer model, neural activity and genetic programs interact to specify the composition and organization of neural circuits during all stages of development [89].

Specific neuronal activity seems to be required to regulate development [90] and for the repair of the human CNS. It seems that activity controls neurogenesis, maturing of neurons, specific migration of neurons, and the development of their axons and dendrites. The specific complexity of axonal and dendritic arborization, and weights of synapses develops through this natural activity, generated by natural movements and functions of body organs. The natural impulse patterns generated by receptors (Figure 6) of muscle-limb mechanics interact with the impulse patterns generated by organs (as for example the urinary bladder (Figure 12) or eyes) in the networks (Figure 59) to develop the neural networks or repair them.

There is indication that genetic programs initiated at the progenitor stage are modulated during development by activity [90]. A critical period for enhanced synaptic plasticity in newly generated neurons in the adult brain has been identified. Adult-born neurons exhibit the same classic critical period plasticity as neurons in the developing nervous system [91].

Redundancy is built into the neural circuits to ensure that perturbations during development can be “corrected en route” and that small network injuries can be repaired in later life.

As shown in Figure 18a, healthy 5-months-old Jürgen has still not developed a proper motor program for walking (Figurc 18b). His premotor spinal oscillators (Figures 13,20,137) will work properly with respect to stimulation (Figure 70) and generate a motor program (Figure 23) by transiently occasionally [1, 92] and oscillatory firing (Figure 12) of the motoneurons (Figure 22) to develop the neuronal networks for walking in the spinal cord and supraspinal centers. Three months later, his motor programs had improved (Figure 18d) with further development. Probably in an earlier stage of development, when he was a fetus, these premotor spinal oscillators were used to generate spontaneous activity for network forming during development. These premotor spinal oscillators consist of the motoneuron and excitatory and inhibitory firing interneurons (Figure 137). It is conceivable that during the early stage of development, when inhibition was not established sufficiently, the premotor spinal oscillators fired already oscillatory with little input. In this way there was activity generated in the networks to coordinate the communication among neurons for integration. Other cell or network oscillators probably contributed to the activity for neuron integration.

Following neural network damage as for example in spinal cord injury the oscillatory firing of the premotor spinal oscillators becomes impaired (Figure 37) and has to be repaired by the movement-induced afferent input induced when the patient is exercising on the special CDT device (Figure 15) and/or is performing rhythmic dynamic stereotyped movements like jumping on springboard (Figure 30). The oscillators become entrained (Figure 64) and improve their Eigen-frequencies.

It is believed that even though many components of cell identity, such as general intrinsic physiological properties and layer position, are likely determined at the progenitor level, other components such as choice of synaptic partners are likely influenced by local environmental cues. These two mechanisms may contribute to different aspects of cell specification.

Abnormal development of cortical interneuron subtypes during late embryogenesis due to environmental perturbations coupled with genetic abnormalities might represent a primary cause for many neurodevelopment disorders [93] as for example cerebral palsy.

In abnormal development and repair following injury, movement-based learning treatment has to be started as early as possible, including vision (Figures 79 and 80). CDT, including visual and auditory input, can be started after birth even in premature born babies (Figures 86 and 140).

4.7. Neurogenesis of Premotor Spinal Oscillators in Myelomeningocele and Following Cancer Treatment in Infants by Movement-Based Learning

There is indication that new motoneurons and probably interneurons can be built by CDT in the adult human spinal cord [3]. A partial structural repair seems possible in spinal cord injury or malformation, especially in infants.

In Figure 135, the three year-old patient Kiki, with a malformation of the caudal spinal cord (spina bifida) is exercising in the sitting (A) and lying position (B). Her higher mental functions are normal. But she is incontinent and cannot move her legs. Therefore, her mother is supporting the legs in Figure 135A and the legs are fixed in 135B. When the Author supported her legs, she liked to crawl. If knees and pelvis would be supported, she could train sky-walking without weight support. It was shown that in a child with spina bifida the CNS could partly be repaired when CDT was administered intensively for a few years [17]. It is therefore likely that the CNS in Kiki could partly be repaired, if the mother could apply CDT to her daughter. However, they do not have the money to pay for devices and mainstream medicine is offering only one to two hours per week instead of the necessary twenty. In twenty years, Kiki will be a young lady who is incontinent and cannot walk. Very often society fails to properly care for its children.

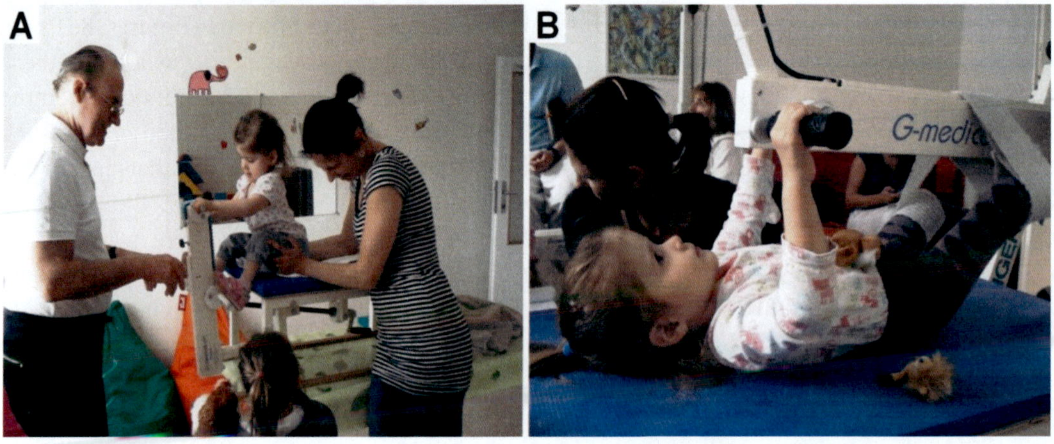

Figure 135. Three-year-old girl Kiki with a myelomeningocele during exercising on a special CDT device in the sitting (A) and lying position (B). For training in the sitting position the mother is needed for support; for training in the lying position only bandages are needed for support.

Following cancer treatment (see below), CDT can help to repair the injured CNS. Since in young infants, the neural circuits are still developing and new neurons can be built, especially when training at limits, movement-based learning can repair damaged neural networks and additionally activate tumor suppressor genes to reduce the probability of reoccurrence of cancer.

The girl of Figure 136 (and 30,32) had cancer. Besides spinal cord removal, she suffered an incomplete lumbar spinal cord injury in the operation. Now she has problems with walking and is incontinent. In ten years, she is a young lady who is incontinent and has problems walking. The non-functioning urinary bladder is the real problem. But it was reported that the urinary bladder could be repaired [1,15,16]. In infants, the repair was faster than in adults.

There is a good chance that structural repair can also be achieved because her networks are still under development. Since neural activity and genetic programs interact to specify the composition and organization of neural circuits during all stages of development [89], there is a good chance that the spinal cord can be repaired and urinary bladder functions and movements can be repaired or established in both girls.

During development and during repair, these premotor spinal oscillators may mediate spontaneous activity for the developing neural networks. The movement induced afferent input probably enhances the building of new circuitry and stimulates its integration into the existing networks.

By including all the movements in the movement-based learning program, which are performed by infants during development, the neural activity and genetic programs interact in the right way to specify the composition and organization of neural network building to become physiologic. In Figure 137 principle premotor spinal oscillator networks are pictured.

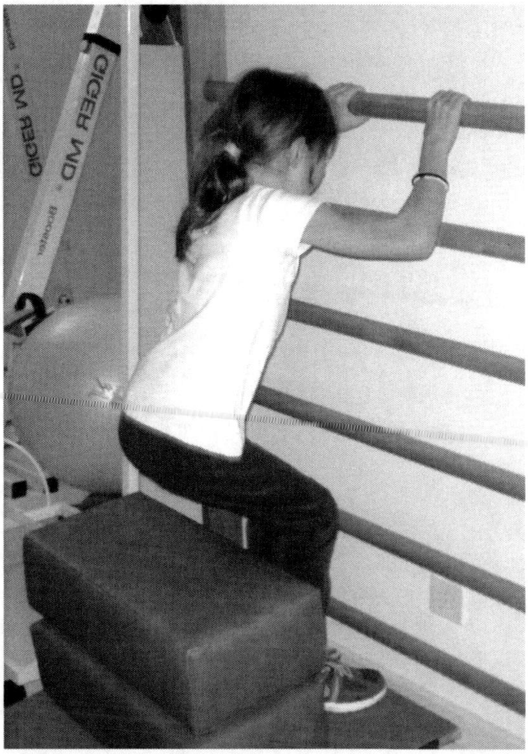

Figure 136. Up-right exercises of a girl following tumor removal and damage of the lumbar spinal cord to repair with other movements the lumbar cord by learning.

4.8. Structure of Premotor Spinal α_2-oscillators and Their Coordinated Firing

It is assumed as a simplified working hypothesis that these spinal oscillators are organized by excitatory reverberatory loops of interconnected interneurons in the form of synfire chains [49,94] to which inhibitory interneurons essentially contribute because their inhibition time, 30 ms (rat) and 150 ms (cat) [95], falls within the range of the 30 ms building time block of the oscillation period and the oscillation period itself (160 ms) (Figure 137). Premotor α_1-oscillators show little rhythmicity generated by small networks, whereas α_3-oscillators show much rhythmicity generated by a larger local subnetwork [30]. The oscillator loops consist of chains of neurons (Figure 137) are therefore only schematized network loops. This network loop seems to be rather continuous for α_3-oscillators, whereas the α_2-oscillator loop may show rather discrete sub-loops (see below) accounting for the firing with different impulse trains length (Figure 137) and therefore different loop periods according to T = 70ms + 30ms • n_{AP}. The network loop of an α_1-oscillator is more in the direction of a single neuron chain loop (Figure 13). The measurements give no direct information about the number of neurons contributing substantially to the self-organization of the spinal oscillators. But since α_2-motoneurons show multi-frequency oscillation and α_3-motoneurons show a rather continuous change of the frequency (indicating rather complex local neuronal networks), one can speculate that the number of neurons contributing substantially to the oscillation may vary between a few (α_1-oscillators) and a few hundred (α_3-oscillators). The characteristic of α_2-oscillators to be able to change their organization when activated additionally by the parasympathetic nervous system division (the number of driving phases changes from two to three) (Figure 59), points towards complex local oscillators consisting of many interneurons.

For more details of the structure of human premotor spinal oscillators in connection to animal data see [1].

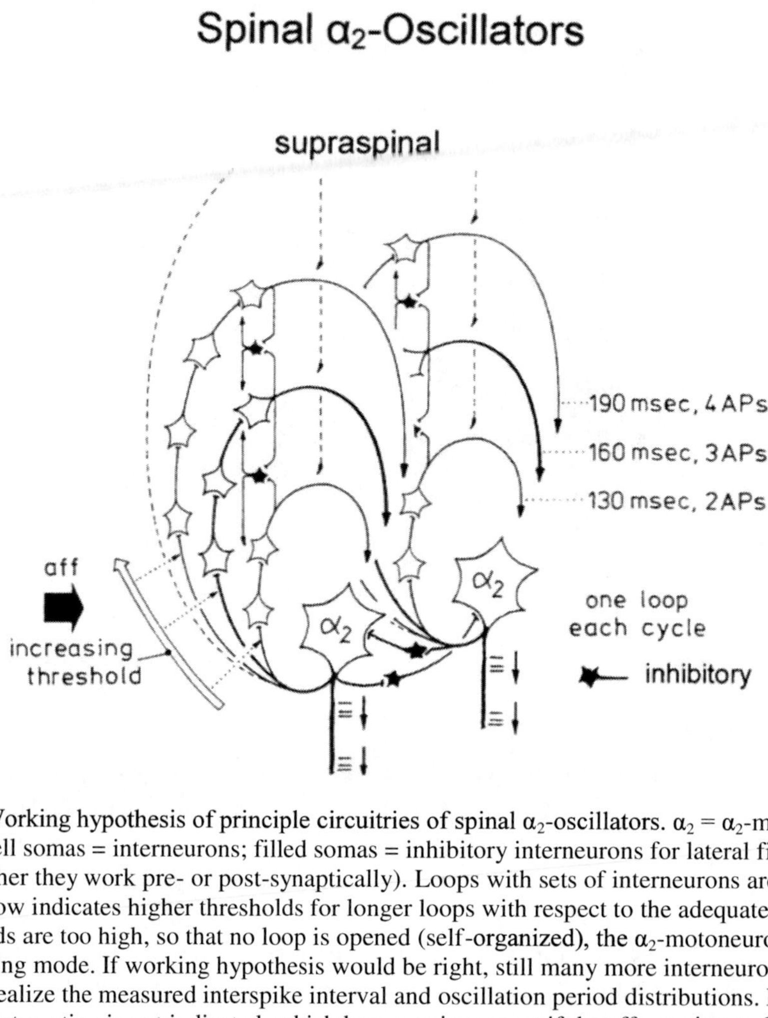

Figure 137. Working hypothesis of principle circuitries of spinal α_2-oscillators. α_2 = α_2-motoneuron soma; open cell somas = interneurons; filled somas = inhibitory interneurons for lateral field inhibition (unclear whether they work pre- or post-synaptically). Loops with sets of interneurons are indicated. Threshold arrow indicates higher thresholds for longer loops with respect to the adequate afferent input. If all thresholds are too high, so that no loop is opened (self-organized), the α_2-motoneuron fires in the occasional firing mode. If working hypothesis would be right, still many more interneurons are necessary to realize the measured interspike interval and oscillation period distributions. For example, the interloop interaction is not indicated, which becomes important if the afferent input changes quickly; more inhibiting interneurons are necessary; the preference for the three AP loop is not explained by the model; the function of the oscillator is not known: Oscillates the α_2-motoneuron in itself (impulse train) and is transiently inhibited by the interneurons (time between the impulse trains) or oscillates the α_2-motoneuron together with the interneurons. One loop each cycle = probably the excitation uses in a first approximation one pathway for an oscillation cycle, but which pathway is taken depends on the probability distribution for a certain afferent input.

The rhythmic firing of the network oscillators is mainly coordinated (Figure 24), similar to the whole neuronal network of the CNS, by the organization tendencies of the network, the descending impulse patterns and the spatiotemporal afferent impulse patterns. The coordinated firing of single motoneurons and single motor units was shown in Figures 22 to 24.

If the three kinds of premotor spinal oscillators would not coordinate their firing and partly synchronize their firing, tremor would occur. Pathologic synchronization was observed in patients with Parkinson's disease (Figures 73 and 74). This pathologic synchronization of the oscillators, giving rise to tremor, can partly be repaired by movement-based learning (Figure 76). Physiologic motor patterns can 'catch' the pathologic premotor oscillator network patterns, entrain it, and repair it by learning transfer. The premotor oscillator network structure seems to be important for understanding motor functioning, development and CNS repair by learning. After emphasizing the importance of premotor neural networks we turn back to clinical considerations.

4.9. Transient Fast Exercising During Deviant Motor Development

The frequency amplitude of transient fast exercising was high during normal development (Figure 131C,D; ~1Hz) and low in aging (~0.2Hz). In deviant motor development, the frequency amplitude of transient fast exercising was very high (Figure 131A,B; ~2Hz). It seems that the CNS tried strongly, by fast exercising, to improve/repair neural network functioning in some similarity of early development (Figures 128, 129E). More data are needed.

4.10. Severe Impairments of CNS Functioning with Aging

To teach the aging or injured CNS to repair itself by trial and error elimination processes, the CNS has, in similarity to the development, to recognize upon CDT, which sub-networks, regulation units or sub-loops are not functioning properly (or are missing) and to repair them by error elimination, including the possibility that other brain parts partly take over functions and sub-networks built anew to a limited extent.

In aging neurogenesis is very limited. To optimize what is left is the main strategy for the improvement of CNS functioning. If dementia or ALS (amyotrophic lateral sclerosis) is in progress, it seems possible to postpone the disease.

In patients with Parkinson's disease or brain injury, it may happen that they have problems initiating the walking pattern. The fast walking or running is normally stated with the hands and arms. If such patients start the walking with arm and legs in the trot gait or pace gait coordination, the walking pattern can be started volitionally. The in-phase movement is easier for the CNS. If one recognizes in aging or diseases which processes are deficient, one may then be able to offer the patient a cue to start other processes similar to learning in deviant motor development.

Often the fight for better CNS function may be lost, especially in aging. Still, it is worthwhile to fight by using movement-based learning. The ancient Greeks put such fight into a nice sculpture (Figure 138). The patient fights with the remaining volitional power, following CNS injury or in aging, against gravity forces and pathologic organization of the CNS. Going to the limits is beautifully represented in the sculpture the Laocoon group,

Compare the faces of the patients in Figures 106A,C and 78 with the faces of Laocoon and his sons.

Figure 138. Laocoon group, made by the Rhodian sculptors Agesandros, Athcnodoros and Pylodoros.

4.11. Movement-Based Learning in Care

In very severe cases of aging the special coordination dynamics therapy device can in principle also be used in the water (Figure 139). However, the human nervous system has no automatisms developed for moving in the water. Athletes, therefore, have to train for different kinds of swimming for a few hours per day, so that the athletes' nervous system learns to swim optimally in the water. Fetuses were in fluid before birth, but were not swimming in it. From the aspect of phylogenies, there was therefore no need to develop automatisms for swimming.

But for an optimal care in severe aging, moving in the water is possible. A good feeling is helpful to keep the patient in a good mood. In the final stage of ALS, when patients can still be very intelligent, some help can be given to them. When they cannot move anymore, because too many motoneurons died already, they have to be passively moved. It is actually a horror for patients with ALS: with full intelligence they recognize that step-by-step even those functions necessary for life deteriorate, for example breathing.

Figure 139. The use of a special coordination dynamics therapy device in the water.

4.12. Stage of Repair/Rejuvenation 6 Years after Cancer Treatment and 5 Years after Reconstruction

In the six years following cancer treatment CDT was performed by the Author and in the last year additionally repeated prolonged fasting:

a. The opacity of the eyes got better, but is still not the same as before chemo and radiation therapy. The eyes are not getting dry any more overnight.
b. Touch and pain afferents (Figure 6) regenerated into the transplanted skin, which had changed into mucosa. The Author cannot distinguish any more by feeling, which mucosa area is new and which one belongs to the surrounding mucosa. Temperature sensitivity returned only little. By the extirpation of the cancer denervated upper lip parts did not or became very little re-innervated, even though innervated tissue was only 1cm away. Also, the denervated smooth muscle of the lip did not become re-innervated by sprouting. Therefore, the power of re-innervation varies strongly in the human head. In animals, the power of regeneration will vary among species and is much higher. Animal data on regeneration, therefore, have first 'translated' to human data before designing clinical studies.
c. The mucosa of the mouth at the place of the former cancer adapted further and still adapts to the tooth prosthesis.
d. Two and a half years after hip replacement, the gluteus muscles of the affected side enlarged and became normal again.

4.13. Cell Replacement Training (Learning)

The strategy of prolonged fasting was to recycle damaged, old, or pathologically functioning cells and replace them with new cells. CDT was administered to give the body additional information by movement-based learning, to optimally place the new cells in the tissue to give rise to physiologic macroscopic organ function. The renewal of the body by cell replacement is the general strategy of the body. By prolonged fasting, however, the renewal or rejuvenation process is enhanced. A repeated prolonged fast trains this renewal regulation. The repeated prolonged fasting and CDT trains the cell-replacement regulation and amplifies this regulation by recycling more cells due to fasting.

At least three benefits take place. First, by repeated fasting we train and optimize the cell replacement regulation. Second, with repeated fasting and longer fasting periods, the amount of cell renewal is larger. Third, with a larger cell replacement and an improved immune system for recognizing better pathological functioning cells, the probability of recycling cancer cells is higher. We may also reach those cancer cells that are in a dormant stage and recycle them. The probability of a reoccurrence of the cancer is reduced and the occurrence of metastasis reduced. For the improvement of the immune system, prolonged fasting and CDT is needed. Therefore, such repeated prolonged fasting, in combination with movement-based learning, has to be repeated. Rejuvenation of nerves and other cells comes about not only by the renewal of cells but also by optimizing the regulation of cell renewal.

Interesting to know is what contributes more to the health, the recycling of old, damaged or pathologically functioning cells or the substitution of new cells according to functional aspects.

The Author administered to himself so far a rejuvenation period of one year (2014). He started with only CDT (coordination dynamics therapy) (Figure 116), and continued with CDT and repeated prolonged fasting (Figures 124 and 126).

It is conceivable that such rejuvenation of the whole body is helpful in other diseases, such as cerebral palsy, CNS injury, or Parkinson.

Chapter V

Learning to Improve Higher Mental Function and Reduce Depression and Anxiety Patterns

Abstract

When adding vision, hearing, and speech to movement-based learning, and coordinating those with movement-based learning, the efficiency of pattern learning can be improved, because the CNS neural networks are activated more integratively, and the learning is more integrative. Especially when moving, vision, hearing, and speech are coordinated up to milliseconds, the neural network learning is more efficient, and goes deeply into the complexity of CNS organization. Higher mental functions can be reached and improved by learning. The repertoire of learning can be enlarged and the rate of learning enhanced when correlating and coordinating the different learning methods exactly with certain movements. Depressions and anxious patterns can be reached and their stability reduced. Learning processes can partly be seen in the facial expressions of patients.

1. Efficient and Integrative Learning When Exercising on the Special CDT Device

In Figures 79 and 80, it was shown that the coordinated movement induced afferent input could be enlarged if, in addition, coordinated input from vision, hearing and speaking is included. In exact coordination with the arm and leg movements, given by the device, numbers, letters, or pictures appear on the display. The patient may also hear words spoken in coordination. When the patient, by himself, has to say the words or letters just when they appear on the screen, the speech is also trained in exact coordination with the movements. This learning procedure is similar to the speech therapy of Figure 78. The advantage to the speech therapy of Figure 78 is that all afferent input processing and organized patterns in the CNS are coordinated. Such coordination of input and simultaneously organized patterns will give rise to efficient learning. This exactly coordinated processing in the CNS is the new step in learning. Since the movements and the other perceptions are complicated, the learning efficiently reaches the deep complexity of CNS organization. In cerebral palsy not only movements, vegetative functions, and vision, hearing and speech are reached for improved functioning, but also the patterns of social behavior and intelligence are reached. The progress in learning is that the different perceptions and patterns are trained simultaneously and in an exact coordination. The damaged nervous system can learn from the device. Learned is a change of the geographical landscape of patterns (pattern learning) and the improvement of phase and frequency coordination of neuron firing. This improved phase and frequency improves the stability of different patterns of CNS organization. Not only the stability of walking (Figures 38 and 39) and other movements is improved but also pattern stability of higher mental functions are improved, such as concentration to a certain task. If one can improve in an autistic child the stability of speaking, walking, thinking, and intention then the child would not be autistic any more. The main problem with the autistic child is that it is not cooperative enough to administer sufficiently long treatment.

Without knowing where certain functions are mainly organized, the CNS of the child learns from the device to improve its functioning. On the other hand, if we understand the organization principles of the CNS, and know with what impulse patterns the neurons communicate, the efficiency of learning can most likely be further enhanced. Since no nervous system is functioning optimally, the healthy CNS can be improved in its functioning and learn to learn more efficiently.

2. Learning to Learn

The Author studied engineering, theoretical physics, and medicine. His memory was always poor. A psychologist told him that it is no problem if his memory is rather poor, because then he has to train the thinking more. During the medical study, anatomists told him often: don’t think but learn by heart. At the end of the medical study after 5 years, the Author had the impression that the learning of facts became easier. He learned to learn better.

3. War Children (Kriegskinder)

As mentioned above, nightmares could be cured during the CDT treatment of scoliosis and in cerebral palsy [9].

However, the CNS of children cannot only suffer damage from a traumatic brain injury, infection, or malformation, but also from the fallout of conflict zones, especially in the absence of parents. They can suffer nightmares that may last for the rest of their lives. Figure 140 shows the ruins of Dresden following the bombing in 1945. Such situations may induce nightmares in those who survived.

Figure 140. The Bombing of Dresden took place in the final months of the Second World War. In four raids between February 13-15, 1945 the bombing and the resulting firestorm destroyed over 6.5 km^2 of the city. An estimated 25,000 people were killed. There was no army in Dresden. A pilot of the United States Air Force refused to bomb Heidelberg, because he had studied in Heidelberg. Dresden was a cultural landmark and is sometimes referred to as "Florence on the Elbe". Culture is only protected by those ones who have culture.
"A people's culture can only be preserved by those who value it themselves."

The father of the Author, an artist in painting, copied the "Sixtinische Madonna" of Raffael (Figures 141A and 142A). In deference to the painting of Raphael (Figures 141B and 142B), he pictured the anxiousness of children and mothers, suffering from the Nazis and bombardments, into the faces (Figure 142A). Neuronal network deterioration due to the war and praying to God for help may be a sufficient pathogenesis of schizophrenia. CDT should help in the treatment of schizophrenia. But drugs are easier to administer.

Interesting is that even in Raphael's painting the Madonna shows anxiousness in the face, whereas the Jesus child not so much.

The Author experienced such situations as a child. To get rid of historic trauma can be difficult. CDT can help to get rid of nightmares. Exercising on the special CDT device is especially helpful in getting rid of nightmares because the complicated coordinations, which are trained, go deeply into the complexity of CNS organization where the higher mental functions are located. The higher mental functions can be reached and improved in their organization. Such therapy may need years. Running, on the other hand, is a very good movement in 'movement medicine', but the complexity of CNS functioning is not efficiently reached.

The destruction of culture, including those of buildings, has consequences and may change the way of life of the people who lost the culture. From the culture of architecture over hundreds of years comes the wisdom: "It is the very building of culture of our ancestors which reveals man's wisdom in all his actions." (in German "Nur angesichts eigenen bauhistorischen Erbes behält der Mensch die Orientierung über die Proportionen seines Handelns").

Figure 141. "Sixtinische Madonna" (Sistine Madonna) is an oil painting by the Italian artist Raphael (right). Commissioned in 1512 by Pope Julius II as an altarpiece. On the left side the copy of Edmund Schalow with some variation; the mark in the painting stem from Stalin soldier who hit the painting. The painting of Raphael was probably modulated in socialistic times, to make the colors lighter and to reduce the "Heiligenschein" (holy light).

Figure 142. Details of the Sixtinische Madonna; left from Edmund Schalow; right from Raphael. Note the different expressions in the faces between Schalow and Raphael. Both reproductions are suboptimal, especially the one of Edmund Schalow.

4. Depression and Repair of Higher Mental Functions

Depression can be counteracted by antidepressant drugs. If depression is inherited, in that it is genetically determined, drug therapy will also work. However, drugs will only mask the symptoms, not improve the function of the neuronal networks. The genetically determined disease may progress in spite of the drugs. This pathologic CNS organization may generalize over a period of years so that other higher mental functions also become impaired or unstable. The drugs can be changed to cover different symptoms, but the disease is still progressing. But applying CDT additionally, the progress of the disease may be stopped or the disease can even be cured. I expect that the repair of the higher mental functions will take years. Similar to motor functions, being euphoric or depressive are organizational states whose stability can be changed and balanced, with respect to each other, and the stability of pathologic emotional states can be reduced. In contrast to the impairment of motor functions, it seems that the low-load and high-load coordination dynamics values do not reflect changes in the higher mental functions, even though exercising on the special CDT devices seems to improve the higher mental functions, since the higher mental functions improve in children with cerebral palsy. The babies or young infants cannot exercise by themselves on the special CDT device but the functioning of their CNS improved following supported exercising (Figures 15 and 86). Therefore, when exercising on the special CDT device, CNS functioning improves, even though it cannot always be measured or the patient cannot exercise independently.

5. Facial Expression

5.1. Facial Expression and Learning

To judge the patient's CNS organization partly by the facial expression, a general understanding of the facial expression and recognition is necessary. This knowledge is then applied to learning for repair. The facial expression is used to partly judge CNS organization especially in coma patients.

Facial expression: A facial expression is one or more motions or positions of the muscles beneath the skin of the face. These patterns of skin muscle activation convey the emotional state of an individual to observers. Facial expressions are a form of nonverbal communication. They are a primary means of conveying social information between humans, but they also occur in most other mammals and some other animal species.

Humans can adopt a facial expression voluntarily (Figures 144 and 145B) or involuntarily (Figures 145A (cerebral palsy), 90C,D,E (normal)), and the neural mechanisms responsible for controlling the expression differ in each case, similar to voluntary and involuntary movement-based learning. Voluntary facial expressions are often socially conditioned and follow a cortical route in the brain. Conversely, involuntary facial expressions are believed to be innate and follow a subcortical route in the brain.

Facial recognition is often an emotional experience for the brain and the amygdala is highly involved in the recognition process. When crawling in interpersonal coordination

(Figures 25 and 26) movement and facial recognition is experienced by the patient's brain and the patient.

The eyes are often viewed as important features of facial expressions. Eye contact is considered an important aspect of interpersonal communication. Beyond the accessory nature of facial expressions in spoken communication between people, they play a significant role in communication with sign language. Many phrases in sign language include facial expressions in the display.

It seems that many facial expressions are innate and have roots in evolutionary ancestors. Also, automatisms like crawling (Figures 25 and 26), walking, and running are innate and have roots in evolutionary ancestors.

Facial muscles: Facial expressions are vital to social communication between humans (Figures 144 and 145). They are caused by the activation and movement of muscles that connect to the skin and fascia in the face. These muscles move the skin, creating lines and folds and causing the movement of facial features, such as the mouth and eyebrows. These muscles develop from the second pharyngeal arch in the embryo.

Neuronal pathways: There are two brain pathways associated with facial expressions; the first one is voluntary expression. Voluntary expression travels from the primary motor cortex through the pyramidal tract, specifically the corticobulbar projections. The cortex is associated with display rules in emotion, which are social precepts that influence and modify expressions. Cortically related expressions are made consciously [101].

The second type of expression is emotional. These expressions originate from the extrapyramidal motor system, which involves subcortical nuclei. For this reason, genuine emotions are not associated with the cortex and are often displayed unconsciously. This is demonstrated in infants before the age of two (Figures 18a and 145A); they display distress, disgust, interest, anger, contempt, surprise, and fear. Infants' displays of these emotions indicate that they are not cortically related. Similarly, blind children also display emotions, proving that they are subconscious rather than learned. Coma patients also seem to display emotions (Figures 121 and 122). In patients with severe brain injury, the improvement of CNS functioning is displayed in the face (Figure 120). The facial expression during CNS repair is emotional, but is modified by the non-optimal CNS organization. This modification of the facial expression can be recognized and judged by the therapist.

Neural mechanisms in face perception: The amygdala plays an important role in facial recognition. Functional imaging studies have found that when shown pictures of faces, there is a large increase in the activity of the amygdala. The amygdala receives visual information from the thalamus via the subcortical pathways. The amygdala may also have a significant role in the recognition of fear and negative emotions. It is believed that the emotion of disgust is recognized through activation of the insula and basal ganglia. The recognition of emotion may also utilize the occipitotemporal neocortex, orbitofrontal cortex and right frontoparietal cortices [102].

Gender and facial cues: More than anything though, what shapes a child's cognitive ability to detect facial expression is being exposed to it from the time of birth. The more an infant is exposed to different faces and expressions, the more able they are to recognize these emotions and then mimic them for themselves. Infants are exposed to an array of emotional expressions from birth, and evidence indicates that they imitate some facial expressions and gestures (e.g., tongue protrusion) as early as the first few days of life.

The same holds for premature born babies. They will also imitate facial expression and gestures if they are able to, according to the developmental stage of their CNS and according to their deviant development. Additional exercising on the special CDT device is needed. The device has to be smaller than that of Figure 135B.

Communication: A person's face, especially their eyes, creates the most obvious and immediate cues, which lead to the formation of impressions [103]. This article discusses eyes, and facial expressions, and the effect they have on interpersonal communication.

A person's eyes reveal much about how they are feeling, or what they are thinking. Blink rate can reveal how nervous or at ease a person is. Stress levels may be revealed by blink rates, which may give information whether a coma patient has stress during CDT.

Eye contact is another major aspect of facial communication. It is hypothesized that eye contact is due to infancy, as humans are one of the few mammals who maintain regular eye contact with their mother while nursing [104]. Eye contact serves a variety of purposes. It regulates conversations, shows interest or involvement, and establishes a connection with others.

Even beyond the idea of eye contact, eyes communicate more data than a person even consciously expresses. Pupil dilatation is a significant cue to a level of excitement, pleasure, or attraction. Dilated pupils indicate greater affection or attraction, while constricted pupils send a colder signal.

Universality hypothesis: The universality hypothesis is the assumption that certain facial expressions are signals of specific emotions (happiness, sadness, anger, fear, surprise, and disgust), which are recognized by people everywhere, regardless of culture or language. The evolutionary basis of these kinds of facial expressions can be traced back to Darwin's "The Expression of the Emotions in Man and Animals".

Evolutionary significance of universality: Darwin argued that the expression of emotions has evolved in humans from animals, who would have used similar methods of expression. Darwin believed that expressions were unlearned and innate in human nature and were therefore evolutionarily significant for survival. Darwin obtained evidence of this through his research on different cultures and species and on infants [105]. Ekman found that people shared many similar facial expressions despite never having met previously before. He was successfully able to confirm Darwin's initial hypothesis that expressions were indeed unlearned behaviors that develop independently from cultured expressions [106].

Innate movement automatisms as crawling, walking and running are trained in CDT to repair CNS functioning, because these automatisms were evolutionarily significant for survival and probably get more genetic support for repair than other movements. The training of facial expressions may be helpful for repair by learning.

5.2. Facial Expression for Social Communication

It was shown above that not only the improvement of phase and frequency coordination is necessary for CNS repair by learning, but also other integrative movement functions have to be trained, so that other parts of the CNS can take over damaged neural network functions. When the patient trains on the special CDT device, his CNS learns to improve the phase and frequency coordination of neuron firing. If he is moving in interpersonal coordination with a therapist, the patient's CNS can also learn from the interpersonal coordination. In Figures 25 and 26, the therapist is crawling in interpersonal coordination with a cerebral palsy child. The vision of the therapist and the patient is involved. With the inclusion of vision the facial expression and the mien of the whole body of the patient and therapist becomes involved in the repair of the CNS by interpersonal coordination and communication.

In Figure 120, it was shown that the improvement of CNS functioning by learning (CDT) over 10 years can be seen in the improvement of the expression of the face. Also, the severance of the brain injury and therefore the deepness of the coma can partly be seen in the face of a patient. The patient Benjamin (Figure 121) suffered a severe brain injury. But his injury was still less severe than that of the patient Manolis. By comparing the face expression of Benjamin (Figure 121) with that of Manolis (Figure 122), both being in a coma, it can be seen that Benjamin's face looks healthier. The organizational state of the CNS of coma patients can partly be judged by the expression of the face.

The young infant of Figure 143 looks during treatment at the Author. His emotional state is expressed in his face. His face expression shows that he is interested in the treatment and concentrates on it, but he is also a bit anxious for what all this is for.

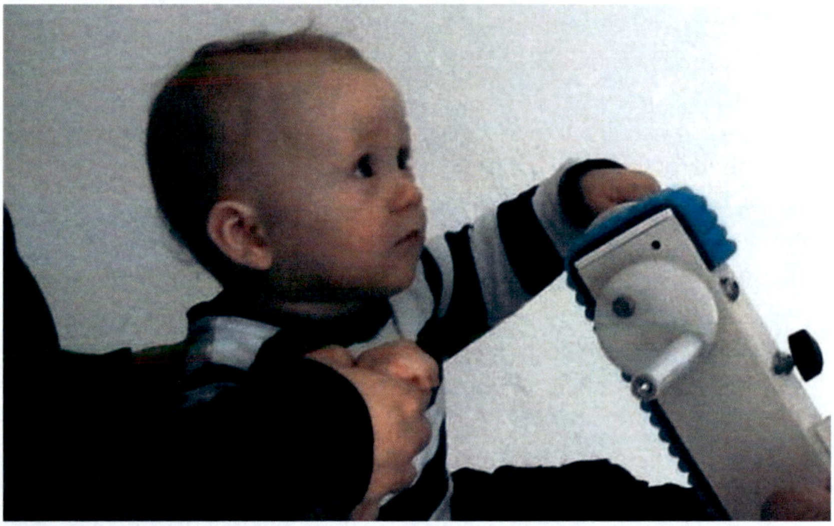

Figure 143. Facial expression of an infant with cerebral palsy during CDT.

If the facial nerve is not damaged, one can see in the face of the patient, what mental pattern he is in. Such emotional patterns are automatisms and are generated by very complicated CNS organizations. But the therapist recognizes automatically the patient's emotional state and can adapt his treatment to it.

When a surgeon visits a patient before the operation there also occurs an important social communication. If the surgeon looks optimistic and friendly, then the patient will become optimistic as well, and be sure that the operation will go well. If the surgeon shows a hard and strong face or even a depressive one, then the patient becomes afraid and thinks the operation might not go well, even though the social communication in that moment has nothing to do with the performance of the surgery of the next day. For social communication and social life, communication through faces is very important.

But what holds for patients, holds for the normal social life in a society as well. To see the head and face is necessary for an advanced social life. Wearing sunglasses can disturb social communication. Children with cerebral palsy need to see the face of the mother to have the feeling of security, to develop properly and not to become aggressive (see above).

In the German fairy-tale film called "Allerleirau" the actress Henriette Confurius expressed beautifully different emotions in her face (Figure 144). From the point of research it is interesting that her face seems to look differently when her head is partly covered by the fur coat. The Author's perception of the expression of her face is influenced by the covering of her head.

Figure 144. Expressions of the face of the actress Henriette Confurius in the fairy tale film "Allerleirau". In A and D she is smiling; in B and E she is angry and in C and F she looks to be in love with the king. My impression of her face expression seems to be different in D, E, F than in A, B, C due to the covering of the head. In G the impression is a bit to be in love, in H very angry and in I she shows a strong and angry face.

In Figure 144A and D the actress shows a friendly face. In B and E, she is angry and anxious, and in C and F her face shows the impression of being in love. The impression of her emotional states, expressed nicely in her face, seems to be different in D, E and F than in A, B and C. The partly covered head seems to change the expression of her face. Also, the hairstyle seems to change the perception of her facial expression. In reality, the perception of the face is modulated. In G, the emotional pattern is to be in love, in H to be very angry and in I she is demonstrating power.

Figure 145 shows twins. In "A" the twins have the same hairstyle and look very similar and in "B" they have a different hairstyle and look different. On the left side of "A" and "B" is one twin and on the right side the other. Obviously, the perception of the facial expression also depends on the hairstyle and not only on the partly covered head (Figure 144).

Figure 145. Twins with different hair sets. One twin is on the left side and the other one on the right side. Note that hairstyle and clothing change the perception of the facial expression. The pictures are out of the film "Das doppelte Lottchen", made in 1950 (Erich Kästner).

The expression of the face can also be used to judge a patient's nervous system organization improvement during a movement-based learning therapy.

Figure 120A show the intelligent patient Dr. Cwienk before his severe brain injury. His face shows not only a friendly expression, but also some power of intelligence. After suffering severe brain injury and partly recovering from it during 10 years of CDT, he shows a friendly face again in Figure 120I, but the power of intelligence is missing. He himself used to say now: "Before the injury I could do two things at the same time, but now only one. Therefore, the mental and emotional states of a patient can partly be seen in the patient's face expression. In coma patients, we may judge a bit from the facial expression, what state the patient's nervous system is in and get some additional information for the prognosis of the outcome. This is important, because we cannot ask the patient. In Figures 121 and 122, it was shown that the facial expression of the patients in a coma gives some information about the brain injury. The patient with a less severe injury (Figure 121) looks healthier than the one with a more severe injury (Figure 122). The wife of Dr. Cwienk reported that her husband looked good while he was in a coma. Conventional treatment was only started three months after the accident. When the patient was out of the coma in Figure 120B-F his face did not look healthy. There are two possibilities, which may have contributed to his state. Firstly, efficient movement-based learning was started too late with the consequence that the injured brain deteriorated in the meantime. Secondly, the patient may have been in a good emotional state while he was in the coma.

The film "Allerleirau" of which the pictures of Figure 144 were taken is really a film about German culture, worthwhile to watch in detail. Also the beauty of a woman's hand is described (poet Schiller) or the classic view of the unity of contents and form (poet Goethe) are touched ("Geist und Kleid" belong together). Mostly the actress was able to express her played emotional state in her face, but not always. The sensitive perception of the facial expressions is astonishing. For the time being, it is not possible to judge someone from the motor programs, which activate the face muscles, what expression a face will have. This indicates how tremendously complex and accurate the organization in the human CNS is. The way of touching the skin of a child stimulates different emotional states in the child. In Figure 6, impulse patterns induced by touching and pin-pricking the skin are shown. Natural impulse patterns of many skin afferents are seen in Figure 6F. It is interesting how sensitive a person's feeling is on the distribution of patterns with respect to time and space. Only stroking the skin gives the feeling of touch. But caressing the skin stimulates additionally an emotional state. Again, small differences in the distribution of the touch afferent input with respect to time and space stimulate completely different organizational states in the human CNS. In Figure 60, it was shown that slightly different skin stimulations changed the organizational states of the CNS from enhancing the continence pattern to a protection reaction.

5.3. Recapitulation Development of Facial Expression

Figure 146A shows a young boy with mild cerebral palsy, and Figure 146B a good actress. The expressions in both faces are similar. It seems that the actress can reactivate automatisms from the childhood. The actress is able to simulate how a certain person feels in a certain moment and expresses this feeling in her face. She is able to activate feeling expressions from childhood. Certain complicated patterns of automatisms are still deep in the complexity of CNS organization and an actress or an actor has the mental strength to recruit them.

It may well be that the human CNS has innate patterns from the evolutions, which are hidden in the complexity of CNS organization of which we do not know. If we would know of some of them, one could use them for pattern-based learning for CNS repair. It may be worthwhile to go backwards in the phylogenies to analyze the behavior, communication, and social life of the ancient humans to learn from them, because our genetics may not have changed very much during the last part of phylogenies.

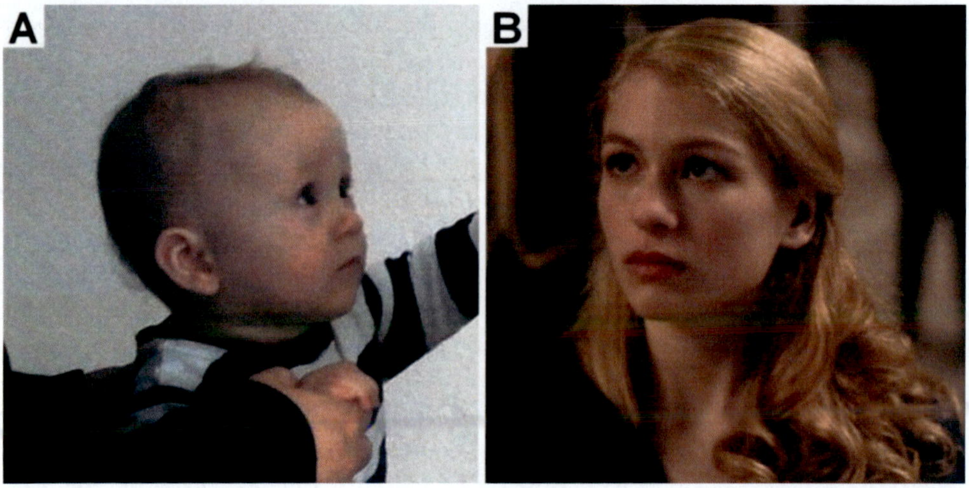

Figure 146. Face expression comparison of a boy with cerebral palsy and an actress. Note the similarity, even though the boy seems to have even a more emphasized expression.

5.3.1. Facial Expression and Perception and Social Communication

Not only can the sympathetic (thoracolumbar) and parasympathetic nervous system divisions (Figure 51) be repaired by learning (Figure 62), but also the enteric division. With the repair of the urinary bladder by CDT also the defecation is repaired. When exercising on the special CDT device or jumping on springboard, bowel movement is achieved passively and actively.

From Figure 120 it can be seen that the facial expression of the patient with severe brain injury improved during 10 years of therapy. It was diagnosed that the cerebellum, the frontal lobe, and the parietal lobe were injured (Figure 147), and he lost a lot of nervous tissue of the brain due to the broken skull. He lost more than 50% of the cerebellum and the pons were also damaged. With the damage of the pons the connections to many nuclei were interrupted. As stated above, the emotional expressions originate from the extrapyramidal motor system, which also involves subcortical nuclei. The improvement of the facial expression documented therefore the repair of the extrapyramidal system and the involved subcortical nuclei.

The cerebral palsy patient Popi of Figure 148 had a severe brain malformation and additional brain damage due to high brain pressure for a few years. The main impairment of CNS functioning, however, originated from the damage of the thalamus. Full urinary bladder continence was achieved within 6 months of CDT and speech was learned in a few years. Broca's and Wernicke's areas were probably not damaged. Further details are given above (Figure 87) and in [1,2]. The proper eye positioning (squint) (Figure 148) was relearned quickly by CDT. The main problem during therapy was the thalamus damage and the connected sensory deficits ("thalamic pain"). The Author was not allowed to touch her and the mother after a few years of CDT because of the sensory deficits. But the thalamus is also involved in facial expression recognition. It seems therefore that some problems of CNS functioning were caused by a disturbed perception of the facial expression. But as her CNS repair showed, her facial expression improved and probably also the perception of facial expressions.

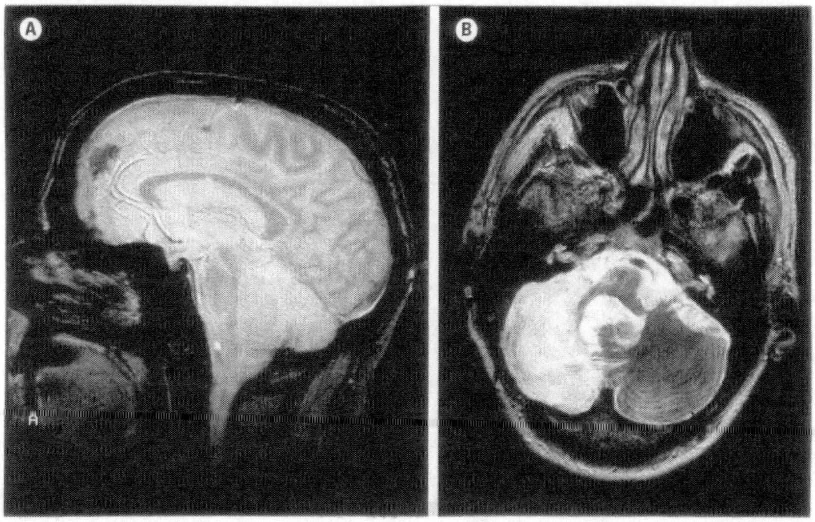

Figure 147. Magnetic resonance imaging (MRI). 63-year-old patient who suffered a severe cerebellar and cerebral injury. The cerebellum has been destroyed to approximately 60% (B, light parts of the cerebellum). There is a loss of brain tissue in the frontal lobe (A, dark areas of the forebrain).

Figure 148. Picture of the 5-year-old patient with very severe cerebral palsy. Note that the positioning of the eyes is not optimal (slightly squint).

This is the great advantage of movement-based learning, even though the treatment effort is enormous; nearly all functions of the CNS improve, including the social functions. As the mother reported again and again, her daughter's social behavior improved strongly. Such improvement of the social functions is very important. How can the mother otherwise survive when living together with her child.

6. How to Continue with Human Neurophysiology, that Means with Medicine, to Improve Health

If one really wants to copy the human brain by electronic circuits, then one has to start with the measuring of impulse patterns elicited by natural stimulation, such as touch of the skin (Figure 6), stimulation of the urinary bladder (Figure 12) or visual stimulation (Figure 79). These naturally induced impulse patterns code the stimulation from the outside world. Together with volitional impulse patterns they give rise to neural network organization changes in the CNS. Measuring simultaneously the natural impulse patterns conducted out of the CNS in efferent nerve fibers (Figures 16 and 17) and into the CNS (Figure 6 and 45); organizational states of the human CNS can be analyzed (Figures 56-61). Single-motor unit recordings from different muscles of the body (Figures 20-24) give further information about CNS output patterns. Penetrating additional electrodes into the brain or spinal cord at specific sites, as possible in brain-dead humans, one additionally obtains impulse patterns for correlation with the impulse patterns running in single afferent and efferent nerve fibers in and out of the CNS. The popular animal anatomy, transmitter distributions and genetics cannot substitute for the knowledge needed to find out how the human CNS modulates its organization at the neuronal level to adapt to changes in the outside world. The recording of the firing of single neurons in the human CNS (Figures 9,49-51) was an essential step forward in the direction of understanding how the neural networks of the human CNS organize themselves to generate certain functions. The achieved progress in understanding neural network functioning under physiologic and pathologic conditions led to substantial progress in curing or repairing nervous system diseases [2-18]. Since the nervous system contributes to most body functions, the health of humans can be enhanced by improved nervous system function. Even a little bit of rejuvenation of the human body seems possible, using movement-based learning to live longer with a better quality of life. As this research hopefully shows, unity of theory and praxis is in the interest of world society to improve its health.

References

[1] Schalow, G., (2013): *Human Neurophysiology: Development and Repair of the Human Central Nervous System.* Nova Science Publishers, Inc., Hauppauge NY, USA, 734 pp.

[2] Schalow, G., (2015): Repair of the Human Brain and Spinal Cord. Nova Science Publisher. In press.

[3] Schalow, G., (2009). Building of New Motoneurons in the Human Spinal Cord upon Coordination Dynamics Therapy to Improve Finger Functions in Motoric Complete Cervical Spinal Cord Injury. In: Berkovsky, T.C. (Ed.*), Handbook of Spinal Cord Injuries,* Chapter 4. pp. 231-264, Nova Science Publishers.

[4] Schalow, G., (2002). Stroke recovery induced by coordination dynamic therapy and quantified by the coordination dynamic recording method. *Electromyogr. Clin. Neurophysiol.* 42, 85-104.

[5] Schalow, G., (2002). Improvement after traumatic brain injury achieved by coordination dynamic therapy. *Electromyogr. Clin. Neurophysiol.* 42, 195-203.

[6] Schalow, G., (2002). Recovery from spinal cord injury achieved by 3 months of coordination dynamic therapy. *Electromyogr. Clin. Neurophysiol.* 42, 367-376.

[7] Schalow, G., (2003). Partial cure of spinal cord injury achieved by 6 to 13 months of coordination dynamic therapy. *Electromyogr. Clin. Neurophysiol.* 43, 281-292.

[8] Schalow, G., Pääsuke, M., Ereline, J. and Gapeyeva, H., (2004). Improvement in Parkinson's disease patients achieved by coordination dynamics therapy. *Electromyogr. Clin. Neurophysiol.* 44, 67-73.

[9] Schalow, G., Jaigma, P., (2005). Cerebral palsy improvement achieved by coordination dynamics therapy. *Electromyogr. Clin. Neurophysiol.* 45, 433-445.

[10] Schalow, G. and Jaigma, P., (2006). Improvement in severe traumatic brain injury induced by coordination dynamics therapy in comparison to physiologic CNS development. *Electromyogr. Clin. Neurophysiol.* 46, 195-209.

[11] Schalow, G., (2006). Hypoxic brain injury improvement induced by coordination dynamics therapy in comparison to CNS development. Electromyogr. Clin. Neurophysiol. 46, 171-183.

[12] Schalow, G., (2006). Cerebellar injury improvement achieved by coordination dynamics therapy. *Electromyogr. Clin. Neurophysiol.* 46, 433-439.

[13] Schalow, G., Jaigma, P. and Belle, V.K., (2009). Near-total functional recovery achieved in partial spinal cord injury (50% injury) after 3 years of coordination dynamics therapy. *Electromyogr. Clin. Neurophysiol.* 49, 67-91.

[14] Schalow, G., (2009). Building of New Motoneurons in the Human Spinal Cord upon Coordination Dynamics Therapy to Improve Finger Functions in Motoric Complete Cervical Spinal Cord Injury. In: Berkovsky, T.C. (Ed.), *Handbook of Spinal Cord Injuries,* Chapter 4. pp. 231-264, Nova Science Publishers.

[15] Schalow, G., (2010). Scientific basis for learning transfer from movements to urinary bladder functions for bladder repair in patients with spinal cord injury. *Electromyogr. Clin. Neurophysiol.* 50, 339-395.

[16] Schalow, G., (2010). Cure of urinary bladder functions in severe (95%) motoric complete cervical spinal cord injury in human. *Electromyogr. Clin. Neurophysiol.* 50, 155-179.

[17] Schalow, G. and Nyffeler, T., (2001). *Koordinationsdynamik-Therapie: Myelomeningozele (Spina bifida). Physiotherapie.*

[18] Schalow, G. and Nyffeler, T., (2000). *Koordinatiosdynamik-Therapie: Skoliose. Physiotherapy.*

[19] Schalow, G., (2005). Tremor in Parkinson's disease patients can be induced by uncontrolled activation and uninhibited synchronization of α_2-motoneuron firing to which α_1-motoneuron firing synchronizes. *Electromyogr. Clin. Neurophysiol.* 45, 393-406.

[20] Kelso, J.A.S., (1995). *Dynamic Patterns.* The Self-Organization of Brain and Behavior. MIT Press, Cambridge.

[21] Haken, H., Kelso, J.A., & Bunz, H., (1985). A theoretical model of phase transitions in human hand movements. *Biological Cybernetics* 39, 139-156.

[22] Schalow, G., (2006). Functional development of the CNS in pupils aged 7 to 19 years. Electromyogr. *Clin. Neurophysiol.,* 46, 159-169.

[23] Schalow, G. and Jaigma, P., (2006). Improvement after severe traumatic brain injury induced by coordination dynamics therapy: Comparison with physiologic CNS development. *Electromyogr. Clin. Neurophysiol.,* 46, 195-209.

[24] Zanone, P.G. and Kelso, J.A.S.: Evolution of behavioural attractors with learning: Nonequilibrium phase transition. *Journal of Experimental Psychology: Human perception and Performance,* 18 (2): 403-421, 1992.

[25] Schalow, G., (1990). Feeder arteries, longitudinal arterial trunks and arterial anastomoses of the lower human spinal cord. *Zentralbl. Neurochir.* 51, 181-184.

[26] Schalow, G. and Zäch, G.A., (1996). External loops of human premotor spinal oscillators identified by simultaneous measurements of interspike intervals and phase relations. Gen. Physiol. Biophys. 15, Suppl.1, 95-119.

[27] Schalow, G. and Zäch, G.A., (1994). Nerve compound action potentials analysed with the simultaneously measured single fibre action potentials in humans. *Electromyogr. clin. Neurophysiol.* 34, 451-465.

[28] Passatore G.M., Fillipi G.M., Grassi G., (1985). Cervical sympathetic nerve stimulation can induce an intrafusal muscle fibre contraction in the rabbit. In: *The Muscle Spindle* (I.A. Boyd and M.H. Gladden, Eds.) pp 221-226, Stockton Press, New York.

[29] Schalow, G., (2010). Phase relation changes between the firings of α and γ-motoneurons and muscle spindle afferents in the sacral micturition centre during

continence functions in brain-dead human and patients with spinal cord injury. *Electromyogr. Clin. Neurophysiol.* 50, 3-27.

[30] Schalow, G., (1993). Spinal oscillators in man under normal and pathologic conditions. *Electromyogr. Clin. Neurophysiol.* 33, 409-426.

[31] Schalow, G. and Zäch, G.A., (2000). Reorganization of the Human CNS, Neurophysiologic measurements on the coordination dynamics of the lesioned human brain and spinal cord. Theory for modern neurorehabilitation (31 case reports). *Gen. Physiol. Biophys.* 19, Suppl. 1, 1-244.

[32] Schalow, G. and Wattig, B., (1993). Recruitment of α and γ-motoneurons in rats, dogs and humans. Electromyogr. Clin. Neurophysiol. 33, 387-400.

[33] Schalow, G., (1992). Recruitment within the groups of γ_1, α_2 and α_3-motoneurons in dogs and humans following bladder and anal catheter pulling. *Gen. Physiol. Biophys.* 11, 101-121.

[34] Schalow, G., (2005). Phase and frequency coordination between neuron firing as an integrative mechanism of human CNS self-organization. *Electromyogr. Clin. Neurophysiol.* 45, 369-383.

[35] Schalow, G., (2009). Relative Coordination between Neuron firing and Generation of a Motor Program in the human CNS. In: Berkovsky, T.C. (Ed.), *Handbook of Spinal Cord Injuries,* Chapter 5. pp 265-295. Nova Science Publishers.

[36] Schalow, G., (2009). Impaired coordination between oscillatory firing FF and FR-type motor units in Parkinson's disease and patients with spinal cord injury. In: Berkovsky, T.C. (Ed.), *Handbook of Spinal Cord Injuries,* Chapter 15. pp 501-517. Nova Science Publishers.

[37] Pavlidis, T., (1973). *Biological Oscillators: Their Mathematical Analysis.* Academic Press, New York.

[38] Blankenship, A.G. & Feller, M.B., (2010). Mechanisms underlying spontaneous patterned activity in developing neural circuits. *Nature Rev. Neurosci.* 11, 18-29.

[39] Schalow, G., Bersch, U., Göcking, K. and Zäch, G.A., (1994). Interspike intervals of secondary muscle spindle and urinary bladder afferents in relation to the oscillation periods of sacral spinal oscillators for continence in man. *Physiological Research* 43, 7-18.

[40] Schalow, G., (1991). Oscillatory firing of single human sphincteric α_2 and α_3-motoneurons reflex activated for the continence of urinary bladder and rectum - restoration of urinary bladder function in paraplegia. *Electromyogr. Clin. Neurophysiol.* 31, 323-355.

[41] Schalow, G. and Lang, G., (1989). Elektrodiagnosis of human dorsal sacral nerve roots by recording afferent and efferent extracellular action potentials. *Neurosurg. Rev.* 12, 223-232.

[42] Schalow, G., (1991). Conduction velocities and nerve fibre diameters of touch, pain, urinary bladder and anal canal afferents and α and γ-motoneurons in human dorsal sacral roots. *Electromyogr. Clin. Neurophysiol.* 31, 265-296.

[43] Brown, P., (2000). Cortical drives to human muscle: the Piper and related rhythms. *Progress in Neurobiology* 60, 97-108.

[44] McAuley, J.H., (2003). The physiological basis of clinical deficits in Parkinson's disease. *Progress in Neurobiology* 69, 27-48.

[45] Koepchen, H.P., (1990). Physiology of rhythms and control systems: An integrative approach. In: Haken H, Koepchen HP editors. *Rhythms in Physiological Systems.* Berlin, pp 3-20.

[46] Lambertz, M. & Langhorst, P., (1998). Simultaneous changes of rhythmic organization in brainstem neurons, respiration, cardiovascular system and EEG between 0.05 Hz and 0.5 Hz. *J. Auton. Nerv. Syst.* 68, 58-77.

[47]. Schalow, G., (2003). Pääsuke, M., Ereline, J. and Gapeyeva, H. Coordination dynamics in Parkinson's disease patients and healthy subjects quantified by the coordination dynamics recording method and sEMG. *Electromyogr. Clin. Neurophysiol.* 43, 473-485.

[48] Schalow, G., Pääsuke, M., Jaigma, P., (2005). Integrative re-organization mechanism for reducing tremor in Parkinson's disease patients. *Electromyogr. Clin. Neurophysiol.* 45, 407-415.

[49] Windhorst, U., (1988). *How Brain-Like is the Spinal Cord?* Springer Verlag, Berlin.

[50] Covic, M., Karaca, E. and Lie, D.C., (2010). *Epigenetic regulation of neurogenesis in the adult hippocampus. Heredity* 105, 122-134.

[51] Deisseroth, K., Singla, S., Toda, H. et al., (2004). Excitation-neurogenesis coupling in adult neural stem/progenitor cells. Neuron 42, 535-552.

[52] Eriksson, P.S., Perfilieva, E., Björk-Eriksson, T., Alborn, A.M., Nordborg, C., Peterson D.A. and Gage, F.H., (1998). Neurogenesis in the adult human hippocampus. *Nature Medicine* 4, 1313-1317.

[53] Praag, van H., Kempermann, G., Gage H., (1999). Running increases cell proliferation and neurogenesis in the adult mouse dentate gyrus. *Nature Neurosci.* 2(3), 266-270.

[54] Ge, S., Yang, C., Hsu, K., Ming, G. and Song, H., (2007) A critical period for enhanced synaptic plasticity in newly generated neurons of the adult brain. *Neuron* 54: 559-566.

[55] Alvarez-Buylla, A. and Lim, D.A., (2004). For the long run maintaining germinal niches in the adult brain. *Neuron* 41, 683-686.

[56] Cao, L., Jiao, X., Zuzga, D.S. et al., (2004). VEGF links hippocampal activity with neurogenesis, learning and memory. Nat. Genet. 36, 827-835.

[57] Palmer, T.D., Willhoite, A.R. and Gage, F.H., (2000). Vascular niche for adult hippocampal neurogenesis. *J. Comp. Neurol.* 425. 479-494.

[58] Zhao, C., Deng, W. and Gage, F.H., (2008) Mechanisms and functional implications of adult neurogenesis. *Cell* 132, 645-660.

[59] Drapeau, E., Montaron, M.F., Aguerre, S. et al., (2007). Learning-induced survival of new neurons depends on the cognitive status of aged rats. *J. Neurophysiol.* 27, 6037-6044.

[60] Schalow, G. and Schmidt, H., (1979). Local development of action potentials in slow muscle fibres after complete or partial denervation. *Proc. R. Soc.* B 203, 445-457.

[61] Song, H., Stevens, C.F., Gage, F.H., (2002) Astroglia induce neurogenesis from adult neural stem cells. *Nature* 417, 39-44.

[62] Blankenship, A.G. & Feller, M.B., (2010). Mechanisms underlying spontaneous patterned activity in developing neural circuits. *Nature Rev. Neurosci.* 11, 18-29.

[63] Wang, C.L. et al., (2007). Activity-dependent development of callosal projections in the somatosensory cortex. *J. Neurosci.* 27, 11334-11342.

[64] Marco Garcia, N.V., Karayannis, T. & Fishell, G., (2011) Neuronal activity is required for the development of specific cortical interneuron subtypes. *Nature* 472, 351-355.

[65] Batista-Brito, R. & Fishell, G., (2009). The developmental integration of cortical interneurons into a functional network. Curr. Top. Dev. Biol. 87, 81-118.

[66] Colvis, C.M., Pollok, J.D., Goodman, R.H. et al., (2005). Epigenetic mechanisms and gene networks in the nervous system. *J. Neurscience* 25(45), 10379-10379.

[67] Batista-Brito, R. and Fishell, G., (2009). The developmental integration of cortical interneurons into a functional network. *Current. Top. Dev. Biol.* 87, 87-118.

[68] Corless, J., (2012). *Powering up. Nature* 485, S62-S63.

[69] Bray, C.W.: Transfer of learning. *Journal of Experimental Psychology,* 11: 443-467, 1928.

[70] Zanone, P.G. and Kelso, J.A.S.: Coordination dynamics of Learning and transfer: Collective and component levels. Journal of Experimental Psychology: Human Perception and Performance, 23: 1454-1480, 1997.

[71] Schmidt, R.A.: *Motor control and learning. Champaign,* IJ: Human Kinetics, 1988.

[72] Tuller, B., Case, P., Ding, M. and Kelso, J.A.S.: The nonlinear dynamics of categorical perception. *Journal of Experimental Psychology: Human Perception & Performance,* 20: 3-16, 1994.

[73] Duus P.(1980): *Neurologisch-topische Diagnostik.* Thieme Verlag, Stuttgart.

[74] Cotta H.(1978): *Orthopädie, ein kurzgefasstes Lehrbuch.* Thieme Verlag,Stuttgart.

[75] Schalow, G. and Pääsuke, M., (2003). Low-load coordination dynamics in athletes, physiotherapists, gymnasts, musicians, and patients with spinal cord injury, after stroke, traumatic brain lesion and with cerebral palsy. *Electromyogr. Clin. Neurophysiol.* 43, 195-201.

[76] Schalow, G., Pääsuke, M and Kolts, I, (2003). High-load coordination dynamics in athletes, physiotherapists, gymnasts, musicians and patients with CNS injury. *Electromyogr. Clin. Neurophysiol.* 43, 353-365.

[77] Piper, M.C. and Darrah, J., (1994). *Motor Assessment of the Developing Infant.* W.B. Saunders Company.

[78] Covic, M., Karaca, E. and Lie, D.C., (2010). Epigenetic regulation of neurogenesis in the adult hippocampus. *Heredity* 105, 122-134.

[79] Corless, J., (2012). Powering up. *Nature* 485, S62-S63.

[80] Chia-Wei Cheng et al., (2014). Prolonged fasting reduces IGF-1/PKA to promote hematopoietic-stem-cell-based regeneration and reverse Immunosuppression. Cell Stem Cell 14, 810-823.

[81] Safdie, F.M et al. (2009). *Fasting and Cancer treatment in Humans: A Case series report.* Aging 1, No 12, 1-20.

[82] Raffaghello, L. et al. (2010). Fasting and differential chemotherapy protection in patients. *Cell Cycle* 9:22, 4474-4476.

[83] Popper, K.R.: *Objective knowledge: An evolutionary approach.* London: O.U.P., 1972.

[84] Touwen, B.: Variability and stereotypy in normal and deviant development. In: Care of handicapped child (Apley, J. (Ed.). *Clinics in Developmental Medicine* No. 67. Philadelphia, 1978, pp. 99-110.

[85] Miledi, R., Parker, I. and Schalow, G. Calcium transients in normal and denervated slow muscle fibres of the frog. *J. Physiol.* 318: 191-206, 1981.

[86] Schalow, G. and Schmidt, H.: Local development of action potentials in slow muscle fibres after complete or partial denervation. *Proc. R. Soc.* Lond, B 203: 445-457, 1979.

[87] Schmidt, H. and Stefani, E.: Re-innervation of twitch and slow muscle fibres of the frog after crushing the motor nerves. *J. Physiol.,* 258: 99-123, 1976.

[88] Schmidt, H. and Stefani, E.: Action potentials in slow muscle fibres of the frog during regeneration of motor nerves. *J. Physiol.,* 270: 507-517, 1977.

[89] Blankenship, A.G. & Feller, M.B., (2010). Mechanisms underlying spontaneous patterned activity in developing neural circuits. *Nature Rev. Neurosci*. 11, 18-29.

[90] Marco Garcia, N.V., Karayannis, T. & Fishell, G., (2011) Neuronal activity is required for the development of specific cortical interneuron subtypes. *Nature* 472, 351-355.

[91] Ge, S., Yang, C., Hsu, K., Ming, G. and Song, H., (2007) A critical period for enhanced synaptic plasticity in newly generated neurons of the adult brain. *Neuron* 54: 559-566.

[92] Schalow, G. (1993). Recruitment of motoneurons in the occasional firing mode in paraplegics. *Electromyogr. Clin. Neurophysiol.* 33: 401-408.

[93] Batista-Brito, R. & Fishell, G., (2009). The developmental integration of cortical interneurons into a functional network. *Curr. Top. Dev. Biol.* 87, 81-118.

[94] Abeles, M. (1982). *Local cortical circuits. An electrophysiological study*. Springer, Berlin Heidelberg New York.

[95] Pavlásek, J.: The binding problem in population neurodynamics: A network model for stimulus-specific coherent oscillations. *Gen. Physiol. Biophys*. 17: 1-18, 1998.

[96] Boulanger, L. and Poo, M., (1999). Presynaptic depolarization facilitates neurotrophin-induced synaptic potentiation. *Nature Neuroscience* 2 (4), 346-351.

[97] Pashler, H., (1997). *The Psychology of Attention.* MIT Press, Cambridge, Massachusetts.

[98] Martinez, A. et al., (1999). Involvement of striate and extrastriate visual cortical areas in spatial attention. *Nature Neuroscience* 2 (4), 364-369.

[99] Quinlan, E.M., Philpot, B.D., Huganir, R.L. and Bear, M.F., (1999). Rapid, experience-dependent expression of synaptic NMDA receptors in visual cortex in vivo. *Nature Neuroscience* 2 (4), 352-357.

[100] Bear, M.F. and Malenka, R.C., (1994). Synaptic plasticity: LTP and LTD. *Curr. Opin. Neurobiol.* 4, 389-399.

[101] Rinn, W.E. (1984). "The Neuropsychology of Facial Expression: A Review of the Neurological and Psychological Mechanisms for Producing Facial Expressions". *Psychological Bulletin* (American Psychological Association, Inc.) 95 (1), 52–77.

[102] Adolphs, R. (2002). "Neural systems for recognizing emotion". *Current Opinion in Neurobiology* 12 (2), 169.

[103] Freitas-Magalhães, A. (2007). *The Psychology of Emotions: The Allure of Human Face. Oporto*: University Fernando Pessoa Press.

[104] Spitz, R.A. and Wolf, K. M., (1946). "The Smiling Response: A Contribution to the Ontogenesis of Social Relations." *Genetic Psychology Monographs* 34, 57-125.

[105] Ekman, edited by Paul (2006). Darwin and facial expression: a century of research in review. Cambridge, MA: Malor Books. p. 12.

[106] Carlson, Neil R. (2012). *Physiology of behavior* (11th ed. ed.). New Jersey: Pearson Education Inc. p. 377.

Author's Contact Information

Dr. Giselher Schalow
Dr. med. habil., Dr. rer. nat., Dipl. Ing.
Untere Kirchmatte 6
CH-6207 Nottwil
Switzerland
Email: g_schalow@hotmail.com

Index